ALFRED TARSKI

Anita Burdman Feferman is an independent scholar and writer. She is the author of *Politics, Logic and Love: The Life of Jean van Heijenoort* (published in paperback as *From Trotsky to Gödel: The Life of Jean van Heijenoort*). She knew Alfred Tarski socially for thirty years.

Solomon Feferman is on the faculty at Stanford University, where he is professor of mathematics and philosophy. He is a recipient of the Rolf Schock Prize in Logic and Philosophy, a Fellow of the American Academy of Arts and Sciences, and has held a Guggenheim fellowship twice. He is the author of *In the Light of Logic* and the editor-in-chief of the multivolume *Kurt Gödel: Collected Works*. He was one of Tarski's students at UC Berkeley in the 1950s.

Alfred Tarski, 1966

Alfred Tarski
Life and Logic

Anita Burdman Feferman
Solomon Feferman

CAMBRIDGE
UNIVERSITY PRESS

CAMBRIDGE UNIVERSITY PRESS
Cambridge, New York, Melbourne, Madrid, Cape Town, Singapore, São Paulo, Delhi

Cambridge University Press
32 Avenue of the Americas, New York, NY 10013-2473, USA

www.cambridge.org
Information on this title: www.cambridge.org/9780521714013

© Anita Burdman Feferman and Solomon Feferman 2004

This publication is in copyright. Subject to statutory exception
and to the provisions of relevant collective licensing agreements,
no reproduction of any part may take place without
the written permission of Cambridge University Press.

First published 2004
Reprinted 2005
First paperback edition 2008

Printed in the United States of America

A catalog record for this publication is available from the British Library.

Library of Congress Cataloging in Publication Data

Feferman, Anita Burdman.
Alfred Tarski : life and logic / Anita Burdman Feferman,
Solomon Feferman
p. cm.
Includes bibliographical references and index.
1. Tarski, Alfred. 2. Logic, Symbolic and mathematical. 3. Logicians – United
States – Biography. I. Feferman, Solomon. II. Title.
QA29.T32F44 2004
199'.438 – dc22   2004045748

ISBN 978-0-521-80240-6 hardback
ISBN 978-0-521-71401-3 paperback

Cambridge University Press has no responsibility for
the persistence or accuracy of URLs for external or
third-party Internet Web sites referred to in this publication
and does not guarantee that any content on such
Web sites is, or will remain, accurate or appropriate.

Contents

1	The Two Tarskis	page 1
2	Independence and University	20

Interlude I:
The Banach–Tarski Paradox, Set Theory, and the Axiom of Choice — 43

3	*Polot!* The Polish Attribute	53

Interlude II:
The Completeness and Decidability of Algebra and Geometry — 69

4	A Wider Sphere of Influence	76

Interlude III: Truth and Definability — 109

5	How the "Unity of Science" Saved Tarski's Life	124
6	Berkeley Is So Far from Princeton	150
7	Building a School	171

Interlude IV: The Publication Campaigns — 189

8	"Papa Tarski" and His Students	196
9	Three Meetings and Two Departures	220
10	Logic and Methodology, Center Stage	246
11	Heydays	257

Interlude V: Model Theory and the 1963 Symposium — 277

12	Around the World	288
13	Los Angeles and Berkeley	310

Interlude VI: Algebras of Logic		334
14	A Decade of Honors	343
15	The Last Times	372

Authors' Note and Acknowledgments   381
Tarski's Ph.D. Students   385
Credits   387
Polish Pronunciation Guide   391
Notes   393
Bibliography   409
Index of Names   417

I

The Two Tarskis

Tarski in His Prime

ALFRED TARSKI was one of the greatest logicians of all time. Along with his contemporary, Kurt Gödel, he changed the face of logic in the twentieth century, especially through his work on the concept of truth and the theory of models. Unlike the reclusive and other-worldly Gödel, Tarski played the role of the "great man" to the hilt, not only through his fundamental work but also by his zealous promotion of the field of logic, his personal identification with the subject, and his charismatic teaching.

Wherever he was, Alfred Tarski was never anonymous. Short, yet somehow grand, he stood out. More than one person thought of him as "kingly"; others used the cliché "Napoleonic" to describe his attitude as well as his size. At lectures he always made an entrance and when walking through a crowded room he never hesitated or shifted from side to side to weave his way around people. Chest out, with quick little steps, he walked straight through the middle, expecting the waters to part.

Far from conventionally handsome, he emanated a physical aura of energy, intensity, and sensuality. His face was mobile, his expressions volatile, reflecting his immediate mood, thought, and appetite. He had a protruding forehead with pulsing veins, giving the impression of having so much brain there wasn't enough room in his head for the whole of it; bright blue eyes, a bulbous nose, a full mouth, seldom quiet; always talking, always smoking, often drinking; he would screw up his face and virtually shudder with disapproval when he shook his head to say he disagreed or that something was not to his taste. He liked to laugh, especially at his own jokes and the gossipy stories he told and retold. In a way, he resembled Pablo Picasso. And, like Picasso, Tarski had supreme confidence in his talent and vision; he conceived of his work in terms of art as well as science and believed in the eros of the intellectual. "The sexiest philosopher I have ever encountered," remembered one casual auditor in his foundations of mathematics class.[1]

He spoke English with a heavy Polish accent – a trilled 'r', the 'h's in the throat, wide open 'a's (he called Sarah Hallam, a mathematics department administrator and long-time friend, Sahrrah Chahlahm) – but his usage was mostly correct and, in writing, the precise word was one of his passions. Using the power of those words, he was a forceful and tireless campaigner for everything he thought was due him, both professionally and personally. He did not like to hear the word "no" in any situation.

Dramatic, theatrical, but always completely controlled, Tarski's lectures were models of elegant clarity, simplicity, and precision. Listeners hung on his every word, mesmerized by the tension and suspense he created through his physical mannerisms as well as the intellectual content at hand. A small, nervous man, a smoker, always in motion, pacing back and forth in a box step, toward the edge of the podium, then to the blackboard, chalk in one hand, cigarette or cigarillo in the other, the ash growing longer as he focused on what he was writing. Would he burn his fingers? Or as he backed away once more to let all regard what he had written, would he take that extra step and fall off the podium? Or moving forward quickly, talking, ready to add to the formula, would he try to write with the cigarette and smoke the chalk? Risk was ever present. According to one account, Tarski did actually set fire to a wastebasket in one of the temporary wooden buildings at the university.[2] Seeing what he had done, he jumped into the basket with both feet and stamped out the flames, blushed a deep red, but then continued lecturing. He was forbidden henceforth to teach in the wooden buildings. His history with cigarettes was terrible. Dale Ogar, a secretary, reported: "He would come to me, all upset, asking 'Where is my cigarette? Where did I put it?' and, after a tense search of his office, I would find it smoldering away, under a pile of papers he had tossed on top of his ashtray."[3]

As a scholar and teacher Tarski set exasperatingly high and memorably rigid standards, but there was a warm, friendly side to him, too. He welcomed and encouraged anyone who asked to attend his classes or seminars: "Oh yes, of course come. You have only to read such and such" he said to many a neophyte applicant, "and then you will have no problem following." However, once the initiate was actually attending and caught up in the systematic Tarskian approach, a maze of hoops and hurdles appeared that had to be negotiated in exactly the right order. Seminar presentations had to be done "Tarski style," with precise definitions and statements of results and clearly presented proofs. Nothing vague, confused, or incomplete was tolerated, and sometimes proceedings would grind to a halt while a hapless student was forced to do things properly. And since Tarski interpreted a student's performance as a reflection of his own abilities, nothing less than the best would do. Depending on the context and the time of day, Tarski commanded admiration, respect, fear and trembling, and affection. A positive result of all this dedicated intimidation was that many of Tarski's ablest students became leading logicians in their own right. On the other hand, in light of his reputation as an authoritarian taskmaster, some very good students – realizing that doing a Ph.D. thesis with him could easily drag on for six or seven years or more – chose another professor as their dissertation advisor. For those who completed the journey, the benefits were everlasting. Fifty years after earning his Ph.D. degree, Bjarni Jónsson, Tarski's first student in the United States, wrote:

I have not yet thought of another person, living or dead whom I would rather have had as a teacher.... He combined an extraordinary mathematical ability with an outstanding talent as a communicator and a willingness to share his ideas with others.[4]

Those who studied with Tarski became part of a school of thought with a tradition of rigor and a set of values that they transmitted to their own students. Even to be known as a student of a student of a student – to be able to trace one's lineage as a descendant – of Alfred Tarski has its cachet. His coterie of graduate and postdoctoral students and assistants at the University of California at Berkeley addressed him reverently as "Professor Tarski"; but when he was not present, his disciples told tales about him and called him "Papa Tarski" with that unique mixture of affection and mockery students use for those they hold in awe.

The distance between the demanding Professor Tarski and the warm "Papa Tarski" was most easily bridged at the frequent parties Alfred and his wife Maria gave, where hospitality seemed limitless. The food and drink reflected their Polish origins: Polish ham and *bigos*, a meat and prune stew, washed down with many rounds of home-made slivovitz – a concoction of fruit from their garden mixed with tequila or some other inexpensive alcohol. Toward the end of the evening when the party had thinned out, a moment would arrive when Tarski announced, "Now, let's drink to *Brüderschaft* [brotherhood]" and he would show the uninitiated how to link arms with him and toss down a glassful of the potent liquor. "Now, you must call me Alfred," he'd say, with an ingratiating smile. Traditionally the *Brüderschaft* ceremony was restricted to those who had just received their Ph.D. but sometimes there were exceptions. At a party, when spirits were high, it was easy enough to call him Alfred, but the next day it was understood by all that he was once again Professor Tarski. It usually took time and many more *Brüderschaft* ceremonies before a young protégé could comfortably call him Alfred, and a few followers never found it possible.

There were six women among the some twenty-five students who worked for a Ph.D. under Tarski's direction, a remarkable number considering how few women study mathematics or logic. This was not by chance. In more than one sense, he was a lady's man: he made it clear that he liked smart lively people and, especially, smart lively women. Although he believed men much more likely to do mathematics at the very highest level, he was strikingly different from most of his colleagues in the way he encouraged and welcomed women students. At least one woman associate claimed that Tarski expressed great anger when he thought she was given an inferior position because she was a woman.[5] Nor did Tarski restrain himself when the attraction went from the intellectual to the physical. Not surprisingly, Tarski's enthusiasm in this direction did not always lead to the happiest results for the women involved. If it was difficult for his male students to measure up, his relations with the women who studied with him, when intertwined with romance, were even more complicated. He was brilliant, charismatic, proud, and ceaselessly persistent; it took an ingeniously tactful woman to resist his advances and remain on good terms with him. And, of course, those who accepted or even welcomed his amorous attentions were negotiating difficult terrain.

Over time, Tarski laid claim to a great deal of territory in the world of logic, mathematics, and philosophy, especially in the areas of set theory, model theory, semantics of formal languages, decision procedures, universal algebra, geometry, and algebras of logic and topology. Between the late 1940s and 1980 he created a mecca in Berkeley to which the logicians of the world made pilgrimage, but he had to push and keep pushing for position, priority, and recognition. For better or worse, it became a habit that he continued long past the point of necessity – or at least so it seemed to others. Even one of his greatest admirers, the philosopher John Corcoran, said: "He was such a glory hound, it was embarrassing. He once confided to me that he considered himself 'the greatest living sane logician'," thus not so subtly avoiding the problem of comparison with Kurt Gödel.[6]

Beginnings

Alfred Tarski was born with the twentieth century, on January 14, 1901,[7] in Warsaw, with the unmistakably Jewish name Teitelbaum (or, in the Polish spelling, Tajtelbaum). He was the first of the two children, both boys, of Rosa (Rachel) Prussak and Ignacy (Isaak) Teitelbaum. Rosa (1879–1942) was from a prominent and wealthy family that had made a fortune in the textile industry of Lodz as the owners of factories and shops. Lodz had been a rural village, a backwater with a few muddy streets, but during the mid–nineteenth-century industrial revolution it became a booming textile center. Factories rose overnight and the population exploded with newly arrived Germans and Jews who came to work in the mills. In this boom town, the Prussaks, an old established family, had been selling and manufacturing cloth since the beginning of the nineteenth century. Rosa's grandfather Abraham Mojżesz Prussak owned one of the first wool factories in the city and was the first to use steam.[8]

Not only was Rosa Prussak an heiress, she was also brilliant, well educated, and willful. One of the rare young women to attend a secondary school, she was the best student in her *gymnasium* and had a gold medal bestowed upon her by the Czar of Russia to prove it. Since 1815 the Congress Kingdom of Poland (also known as Vistulaland), which included Lodz and Warsaw, owed allegiance to the Russian Empire; that was still the situation at the time of Alfred's birth and would continue to be the case until the middle of World War I.

In his historical novel *The Brothers Ashkenazi*, Israel Joshua Singer (older brother of the more famous Isaac Bashevis Singer) depicts the rise of the bourgeoisie in Lodz and the conflicting ideals, goals, and passions of the Jewish population. The Prussak family might well have been the model for some of his main characters. Dinele, the heroine, the educated

Rosa Prussak, Alfred's mother, before her marriage, c. 1899.

daughter "interested in literature and disdainful of common ways," is a perfect picture of the flesh-and-blood Rosa, an imposing woman of high intellect and high standards, demanding of others, and never easily satisfied. In later years, Tarski explicitly credited his own mental rigor and astounding memory to her. By contrast, his father was kind and gentle, a man of the heart.[9] Setting aside the "gentle," which Tarski decidedly was not, the origin of his own generosity of spirit and warmth begins with Ignacy Teitelbaum.

Ignacy (1869–1942), one of five children of Berek Teitelbaum and Niute Weinstock, was born in Warsaw. After Niute's death, Berek married Niute's sister, and a daughter (Ignacy's half-sister) was born. Beginning with that generation, a series of "double cousins" complicates the family tree. One branch includes the Swiss mathematician Joseph Hersch and his sister, the philosopher Jeanne Hersch, who were Tarski's cousins once removed. Complicating the genealogy even further, Ignacy's brother Stanisław married Rosa's younger sister, making her Tarski's "double aunt." The phenomenon of brothers marrying sisters of their deceased wives was a common occurrence in those years.[10]

Ignacy Teitelbaum, Alfred's father (date unknown).

The Teitelbaums, like the Prussaks, were business people whose lives were mostly but not exclusively oriented toward Jewish society. Unlike the large percentage of Warsaw Jews who lived their lives entirely within the self-contained Jewish neighborhoods and community – with their own religion, language, and schools – the Teitelbaums participated in the broader Warsaw scene.

Little is known about the specific occupation of Tarski's paternal grandparents, or of the nature of Ignacy's education, or of how it came about that Ignacy Teitelbaum of Warsaw met and married Rosa Prussak of Lodz. In the Singer fiction, a bold heiress from Warsaw falls in love with a young man at Dinele's wedding and begs her father to send a matchmaker to the appropriate parties in Lodz. With the cities reversed and life imitating art, a similar arrangement might have led to the marriage of Rosa and Ignacy. Since the Prussaks were people of consequence in Lodz – and since daughter Rosa was endowed with intelligence, good looks, and a sizable dowry – Ignacy, too, had to have been in some way alluring. One attraction was that he lived in Warsaw, an interesting, cosmopolitan, glamorously European city whose Parisian flavor eclipsed the gritty

Koszykowa Street, c. 1910. (Number 51, Alfred's birthplace, no longer exists – all the buildings in the immediate area were replaced by Constitution Square several years after World War II.)

frontier town of Lodz. He was undoubtedly a man of some sophistication and culture, and surely his personal warmheartedness was in his favor.

In *The Brothers Ashkenazi*, when the heroine Dinele is betrothed and her trousseau needs to be made, her mother considers it a triumph that Mademoiselle Antoinette, a French dressmaker, has accepted the commission and convinces her husband that it is a privilege to hire a person whose fee is double or triple the normal price, because of the status it will confer. "The wedding will be the talk of Lodz." Something similar happened for Rosa Prussak. A portrait taken by a photographer to the court of St. Petersburg shows a fair young Rosa, her ample hair piled high, wearing an elegant dress trimmed in sculptured lace. There is every indication that she was a dazzling bride ready for a dazzling wedding.

Rosa and Ignacy married on 16 January 1900; their son Alfred was born a year later and his younger brother Wacław in 1903. Although the Teitelbaums identified themselves primarily as Jews and observed the traditions, they resided in an integrated part of the city, not in the ghetto near the old town where the majority of the poorer, most *Jewish* Jews resided. Their home was in the heart of Warsaw, in a large apartment on Koszykowa Street, close to Marszałkowska, one of the grand commercial boulevards of Warsaw, a lively thoroughfare lined with shops and cafés. It was a fine neighborhood, near the handsome foreign embassies on Ujazdowski Boulevard, which runs into Nowy Świat [New World Avenue] and then changes its name to Krakowskie Przedmieście [the Cracow Way] as it approaches the Castle Square and Old Town. From Koszykowa Street it is an easy walk to Ujazdowski Park, to the Botanical Garden and to Łazienki Park – the most elegant and beloved of the Warsaw public gardens, with its summer palace, the Belvedere, its statues, its outdoor theatre on the lake, and the rose garden where Chopin and Paderewski and other great artists had performed.[11] Taking advantage of their surroundings, the Teitelbaums lived a comfortable bourgeois life: they dressed well, entertained friends and family, had servants to help, played bridge, went to cafés, theater, and concerts, and sent their children to excellent schools nearby. The Botanical Garden was one of Alfred's favorite places; there he would develop a lifelong love of plants of all kinds, but especially the most exotic ones.

From 1900 to 1939, Warsaw had the second-largest Jewish community in the world, comprising thirty to forty percent of its population. (The largest, following the enormous waves of migration of European Jews to the United States in the early 1900s, was in New York City.) In the Warsaw ghetto, those who chose to or had no other options could easily conduct a life entirely within the Yiddish-speaking community. It was possible to grow up, go to school, work, do business, and shop for everyday goods without speaking or even understanding more than a few words of Polish.

The Teitelbaums, however, were part of mainstream Warsaw, assimilated to the degree that they spoke Polish as their primary language and sent their children to schools where instruction was either in Russian or Polish. On the other hand, after school the boys went to temple to learn Hebrew and study the Torah. The family celebrated the Jewish holidays and traditions. However, by the time he was in his twenties, Tarski's identification became (with some exceptions) very decidedly Polish rather than Jewish, and in his own home he gave Easter and Christmas parties for his family and friends. Even so, he would always recall with pleasure the details of the Jewish holiday celebrations of his youth in Warsaw, and particularly the gilt-wrapped chocolate Hanukkah coins he was invited to search for in his uncles' pockets.[12]

Ignacy Teitelbaum was in the lumber business, although accounts vary as to precisely what he did: some say he owned a factory which produced lumber, others that he owned a store, others that he was a trader. Perhaps at one time or another he did all three. He is also remembered as having been an inventor whose creations were never put to practical use – in other words, a dreamer. One thing seems quite clear: he was not a good businessman. As a "man of the heart" he was, by implication, not hard-headed, not tough enough in his commercial dealings and therefore only intermittently successful. Financially, he did not live up to the standards to which his wife had been accustomed. She had expensive tastes in furniture and in clothing; she liked to live well and made no secret of her opinion that her husband never provided her with enough money. Like most women in her milieu, Rosa did not work outside the home, but she managed the household and (according to later reports from Tarski's wife) gave her servants a hard time, frequently upbraiding them publicly about their inadequacies. One can imagine that, along with her husband and her domestic help, the boys Alfred and Wacław had their work cut out for them when it came to pleasing mother.[13]

Troubles

The mature Tarski was a raconteur. He liked talking about his past experiences and adventures to friends and students, but aside from the Hanukkah stories, he had surprisingly little to say, even to his close friends, about his early childhood. He did tell stories of his interests as a young man – about his long hikes in the Tatra mountains, his love of nature, his passion for botany and biology, his fascination with language, the general excellence of Polish education, and his political concerns – but he revealed almost nothing of his inner life and feelings as a child.

Of course, material for conjecture exists. Alfred was a gifted child, quick and hardworking, and eager to perform. With his intensely bright blue eyes and golden curls, he surely satisfied most if not all of his parents' expectations and was a favored child. The mature Tarski had winning ways; he had learned how to be charming and anticipated that both women and men would respond to his lively conversation and courtly manners. From long experience, he was accustomed to being the center of attention and was unhappy if he was not.

The younger Wacław, who became a lawyer, was the more practical, less intellectual of the two brothers, but they were close and enough alike

Little Alfred, c. 1906.

to have had similar thoughts about assimilation. In their twenties, they would work together on the idea of changing their name from Teitelbaum to Tarski.

As a young adult and until he left Poland, Tarski maintained close ties with his family although, given his growing nationalistic identification as well as his sense of his own importance, they would naturally have had many disagreements. Still, aside from a beautiful photograph of a soulful, sensitive boy of perhaps ten years, the details that would yield a deep understanding of personal relations within the Teitelbaum household are scanty. Why?

Professor Peter Hoffman, a younger Polish-born colleague who often visited Tarski in Berkeley in the 1970s and had long conversations with him about everything under the sun, offered this comment about Tarski's silence on personal matters: "Listen, he was powered by his ideas. I don't think human relations were very important to him. He was interested in ideas, in mathematics and philosophy, and in politics – fascism, communism, democracy."

Alfred, c. 1911.

Perhaps so, but there was at least one strong exception to Tarski's silence on human relations: the half-personal, half-political subject of anti-Semitism in Poland, a topic never far from the center of his consciousness, a constant thorn in his side, and *always* on the agenda whenever Tarski and Hoffman met. Hoffman, a philosopher, had also been born a Jew and, like Tarski, he too eventually came to feel more Polish than Jewish. Confident that Hoffman would understand, Tarski unburdened himself and told Hoffman about an early emblematic experience of persecution that had affected him profoundly and stayed with him forever.

As Alfred and his brother walked to school or to Łazienki Park to play or to the Botanical Gardens to examine the great collection of plants and trees, somewhere along the way they would be accosted by the neighborhood toughs, who would block their way and taunt them, asking "Where are you going, little Jews, dirty little Jews?" – abusing them verbally and physically. Tarski was sturdy but small for his age, a *little* boy against a gang of ruffians that would intimidate anyone. It is easy to imagine him calculating the risk of such a frightening encounter against the pleasure of doing what he loved to do. The hazing, and the frustration of being unable to stop it or to retaliate, left its mark. Sixty years later, when he told Hoffman of these incidents along with tales of other injustices he had suffered, he sputtered with anger.[14]

Revenge

The combination of being Jewish and small made Alfred vulnerable, but at least he knew he had other resources. From an early age he was aware that his mental prowess and imagination could work in his favor. At a later time, in another country but in a similar situation, the American writer James Baldwin recalled his own state of mind as a child:

> I knew I was black – of course, but I knew I was smart. I didn't know how I was going to use my mind but I was going to get whatever I wanted that way.... I was going to get my revenge.

In this instance, it was vicarious revenge that had to satisfy Tarski.

When he was twelve he translated a short story, *In Letzer Stunde* [*The Last Hour*] by Hugo Gerlach,[15] from German into Polish and presented it to his parents on their thirteenth wedding anniversary. The choice of story may not have been appropriate for that occasion, but it obviously fit very well with what was happening in little Alfred's mental life. The gift is remarkable for its content, for the glimpse it gives us into his preadolescent psyche and, most miraculously, for the fact that it was saved and preserved first by his parents and later by his wife, who eventually brought it from Poland to the United States, where it now resides in the Tarski Archive in the Bancroft Library of the University of California at Berkeley.

Alfred's translation of *In Letzer Stunde* is written in a clear, careful hand; the title page is decorated with a violet-colored flowery motif, and a separate dedication page reads: "Warsaw, January 16, 1913. For my beloved and respected parents on the thirteenth anniversary of their wedding day from their beloved son who has translated this short story from the German." The cover design suggests romantic poetry – which may well have been what his parents were expecting, because Alfred loved poetry and memorized great chunks of verse that he remembered until his dying day. But *The Last Hour* is about something quite different. Written in a vernacular Berlin dialect, it is about poetic justice and taking charge of one's own life. Here is a condensed version of the story Alfred presented to his parents.

The scene is a prison; death row. A guard hands a prisoner his breakfast and says, "Tomorrow at this time, you will not be here." As the

Alfred's gift to his parents on their thirteenth wedding anniversary, a translation of *The Last Hour*.

condemned man swallows a large piece of bread, the jailer says, "I am surprised at how calm you are. You hardly seem upset."

The prisoner responds: "Well, what good would it do to kick and scream? That wouldn't add another hour to my humble life."

Impressed, the guard asks: "Would you like to see the priest, to confess?"

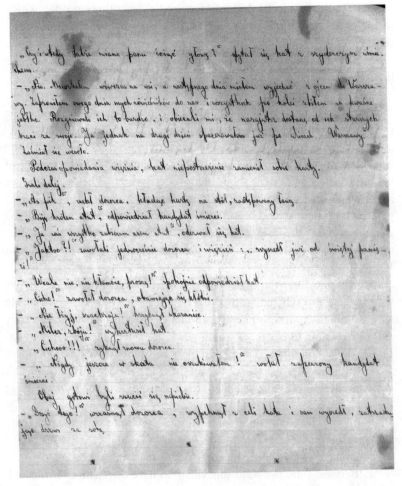

A page from Alfred's handwritten translation of *The Last Hour*.

"No," says the prisoner, "I'm fine. I do not want to confess my sins to a priest. I'd rather be happy."

"Well, what *would* you like to do?"

"I'd like to eat a nice dinner... pork because it is the tastiest. And I'd like some beer and then, as a last wish, I'd like to play a game of cards. I'd like to play 'Skat'."

Skat requires a special deck of cards and at least three people, which the jailer says is a problem. The prisoner proposes the executioner as a third hand.

"Surely you don't want to play with *him*," says the guard, to which the prisoner replies, "Why not, if he is a decent fellow?"

So the guard brings the executioner, the cards, and a stool to the cell. Play begins and, not surprisingly, the prisoner wins hand after hand while the others grumble, annoyed that they won't have a chance to win back.

As the cards are being shuffled in preparation for another round, the prisoner says, "This reminds me of the time when I was a boy of ten and my father and I were were about to move to Berlin.[16] The day before we left my village, I called all the little boys to my house, one by one, and gave each of them a good beating. They were all very angry and told me that the next day their big brothers would give me a sound thrashing. But of course the next day, I was gone, happily walking along the streets of Berlin."

Play resumes, but the prisoner's good cheer evaporates as he catches the executioner exchanging the cards he's been dealt for better ones. He accuses him of cheating. The executioner denies it. They quarrel loudly and nearly came to blows, at which point the jailer, berating the executioner, drags him out of the cell and locks the door and leaves.

Half an hour later he returns to find that the prisoner has hung himself with his suspenders which he had attached to the window bars by using the stool the jailer had brought to sit on during the game.

On his cot the condemned man left a suicide note with the following explanation.

"I have taken my own life for three reasons:

"1. Because I did not want this villain to have the chance to win back 50 rubles from me.

"2. Because it is certainly not right for a cheater like him to cut off my head.

"3. Because it is a sheer pleasure to me to know that nobody can get back at me for what I have done."

A tale of revenge and triumph, not over death but over authority. The hero implements his master plan to perfection, dictates how he will spend his last hour and also has the last word, which is certainly the part Alfred would have liked best.

Of all the stories a twelve-year-old might have chosen to present to his parents, why this one? And where would he have come across this German short story written in a Berlin dialect? Certainly not in his public school classroom, where Russian was the language of instruction. The macabre tale obviously appealed to him, and his parents, surprised though they may have been at the choice, appreciated it enough to have saved it for posterity. Identifying with the crafty prisoner who outwits his executioner and with the prisoner as the youth who beat up the boys in his village and got away with it, Alfred, victim of neighborhood bullies, must have wished he could have done likewise. Here was a story that could offer him at least literary revenge against his tormentors. With its themes of defiance of authority, refusal of confession – "I don't want to tell my sins to a priest. I'd rather be happy" – calm in the face of death, and most of all the triumph of controlling one's final destiny, one's *final hour*, it becomes obvious that the gift to his parents was a joy to create. Incidentally, it enabled Alfred to demonstrate how well he knew German and what a clever translator he was as well as how much understanding he already had about the ways of the world and, slyly, to foreshadow his growing independence from his family. And, as to where he found Gerlach's stories: close to the Teitelbaum home on Koszykowa Street there was (and still is) a fine public library. It would be surprising if Alfred, an intellectual omnivore and avid reader, had not been one of its most active borrowers.

Gymnasium

In 1915, a year after World War I had begun and just before Alfred was ready to enter high school, the German army invaded the Russian sector of Poland, forced the Russians out, and took control of Warsaw. So, when he began his college preparatory studies, for the first time in thirty years the language of instruction was Polish, not Russian. The Szkoła Mazowiecka on Klonowa Street was a well-known *gymnasium* for children of the intellectual elite, which in that era almost inevitably meant people of some means. The new curriculum at Mazowiecka now included Polish language and literature, Latin, German, French, history, religion, law, logic, mathematics (including arithmetic, algebra, geometry, trigonometry, and analytic geometry), physics, cosmography (astronomy, geography, and geology), and hygiene. His final school record does not indicate that he studied Greek, but somewhere along the line he learned that language, too; Tarski was also perfectly fluent in Russian, since all his previous schooling had been conducted in that language.

As a *gymnasium* scholar, Alfred did brilliantly. "A student of extraordinary ability," his teachers wrote, "not to be compared with others." Picture an energetic boy, small for his age, sitting in the front row, fidgety – a classmate recalled the odd detail that he was constantly twisting his handkerchief – driven by an enthusiasm for learning *everything* and bursting with desire to speak up and demonstrate how much he knew. Tarski excelled in all the subjects he studied, getting the highest marks possible. In later years he enjoyed confessing to an audience that "the reason I studied logic in the university was because it was the only subject in which I had gotten a B grade. This made me realize that I had to do better, to learn more." A good story but not true, even if he himself told it. His high-school records show he received the highest mark in logic, too. This is not the only example of Tarski building his own apocrypha.

≥∙ ≥∙ ≥∙

Although World War I was being waged all during Alfred's high-school career, it made little difference to his studies. Most of the fighting between Germany and Russia was on the Eastern front, much of it done by Polish recruits who suffered heavy losses. Warsaw was damaged in 1915 when the retreating Russians destroyed the city's bridges, stations, and some factories; but as the war progressed under the German occupation, the main effect on the city was a volatile situation marked by rising prices, unemployment, and repeated strikes.[17] The Teitelbaum family was not significantly affected in a personal way, and Poles in general were buoyed by the feeling that the result of the war might finally lead to independence. According to the historian Norman Davies: "The spontaneous celebration of the Polish National Day on 3 May 1916, for the first time in fifty years, betrayed the nationalist feelings which were now about to surface."[18] It was to be the independence of Poland at the end of the war that would be most important to Alfred.

At the head of the class when he finished high school in June 1918, Tarski felt on top of the world. His graduation picture shows a serious-looking young man in his school uniform: high buttoned collar, buttoned pockets with a medal pinned to the left one, and buttoned knee britches. The photographer did a fine job of capturing the seventeen-year-old youth, well on his way to manhood. Hair perfectly groomed – as

Alfred ready to enter the university, 1918.

it would be always – a trim little mustache to show he was really a man, Alfred, sitting in a low-backed studio chair, looks composed, relaxed but serious, and very sure of himself.

If the adult he became is any indication, he was already a romantic, a poetry lover, an idealist, passionately opinionated about politics, and a gregarious person who appreciated and needed the stimulation of other intelligent minds to hone his own competitive edge. Poland was about to experience a cultural and intellectual explosion and, as he prepared to enter the university, Alfred was ready and eager to take part.

2

Independence and University

> *The achievements of Alfred Tarski were the most brilliant result of a favorable cultural entanglement which took place in Poland in the first half of the twentieth century.*   Andrzej Grzegorczyk[1]

NINETEEN EIGHTEEN was a momentous year in world history, in Polish history, and in the personal history of Alfred Teitelbaum – on his way to becoming Alfred Tarski. World War I was finally over and Poland, after 150 years of partition and foreign rule, had emerged as an independent nation. It was at this point that Alfred entered the University of Warsaw, now a *Polish* university.

As the nation was defining and redefining itself in terms of political and cultural goals, Tarski was engaged in the debate on both a general and a personal level, trying to come to terms with where he fit in as an individual Polish Jew. Although the center of his existence was always scholarship, Tarski was also passionately interested in other matters. His earliest memory, he claimed, was an image of himself as a very little boy, standing on the balcony of his apartment on Koszykowa Street watching a long line of workers marching down Marszałkowska Boulevard in support of the 1905 revolution in Russia. Alfred was four years old then, but he never forgot that night and the feeling of being in sympathy with the marchers.[2] It was his first taste of what he would later understand about the forces of history and the deep emotions that move a people suffering under repression. From that and other early experiences, politics – especially the politics of Poland's tortuous ups and downs as an occupied or independent nation – remained a major interest for the rest of his life, and he developed very definite ideas about what was moral and just.

"Every Polish schoolboy..."

Beginning in 1772, Poland was ignominiously partitioned between its three most powerful neighbors: Russia, Prussia, and Austria.[3] For hundreds of years long past, Poland – in alliance with Lithuania – had enjoyed a golden age as an empire that stretched from the Baltic to the Black Sea. But the Polish–Lithuanian Commonwealth had been in serious decline since the seventeenth century, and it was only after showing signs of resurgence under the relatively liberal and enlightened rule of its last king, Stanisław-August Poniatowski, that the three powerful neighbors swallowed it up. Hopes for independence were briefly revived following Napoleon's stunning military successes, first against the Russians and the Austrians at Austerlitz in 1805 and then against the Prussians at Jena in 1806. That year, Napoleon himself entered Warsaw, "greeted with triumphal arches and delirious crowds."[4] But the enthusiasm was dampened when, of the almost 100,000 Polish troops who enlisted into Napoleon's army for its invasion of Russia in 1812, fewer than a quarter returned from the disastrous campaign. With Napoleon's defeat at Waterloo in 1815, Poland's fate was sealed at the Congress of Vienna and once more it was torn apart. Prussia took over the western provinces, Austria absorbed the southern province known as Galicia, and Russia annexed the eastern provinces. What remained was called the Polish Kingdom, or "Congress Poland," with Warsaw as its capital; it was nominally independent but in fact was completely under the thumb of the Russian czar. The Kingdom lasted until 1874, when it was absorbed entirely into the Russian empire as "Vistulaland."

Under the increasing repression that followed the 1815 partitions, the Polish masses had no choice but to continue their daily lives as best they could. Various forms of "realistic" accommodation and cooperation with the ruling powers were pursued by those with civic and governmental responsibilities, but little was given back in return. For the romantic idealists – the passionate nationalists – the only viable future lay in the eventual overthrow of the occupiers. There was no lack of humiliating provocations, both in and out of Congress Poland, and these led the nationalists to mount three major revolts between 1830 and 1863. Each of these uprisings was put down brutally and led to increasing Germanization in Prussian Poland and Russification in the Russian-dominated parts; only Austrian Galicia retained some autonomy. The repressions intensified feelings of Polish identity, especially in the Congress Kingdom; in contrast to the generally liberal attitudes with respect to ethnicity, language, and religion that had been the norm throughout the times of the Polish–Lithuanian Commonwealth, for many the spirit of national restoration now narrowed to the view that to be Polish meant one had to speak Polish and be Catholic.

Increasingly during the nineteenth century, nationalistic feelings were stirred by the romantic historians, dramatists, novelists, and poets, who featured such tales as the heroic ancient struggles of the Poles against the Teutons on one side and the Cossacks on the other. Even more than literature, according to Adam Zamoyski, "the painting of this period created a gallery of images which embalmed for all time the myths and heroes of a bygone age. King Bolesław entering Kiev; Batory taking the surrender of Pskov; Sobieski at Vienna; Kościuszko at Racławice – these are images in the mind of every Polish schoolboy." Defeat itself was glorified in these paintings: "The stress was on the enduring and transcending nobility of their actions at the moment of failure, to suggest that failure itself was not important."[5]

Answer to a Prayer

The turning point for Polish nationalism in 1905, the year of Alfred's powerful childhood memory, came as a result of revolutionary events in Russia that erupted over its disastrous and unpopular war with Japan. In St. Petersburg on January 22, troops fired on a crowd of workers who were marching on the winter palace of Czar Nicholas II to petition for an end to the war. This "bloody Sunday" turned into the Russian Revolution of 1905, an upheaval that reverberated throughout the empire. The revolution spread to Russian-occupied Poland, where the increasingly strong socialist party (PPS) organized mass strikes and demonstrations which so worried the Russians that 300,000 troops were sent to suppress the insurgent movements. The leader of the PPS was the fiercely patriotic Józef Piłsudski, a young noble from Vilna in the former Lithuanian provinces. Although others hated Germany and tried to cooperate with Russia, for Piłsudski the main enemy was the Russians; at age twenty he had been arrested and exiled to Siberia for five years for alleged participation in an attempt to assassinate Czar Alexander III.

The second great turning point in the efforts to regain Polish independence came with the outbreak of World War I in August 1914, which pitted the Allies (principally Great Britain, France, and Russia) against the so-called Central Powers (Germany, Austro-Hungary, and the Ottoman Empire). Piłsudski calculated that the Germans, for whom he had no love either, might well be able to bring down the Russians. If the Central Powers then could be defeated by the Allies, that would be an opening once more for Polish independence. For Piłsudski, "August 1914 was the answer to the prayer which the famous poet Adam Mickiewicz ... had once addressed to God for a universal war to free the oppressed Poles."[6] Seeing this on the horizon, in the years before the war Piłsudski had organized clandestine "shooting clubs" in Austrian Galicia; when war came, these formed the core of his Polish Legions. They fought under his command as independent units alongside Austrian and German troops.

Piłsudski's strategy had its first rewards in 1915, when the Germans drove the Russians out, reinstated the Polish Kingdom under their control, and almost immediately permitted a process of Polonization to take place. This was the occasion for restoring Polish as the language of instruction in the Mazowiecka *gymnasium* that Alfred attended and in other schools. The new Polish-language University of Warsaw that he would enter three years hence was established in November 1915 to take the place of the no-longer-functioning Russian Imperial University.

Independence

The Russian Revolution of 1917 brought the communists to power; under serious internal challenges and in desperate circumstances, they sued for peace with the Germans, thus taking Russia out of the war. That left the English and French, now joined by the Americans, to defeat the exhausted Central Powers in 1918. In addition to the millions of soldiers that had perished on both sides, some 450,000 Poles – who had been conscripted into the Russian, German, or Austrian armies – lost their lives, and twice that number came back wounded.[7] Before the end of the war, unofficial emissaries from Poland had been putting pressure on the West to help resolve the "Polish question" once and for all as part of a final settlement. The celebrated pianist Ignacy Jan Paderewski was in the United States, concertizing and propagandizing for Polish independence. He had President Woodrow Wilson's ear, and in an address to the U.S. Senate in 1917, Wilson declared that "statesmen everywhere are agreed that there should be a united, independent and autonomous Poland." Thus it was that the victory of the Allies in 1918 finally brought true independence to Poland, formally proclaimed on November 6 of that year. According to one description:

The Poles were delirious with their new freedom. People took to the streets of Warsaw in wild celebrations. The city was festooned with white-and-red Polish flags. They were raised over the Town Hall, they appeared over Poniatowski's Belvedere Palace, above Warsaw University and the Polytechnic Institute, and on the rooftops of hundreds of buildings. Sidewalks and cafés were packed with milling people.[8]

Under the Treaty of Versailles dictated by the victorious Allies, the country's Prussian regions were returned to Poland, which also regained Western Galicia, including Cracow. For the rest, Poland had to reclaim what it regarded as its former territories prior to the partition of 1772 through a series of what Norman Davies called "nursery brawls."[9] In the period 1918–1921, Poland fought *six* wars: with the Ukrainians, the Germans, the Lithuanians, the Czechs, and – most seriously – the Soviet Union. The war with the Ukrainians returned Eastern Galicia, including the city of Lvov, to the Poles. The most serious threat to Polish independence came in 1920, when Russian troops reached almost to Warsaw on the right bank of the Vistula. In the ensuing battle, Piłsudski's men pushed back the Red Army, and the Poles ended up recapturing a good part of the claimed territory. Thus the new Polish Republic found itself once more with substantial minorities of ethnic Germans, Ukrainians, Byelorussians, Lithuanians, and Jews, all of whom attracted renewed hostility from the Polish Catholic majority. The situation of the Jews was special. Despite considerable migration to America and other parts of the world in the prewar years, there were three million Jews in Poland, about ten percent of the entire population; but in Warsaw, the capital city, a third or more of the nearly one million inhabitants were Jewish. In some quarters they were perceived as a threat not only because of their numbers but also because many of them, particularly in the Eastern provinces recaptured from Russia, were suspected of being sympathetic to communism.[10] Furthermore, the ever-present undercurrent of anti-Semitism was fueled by the country's serious economic stresses of high unemployment and uncontrolled inflation during the 1920s. Once more, the "Jewish question" became a political football and a source of daily concern for the Jews.

How to Be Jewish in Poland

The history of the Jews in Poland dates from the Middle Ages, when the first settlers arrived, but the greatest influx came in the fifteenth and sixteenth centuries following years of persecution and periodic expulsions in Western Europe, especially from Spain and Germany. The Polish–Lithuanian Commonwealth proved to be a haven where the rights of the Jewish community were established by royal charter and Jews were given considerable autonomy and self-governance. This was especially broad during the sixteenth and seventeenth centuries, when the Commonwealth was called "heaven for the Jews, paradise for the nobles, and hell for the serfs."[11] The situation of the Jews in Poland was unique among European countries because they were allowed to participate in a wide range of trades and crafts as well as to manage the estates of the nobility. Their religion continued to be orthodox Judaism, and Poland became a center for Talmudic study and then Hasidic revivalism. The majority of Jews were separated from the Polish Catholics and other ethnic and religious groups in the Commonwealth by their traditional garb and appearance – long black coats and black hats, beards and side curls for the men – and by the use of the Germanic dialect of Yiddish as their everyday language. Immigration from Western Europe continued to such an extent that, by the end of the eighteenth century, three quarters of the world's Jews lived in the Polish–Lithuanian Republic.[12]

Things changed drastically following the Congress of Vienna. As a result of Russia's annexation of the Eastern territories, the majority of Jews came under direct czarist rule. They were then required to live in special cities and *shtetls* called the Pale of Settlement, and they were subject to many restrictions; they could not buy land nor serve at higher ranks in the military and the bureaucracy, and they had limited access to secondary and higher education and the professions.[13] Worse were the frequent pogroms – often officially encouraged. These constant harassments caused many Jews in the Pale to emigrate, some back to Congress Poland, others to Western Europe, and in ever-increasing numbers to America. Spurred by economic growth and industrialization, Warsaw's population tripled in the latter part of the nineteenth century, swelling the numbers of Jews in the north-central part of the city.

While most Jews maintained themselves as a separate community socially and politically, the assimilated ones had several options at the turn of the twentieth century. They could follow Piłsudski's socialists, with its primarily nationalistic emphasis;[14] or they could follow Rosa Luxemburg's socialists, which had an internationalist orientation and was more to the left than the PPS. Theodor Herzl's Zionism, which sought a return to Palestine as the Jewish homeland, was yet another movement that attracted Jews. The first premier of the eventual state of Israel, David Ben Gurion, started out as a socialist in Poland and then became an ardent Zionist.

University Years

By the time that Alfred Teitelbaum was ready to enter the university in 1918, his social identity had been transformed from that of a moderately assimilated bourgeois Jewish boy to that of a Polish patriot. He was neither religious nor a Zionist; politically his leanings were socialist. He may have thought of himself as one with the times, but in reality there was no such thing because the times were highly unsettled and there were many conflicting political and economic forces at work. Although later there would be quotas for Jews in the University of Warsaw and other Polish universities, there were none in 1918 and, as in the population of the city as a whole, a third or more of the students were Jewish.[15] The date of Alfred's matriculation was 15 October 1918, barely a month before Polish independence was declared on the 10th of November. At that time the university was divided into four schools: Theology, Law, Medicine, and Philosophy, the last of which included among its divisions the Exact Sciences. Alfred's plan was to study biology, his first love. Indeed, his student documents show him enrolled for courses in zoology, botany, anatomy, chemistry, and physics, with the names of the professors listed for each. But in his student book these enrollments are crossed out, without the requisite signatures verifying completion of the courses, because, owing to the still unsettled political and military situation marked by Poland's territorial wars with its neighbors, the university suspended all lectures for the academic year 1918/19. Young Alfred was called up for duty in that period, but he managed to avoid action as a soldier and instead worked in a unit that provided food, equipment, and medical help.

When the university reopened in 1919, Alfred abandoned biology as his major subject in favor of mathematics.[16] There would be nothing remarkable about this, since students frequently switch their majors, except that in later years he repeatedly made a "big story" of how this happened. He had, he said, solved an open problem in set theory that had been raised in a logic course by the professor, Stanisław Leśniewski; the latter was so impressed, he encouraged Alfred to switch from biology to logic. A

Entrance to Warsaw University on Krakowskie Przedmieście.

dramatic tale, but the evidence does not bear it out: Leśniewski did not join the faculty until the winter of 1919. By then, Alfred was already fully into his new direction of studies. Furthermore, the result of his solution of the problem in set theory was the subject of his first published paper (in Polish), "A Contribution to the Axiomatics of Well-Ordered Sets," which did not appear until 1921 – a full three years after he had entered the university.[17]

From the beginning, Alfred's professors were among the leaders in the areas of mathematics and logic in Poland, soon to be recognized internationally for their outstanding contributions. It is more likely that given his diverse talents, especially in mathematics (already marked at the Szkoła Mazowiecka), Alfred decided on his own to switch from biology when he became aware – during the unsettled year 1918/19 – of the brilliant assemblage of faculty in mathematics at the university and developed a sense of what he could achieve in that subject. Since the number of students at the University of Warsaw was low to begin with, classes in general were small, especially in subjects like mathematics. Pure mathematics was traditionally a bastion for the smartest, problem-solving, chess-playing boys, whereas classes in the biological sciences were crowded with students gearing up for a career in medicine or pharmacy. From enrollment figures for the University of Warsaw in 1918/20,[18] one can estimate that there were at most twenty-five or thirty students in mathematics, all told. Thus, a student in mathematics could expect to benefit from individual attention from top researchers in the field.

Mathematics in Warsaw

Alfred's choice of studies in mathematics and logic was opportune for another reason – namely, the phenomenal intellectual explosion of these subjects in Poland following its independence. On the mathematical side, the grounds had been laid by a young professor, Zygmunt Janiszewski. He had obtained a doctoral degree in the newly developing subject of topology in Paris in 1912 and was appointed, along with the topologist Stefan Mazurkiewicz, to the faculty of mathematics at the restored University of Warsaw in 1915. Janiszewski had joined Piłsudski's Polish Legions at the outbreak of World War I but left a year later, disheartened by having to fight on the side of the Austrians when they demanded an oath of allegiance. Janiszewski's brilliant idea was to establish a distinctive Polish school of mathematics and to make an impact on the international scene by founding a new journal, *Fundamenta Mathematicae*, devoted entirely to a few subjects undergoing active development. It was to concentrate on the modern directions of set theory, topology, mathematical logic, and the foundations of mathematics that had begun to flourish in Western Europe early in the twentieth century. Moreover, the articles were to be written in international languages – which at the time meant French, German, or English – and to break the tradition of publication in Polish, a language practically no one but Poles could read.

Unfortunately, Janiszewski died in the flu epidemic of 1920 and did not live to see the appearance of the first volume of *Fundamenta* that same year. The inaugural volume consisted entirely of articles by Poles, but later volumes would widen the circle of authors considerably to include many foreign contributors, and the topics were expanded to include abstract algebra and the modern theory of functions of real numbers. Two of the authors in that first volume were the brilliant Stefan Banach (with whom Tarski was later to write a famous paper on what has come to be called the Banach–Tarski Paradox) and Hugo Steinhaus. Banach and Steinhaus were leaders in the illustrious Lvov school of mathematics, which had developed independently of the Warsaw school. Steinhaus's "greatest discovery" (a frequently told story) was of Stefan Banach himself: while walking in the park in Cracow one summer evening, Steinhaus was astonished to hear a young man explaining some of the most modern and exciting concepts of mathematics to a friend. Following this chance encounter, he met frequently with Banach and other young mathematicians at a café to discuss ideas and problems. In 1920 Banach joined Steinhaus in Lvov, and from then on his career took off meteorically. He was soon recognized worldwide as one of the creators and leaders in the subject of functional analysis, an area of mathematical research that applied notions from abstract algebra and set theory to the theory of functions. When Banach and Tarski published their joint paper in 1924, Banach, who was nine years older, was far more established.

Alfred's professors in mathematics at the University of Warsaw were the young and vital Stefan Mazurkiewicz and Wacław Sierpiński; they were joined a few years later by Mazurkiewicz's student Kazimierz Kuratowski. In 1918, the old man of the group was Sierpiński, aged thirty-six; Mazurkiewicz was thirty, while Kuratowski at twenty-two was the "baby." During the interwar years at Warsaw University, Mazurkiewicz, noted for the brilliance and intelligence of his lectures, was the chair and central figure in the mathematics department.[19] He and Sierpiński were the initial co-editors-in-chief of *Fundamenta Mathematicae*. Besides his editorial work, Mazurkiewicz chaired the meetings of the Warsaw division of the Polish Mathematical Society. In Kuratowski's volume of "remembrances and reflections," *A Half Century of Polish Mathematics,* he writes that official meetings continued to be held at a café, where Mazurkiewicz was the life and soul.[20] This was typical: in both Warsaw and Lvov, café tables became the place of mathematical inspiration, and everyone had his favorite café and favorite table. Leśniewski, Tarski's professor of logic, defined an intellectual as "a man who after leaving home does not return without visiting at least one *kawiarnia* [coffee house]." His favorite was Lourse on Krakowskie Przedmieście, near the university.[21]

The senior member in the Warsaw mathematics department, Wacław Sierpiński, was especially noted for his work in set theory, a subject that was to become one of Tarski's main areas of research. Set theory had been created in the late nineteenth century by the German mathematician Georg Cantor. Because of its novel and strange concepts concerning transfinite cardinal and ordinal numbers of different infinite sizes and orders, it was greeted in some quarters with much suspicion and hostility.

But Sierpiński and others transformed it into a systematic field that could be pursued with as much confidence as more traditional parts of mathematics. Sierpiński was extraordinarily prolific: all told he wrote some seven hundred papers and books, including the *Outline of Set Theory* (1912), one of the earliest systematic texts on that subject. Much later – through the work of Tarski, Gödel, and others – axiomatic set theory would become the subject of intensive logical examination.

The Lvov–Warsaw School of Logic and Philosophy

In logic, Tarski's professors were the philosophers, Jan Łukasiewicz and Stanisław Leśniewski, who were the leading lights in Poland in that field. Both had studied in Lvov with Kazimierz Twardowski, who in turn had been a student in the latter part of the 1880s of the influential and charismatic philosopher Franz Brentano, in Vienna. Sigmund Freud, Edmund Husserl, and Alexius Meinong were among Brentano's other notable students.[22] Twardowski was appointed to the Chair in Philosophy at the University of Lvov in 1895 and – through his impressive teaching, organizational abilities, and appealing personality – built up from scratch what was to become the Lvov–Warsaw School of Logic and Philosophy. At first he lectured to practically empty halls, but as news got around about what an exciting speaker he was, the halls gradually filled up until, according to one of Twardowski's students, the lectures finally "had to be transferred outside the university because no university hall could accommodate the listeners who from the early morning hurried to secure themselves a place."[23]

Łukasiewicz was one of Twardowski's first students; he obtained his Ph.D. in 1902 and then taught in Lvov until 1915, when he took up a position in Warsaw. Twardowski attracted a large number of doctoral students, some thirty in all by 1918.[24] Notable among this next generation were Stanisław Leśniewski and Tadeusz Kotarbiński, both of whom completed their work in 1912. Not all of the Lvov group continued to pursue philosophy, though the majority did, and they carried the tradition established by Twardowski to other universities – including Warsaw, Cracow, Vilna, and Poznan. In particular, when Łukasiewicz, Leśniewski, and Kotarbiński later moved to Warsaw, the principal axis of the Lvov–Warsaw school was established. What characterized this school was a *way* of doing philosophy rather than a simply identifiable program. The

Kazimierz Twardowski, professor of philosophy at the
University of Lvov – the teacher of Alfred's teachers.

work of this group and its principal figures are elucidated at length by
Jan Woleński in his *Logic and Philosophy in the Lvov–Warsaw School*. As
described succinctly by Francesco Coniglione, "the philosophy practised
by the [Lvov–Warsaw] School was based on Brentano's thesis that 'vera
philosophiae methodus nulla alia nisi scientiae naturalis est' [the method
of true philosophy is nothing other than that of natural science]."[25] In
general, the school shared with natural science an adherence to the classical concept of truth and some form of realism. In addition, with the move
of Łukasiewicz and Leśniewski to Warsaw, logic began to take on a prominent role. Again, as Coniglione puts it, "[this] was to make contemporary
formal logic ... a fundamental tool with which to reform philosophy and
eliminate the semantic misunderstandings and terminological confusion
it was constantly afflicted by."[26]

Course Work and Seminars

Alfred's studies at the University of Warsaw from 1919 to 1924 were devoted almost entirely to the subjects of mathematics and logic. In his first
year, besides taking standard courses on analytic geometry and the dif-
ferential and integral calculus, he attended Sierpiński's lectures on the
theory of sets and the theory of measure as well as Kotarbiński's lectures

on elementary logic and philosophy. From Leśniewski, he took exercises in the theory of sets and in the foundations of mathematics. That year he also took a course in experimental physics and one in sociology. Some of these courses were quite advanced, especially on the theory of sets and the theory of measure (the very subject on which Banach was expounding when "discovered" by Steinhaus). He did not take a course from Łukasiewicz until the following year because between 1918 and 1920 the latter was on leave as minister of education in the new government headed by Paderewski. His position carried the imposing title of Director of the Higher Education Division in the Ministry of Religious Denominations and Public Education.[27]

Repeatedly we see, in Alfred's student record book for the years 1920–1924, the listings "Logika" and "Seminarium" with Łukasiewicz and "Podstawy arytmetyki," "Podstawy geometrji," and "Podstawy logistyki" (foundations of arithmetic, geometry, and logic, respectively) with Leśniewski. Another course with Łukasiewicz was on the theory of relations, and still another was "O wolności i konieczności" [On freedom and necessity]. At the same time Alfred continued to take many courses with Mazurkiewicz on calculus and higher analysis and with Sierpiński on the theory of sets, as well as higher algebra. Beginning in 1921 and extending to 1924, he also attended Kuratowski's lectures on topology. Kuratowski was noted for the clarity and elegance of his expositions; his teaching and work in set-theoretical topology and the theory of functions of a real variable directly and indirectly influenced many students and followers, and Alfred was fortunate to be among the earliest to profit from his lectures. In 1931, Kuratowski and Tarski were to collaborate on a fundamental paper about the notion of definability in the real numbers.[28]

Alfred's studies and personal relations with Leśniewski and Łukasiewicz were of central importance to him. Łukasiewicz was a small, tidy man with a generally distracted air.[29] Preoccupied mainly with abstract thought, he was concerned as much with its beauty as its truth. His writings and lectures were characterized by their extreme precision and rigorous formulation. In 1910, Łukasiewicz had published an important book entitled *The Principle of Contradiction in Aristotle,* which served among other things to bring the new logic then being developed in Germany and England to the attention of scholars in Poland. During his period in Lvov and then in Warsaw until 1918, he considered himself a philosopher with a strong interest in logic, but he frequently criticized the general state of

Jan Łukasiewicz, professor of philosophy and twice rector at the University of Warsaw.

philosophy. After 1918 Łukasiewicz considered himself to be primarily a logician, and he advanced logic as one of the principal tools for doing philosophy in a proper scientific way. He became most famous for his construction of many-valued logics,[30] beginning with the three-valued logic of "true," "false," and "possible." According to this conception, statements about the future are neither true nor false but only possible, and hence they are neither to be accepted nor rejected. Łukasiewicz's idea was that this could be used to refute determinism; one argument for determinism is that it follows from the classical two-valued logic going back to Aristotle, according to which statements are either true or false but not both. Traditionally, the consequence of two-valued logic for determinism was illustrated by the ancient "sea battle" argument given by Aristotle, according to which it is determined today whether or not there will be a sea battle tomorrow, since it is either true or false that there is a sea battle tomorrow. But if it is true today that there is a sea battle tomorrow then that outcome is already determined, and if it is false today that there is a sea battle tomorrow then that eventuality is also already determined. It thus seems, on the basis of this argument, that the future is completely determined in either case! With his replacement of the classical two-valued logic by three-valued logic, Łukasiewicz believed that

Stanisław Leśniewski, professor of the philosophy of mathematics at the University of Warsaw. Tarski's dissertation advisor.

he had freed the philosophical position of determinism from the shackles of Aristotelian logic. Some regarded Łukasiewicz's discovery of this so-called non-Aristotelian logic as comparable in importance to the discovery of non-Euclidean geometry in the nineteenth century.[31]

In physique and personality, Leśniewski was the antithesis of Łukasiewicz. He was a large, strong man, with a stolid, rather impassive face; he favored smoking large pipes and drinking strong coffee from a big pot.[32] Born in Moscow in 1886, the son of a Polish engineer, Leśniewski came to Lvov in 1910 to work on his Ph.D. with Twardowski after having wandered around German universities studying a variety of subjects. Łukasiewicz recalled his first meeting with Leśniewski in Lvov in 1912:

I lived then with my uncle in Chmielowski Street 10. One afternoon someone rang at the entrance door. I opened the door and I saw a young man with a light, sharp beard, a hat with a wide brim and a big black cockade instead of a tie. The young man bowed and asked kindly: "Does Professor Łukasiewicz live here?" I replied that it was so. "Are you Professor Łukasiewicz?" asked the stranger. I replied that it was so. "I am Leśniewski, and I have come to show you the proofs of an article I have written against you." I invited the man into my room. It turned out that Leśniewski was publishing in *Przegląd Filozoficzny* [*Philosophical Review*] an article containing criticism of some views of mine in the "Principle of Contradiction in Aristotle." This criticism was written with such scientific exactness, that I could not find any points which I could take up with him. I remember that when, after hours of discussion, Leśniewski parted from me, I went out as usual to the Kawiarnia Szkocka [the Scottish Café], and I declared to my colleagues waiting there that I would have to give up my logical interests. A firm had sprung up whose competition I was not able to face.[33]

Of course, Łukasiewicz did not give up at all, but this intellectual – and perhaps even physical – intimidation by Leśniewski, the upstart genius, was characteristic. Leśniewski completed his doctoral work with Twardowski in 1912, with a contribution that was one source of the basic language–metalanguage distinction that was to prove so important in Tarski's later work on the definition of truth. After more years of travel and specialized lectures in Poland and Western Europe, Leśniewski returned to Moscow in 1915, where he taught in a Polish school for girls for a few years; during that same period he was an activist in socialist causes and an assistant to Rosa Luxemburg. He returned to Warsaw in 1918 and, after a year in the Ministry of Education with Łukasiewicz, he was appointed professor of the philosophy of mathematics in the faculty of mathematical and natural sciences. From that point on, he and Łukasiewicz represented logic in Warsaw, with markedly different aims, though both were equally dedicated to exceptional rigor and precision in their work. Whereas Łukasiewicz was not bound to any particular point of view or approach within the subject, Leśniewski began to evolve what he regarded as *the one true system* of logic and foundations of mathematics, a project to which he devoted the rest of his life. Unlike Łukasiewicz, he published little and instead made his influence felt primarily through his lectures and personal discussions – the latter with such intensity, absolute conviction, and scorn for others' views that he frequently terrorized his listeners. Kotarbiński once reported to Twardowski that

in general he [Leśniewski] now plays the part of the first violin here and he gives the same impression he gave in the Lvov milieu: people are terribly afraid of him in discussions; at first they do not understand, then they pout, mock him horribly, insult "the formalists" and so on.[34]

Personalities aside, there was a rare environment of mutual respect and appreciation in Warsaw: the mathematicians had a positive attitude toward logic and encouraged their students to attend the lectures of both

Tadeusz Kotarbiński, professor of philosophy at the University of Warsaw. The teacher Tarski revered.

Łukasiewicz and Leśniewski. Though they were no longer philosophers in the traditional sense, neither could they be counted as mathematicians; they devoted themselves totally to logic and the foundations of mathematics. They also continued to have good working relations with the mainline philosophers – especially Kotarbiński, with whom they traded students. The connection ran deeper, since Kotarbiński developed a philosophy of "reism" that granted reality only to concrete objects, whose source lay in Leśniewski's extreme nominalism. Kotarbiński's philosophy took reism as a point of departure for the justification and development of ethical principles relying only on human relationships and not on any theological belief. His leading role in Polish philosophy was long and important, extending well beyond the Second World War. He was revered by many, including Tarski, as a person and as a teacher.[35]

Teitelbaum/Tajtelbaum-Tarski

Brimming with self-confidence, Alfred made extraordinary progress with this stellar group of teachers and researchers, whose high demands and expectations inspired rather than intimidated him. Even the notorious

Leśniewski did not break his stride. A near contemporary, Kazimierz Pasenkiewicz, recalled that only a few students regularly attended Leśniewski's lectures on ontology; Alfred attended irregularly. "He sat in the last row and read newspapers. After a lecture or during intermission he conferred with the professor, but he did not enter into conversations with fellow students."[36] But when two new students, Adolf Lindenbaum and Mordechaj Wajsberg, showed up for Leśniewski's seminars in 1922, Alfred took a greater interest in them and even helped evaluate their presentations. One subject of the seminar was the book by Louis Couturat, *The Algebra of Logic*, which was being subjected to severe criticism. Pasenkiewicz said that "at one point, Tarski stood up and asked whether it is at all worthwhile to busy oneself with [it]. Leśniewski felt a bit slighted but asked with humor, 'Do you think that my seminar is a waste of time?' Tarski sat down."[37]

Already in 1922, a colleague wrote Twardowski about how fearlessly Alfred was taking on Leśniewski full tilt: "*Przegląd Filozoficzny* doesn't have many papers of any real worth in the editorial office. Warsaw coryphaes write little, being afraid of Leśniewski, although the 'scourge of God' has also risen upon him, in the person of his pupil – Tajtelbaum."[38]

In documents from the Central Archives of Warsaw and from the university, the names 'Alfred Teitelbaum' and 'Alfred Tajtelbaum' are used interchangeably throughout, with no apparent rule being applied. 'Teitelbaum' is the German (or "Jewish") spelling and 'Tajtelbaum' the Polish spelling, pronounced in the same way. The latter version was attached to Tarski's first two papers to appear in an international language. In 1923, four years after embarking on his university studies, he obtained surprising results on a problem posed by Leśniewski concerning the latter's fundamental system of logic. These results constituted his doctoral dissertation and were published the same year in Polish in the philosophical review *Przegląd Filozoficzny*. But they were also translated into French forthwith and published in two parts in *Fundamenta Mathematicae*. The first part, "Sur le terme primitif de la logistique" [On the Primitive Term of Logistic] appeared in 1923 under the authorship of Alfred Tajtelbaum; the second part, "Sur les *truth-functions* au sens de MM. Russell et Whitehead" [On Truth Functions in the Sense of Russell and Whitehead] appeared in 1924 under the name Alfred Tajtelbaum-Tarski. These were the last papers to bear his family name; after that, all his papers were published under the name Alfred Tarski because, shortly before receiving his Ph.D. degree in 1924, he officially changed his surname and concomitantly converted to Catholicism.

How did that come about? And why "Tarski"? The second question is simpler, although there are three different answers, two of them offered years later by Tarski himself. One account is that he and his brother Wacław, who by that time had become a lawyer, decided to "Polonize" by abandoning the Jewish-sounding name Teitelbaum (in either spelling) and replacing it with a Polish name. This was legal unless an objection was pressed by any party bearing that name. In order to minimize the chance of this happening, Alfred and his brother searched the Warsaw registry and found the name "Tarski" listed only once. By doing some detective work they discovered its owner was an old woman, not likely to object, they reasoned, and successfully petitioned for the change.[39] Another story, also Alfred's version, is that he made up the name himself as one sounding suitably Polish yet not already existing, so far as he knew. Much to his surprise, when residing in Berkeley some fifty years later, the Post Office mistakenly delivered a letter to him that was addressed to an Alfred Tarski in the neighboring city of Oakland, California. At this point Tarski$_1$ called the Oakland Tarski$_2$ and discovered that the latter was a commercial airline pilot of Polish origin who had always had that name.[40] The third version was told by Tarski's old friend Bronisław Knaster: some time after the Second World War, Knaster met Tarski's niece Anna (Wacław's daughter). In the course of their conversation, he said – proudly – "You know, *I'm* the man who gave Tarski his name!"[41] Perhaps Alfred had consulted Knaster before settling on the choice. Aside from these conflicting stories, one can also speculate that the similarity in the key sounds of "Tarski" and "Twardowski" was considered by Alfred to be an asset; the simplicity of its spelling and relative ease of pronunciation by foreigners was also important to him.

Returning to the more important question: What impelled him to change his name and convert to Christianity? In the first place, it was not such an uncommon thing for Jews to do in Europe since at least the nineteenth century. Most often the change was a career move because in many countries Jews were not allowed to enter the professions. In Alfred's case, his teachers, Łukasiewicz and Leśniewski, strongly encouraged him to change his name if he planned to have an academic position, because there were too many Jewish names on the rosters of educational institutions. Łukasiewicz, in his memoir, recalled his time as minister of education:

The autumn of 1918 had come ... the next issue that concerned me particularly was the stabilization of Warsaw University. The University did not have professors so far, only temporary lecturers.... [T]he Council of Ministries voted to create a commission composed mainly of professors from Cracow and Lvov, which had to elaborate proposals for the appointment of chairs. Stanisław Michalski forced me into not inviting any Jew to join the commission.[42]

Certainly part of taking a new name was to make professional life possible, but there was more to it than that. During the period of renewed nationalism, pride in being Polish and ideologies about assimilation were gaining strength among certain groups of Jews, particularly those who considered themselves part of the intelligentsia. Some saw this as the only rational solution to the "Jewish question," which was considered to be one of the major political problems of the time. Most of the Socialist Party members were also in favor of assimilation, and Tarski's political allegiance was socialist at the time. So, along with its being a practical move, becoming more Polish than Jewish was an ideological statement and was approved by many, though not all, of his colleagues. As to why Tarski, a professed atheist, converted, that just came with the territory and was part of the package: if you were going to be Polish then you had to say you were Catholic.

Doctor of Philosophy Alfredum Tarski

The actual procedure for changing a name required governmental approval and the inevitable bureaucratic steps. It must have been a frantic experience to make everything happen in time for "Tarski" to appear on the announcement of his doctoral examination. Official governmental approval was granted on 19 March 1924, and the change was duly registered at the university two days later, only three days before the examination and granting of the doctoral degree on 24 March 1924. Perhaps there was help from his mentors. The diploma, headed Summis Auspiciis Serenissimae Rei Publicae Polonorum, declares ALFREDUM TARSKI to be a Doctor of Philosophy for his work "O wyrazie pierwotnym logistyki" [On the Primitive Term of Logistic], having passed an examination in mathematics, Polish philology, and philosophy. The listed officials for this document are Ignacy Koschembahr-Łyskowski (Rector), Wiktor Lampe (Decan), and Stanislaus Leśniewski (Promotor).[43]

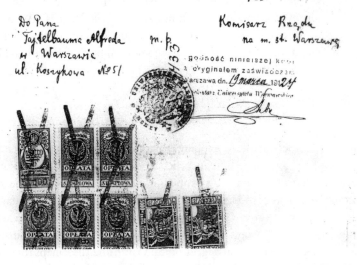

Document for name change from Alfred Tajtelbaum to Alfred Tarski, 19 March 1924.

In the following years, Leśniewski went around saying: "All of my students are geniuses," though everyone knew that Tarski was his one and only Ph.D. student. Also, everyone knew that he was the only one who had the privilege of being invited to Leśniewski's house. They met once a week, on Thursdays, for a private discussion.[44] Yet the dedication in Tarski's 1956 collection of fundamental papers, *Logic, Semantics, Metamathematics*, reads:[45]

> To his teacher
> TADEUSZ KOTARBIŃSKI
> The author

What a retrospective slap in the face to Stanisław Leśniewski, his real teacher! Several things explain Tarski's disaffection.

Almost immediately following his dissertation, Tarski turned to Sierpiński-style set theory as his primary area of research and produced a remarkable number of new results, including the stunning paradoxical decomposition of the sphere in his 1924 paper with Stefan Banach. When he returned to logic, instead of concentrating on Leśniewski's "one true system" he worked with Łukasiewicz on a variety of many-valued systems. Moving beyond that, he adopted the new metamathematical approach that had been gaining ascendancy, under which *any* system of logic could be a candidate as a legitimate object of study. Though Tarski was punctilious in his papers into the mid-1930s about crediting Leśniewski for various specific contributions, the methodological gap between them widened almost from the beginning. Leśniewski himself contributed to that separation in a 1927 article in which he scorned the "happy-go-lucky" set theory practiced by Sierpiński and others. When Leśniewski published the first part of his new system for the foundations of mathematics in *Fundamenta Mathematicae* two years later, Sierpiński responded in kind. Things got so out of hand in their dispute that Leśniewski withdrew as an editor of *Fundamenta* and got Łukasiewicz to do the same, rending the once excellent relationship between the logicians and mathematicians in Warsaw.[46] Tarski took no public stance on this, but in spirit and in his own work he clearly sided with the mathematicians. Even so, the Professors "L" had stamped him indelibly with an unending concern for clarity, precision, and rigor, providing the ground for his fundamental view of logic as the essential cornerstone for the methodology of the deductive sciences and, even more widely, for the sciences in general.

For Tarski, more traumatic than their intellectual separation was the breakdown of personal relations between him and both Leśniewski and Łukasiewicz. This was due to their increasing anti-Semitism, which reached such a point in the mid-1930s that they would not invite him to sit at *their* table in their usual café. Understandably, Tarski was hurt and incensed. How ironic this was, since it was they who had advised him to assimilate and change his name. Kotarbiński, on the other hand, held his place as a hero for his vocal and active opposition to anti-Semitism and for his general humanity.[47]

Years later, when Tarski's doctoral students in Berkeley asked him who *his* teacher was, he replied, "Kotarbiński," whose photo had a privileged position on his desk. Leśniewski's name was never mentioned. Following his death in 1939, Leśniewski disappeared from the logical scene except for a few scattered disciples, although later there was renewed interest in his work.[48] Kotarbiński, on the other hand, survived the war to become the leader of resurgent Polish philosophy. Universally admired, who better to claim as one's *Doktor-Vater* [doctor-father]?

INTERLUDE I

The Banach–Tarski Paradox, Set Theory, and the Axiom of Choice

IN 1924, Stefan Banach and Alfred Tarski published an article in *Fundamenta Mathematicae* entitled "Sur la decomposition des ensembles de points en parties respectivement congruents" [On the Decomposition of Sets of Points into Respectively Congruent Parts]. For Tarski, it was the third paper on the theory of sets to appear in the remarkably productive year directly following his Ph.D.

One of the amazing results in the Banach–Tarski paper implies that a solid ball of any size can be cut up – in theory – into a finite number of pieces that can be reassembled to make a ball of any other size. In other words, "a pea can be cut up to make the sun"! Since this flatly contradicts normal intuition, their result came to be known as the Banach–Tarski Paradox.[1] But unlike other paradoxes that entail actual contradictions, such as the Liar Paradox or Russell's Paradox, the Banach–Tarski Paradox is perfectly consistent with the generally accepted assumptions of modern mathematics – namely, the axioms for set theory. These axioms were first formulated by Ernst Zermelo in 1908 in order to codify and put on a firm basis the groundbreaking but controversial work of Georg Cantor in the last decades of the nineteenth century that dealt with newly discovered relationships between infinite sets. All but one of Zermelo's axioms assert the existence of sets given by explicit defining conditions. The exception is the so-called Axiom of Choice, a pure existence statement without defining conditions. That axiom was necessary to provide a foundation for Cantor's theory of transfinite cardinal and ordinal numbers, but instead of bolstering confidence in such numbers, its blatant nonconstructive character engendered further heated controversy. As it turned out, the Axiom of Choice was essential to the proof of the

Banach–Tarski Paradox. In 1938, Kurt Gödel proved that the Axiom of Choice is consistent with the other axioms of set theory, thus showing that the Banach–Tarski Paradox does not lead to a contradiction provided the axioms of set theory are consistent.

Puzzles about infinite sets go back to antiquity, when the first paradoxes were articulated. Zeno's paradoxes were designed to show that motion is impossible – most vividly in the paradox of the race between Achilles and the tortoise, wherein Zeno purported to prove that Achilles could never catch up with the tortoise if the latter were given a head start. For, by the time Achilles has reached the tortoise's starting point, the tortoise has moved ahead to a new point, and by the time Achilles has reached *that* point, the tortoise has moved on to still a further point, and so on. In modern terms, the paradox is resolved by saying that the described sequence of successive points converges to a point p at which stage Achilles has caught up with and then can pass the tortoise. The problem here is related to the distinction that Aristotle made between the *potential infinite* and the *actual infinite*. For example, the *sequence* of whole numbers $1, 2, 3, \ldots$ is considered to be potential or unending, while the *set* of all these numbers, indicated by $\{1, 2, 3, \ldots\}$ is thought of as being actual or completed. Aristotle rejected any use of notions involving the actual infinite, but the modern resolution of the paradox of Achilles and the tortoise requires conceiving of the sequence of successive points described above as one that can be completed in the sense of convergence to a point.

The Aristotelian horror of the actual infinite dominated mathematical thought for two millennia. Even Galileo, who had freed himself from many Aristotelian dicta, remained suspicious of the actual infinite. In his famous book *Dialogues Concerning Two New Sciences* it is pointed out in the discussions between Salviati, Sagredo, and Simplicio that there are just as many squares of numbers $1, 4, 9, \ldots$ as there are numbers $1, 2, 3, \ldots$, while at the same time, paradoxically, there are fewer of the former than of the latter. Similarly, from one perspective there are just as many points on a line segment one unit long as there are on a segment that is two units long. For the latter segment can be considered as the base of an equilateral triangle in which the former segment may be drawn parallel to the base midway between it and the vertex of the triangle; every line drawn from the vertex to a point of the base passes through a point of the shorter segment, and these points are in exact correspondence. (This is indeed a matter of perspective in the geometrical sense.) On the other hand, from the usual point of view it is clear that there are fewer points in the shorter segment than in the longer one. Salviati in Galileo's dialogue concluded that it is inadmissible to apply the notions "more than" and "fewer than" to infinite sets.

Cantor Tames the Actual Infinite (Almost)

It was only with Georg Cantor three centuries later that the distinctions needed to resolve the Galilean paradoxes were made precise by some simple definitions concerning any two sets A and B; these provide two senses of what it means for A to have fewer elements than B. First of all, A is said to have *the same number of elements* as B if there exists a *one-to-one correspondence* between the elements of A and those of B; in this case we also say that A and B have the *same cardinal number*. Then, following Cantor, A is said to have *fewer elements* than B if it has a *smaller cardinal number* than B, that is, if it has the same number of elements as a subset of B but not as B itself. In the special case that A is a subset of B, the other sense of "fewer than" is that A is a *proper subset* of B. In this second sense, the set $A = \{1, 4, 9, \ldots\}$ of squares of numbers has fewer elements than the set $B = \{1, 2, 3, \ldots\}$ because it is a proper subset of B, but it does *not* have fewer elements in the first sense of having a smaller cardinal number. Similarly, a straight line segment $A = [0, 1]$ from 0 to 1 (in any unit of measurement) has fewer elements than the segment $B = [0, 2]$ from 0 to 2 when considered as a subset, but these sets have the same number of elements in the cardinal number sense because they can be put in one-to-one correspondence.

Though all infinite sets are lumped together in a naive view of the infinite, one of Cantor's first surprising results was that, once the idea of the actual infinite is accepted, one must distinguish *different sizes of infinity*. Namely, he showed that the set of positive integers $\{1, 2, 3, \ldots\}$ has a smaller cardinal number than the set of points on a straight line or any line segment such as $[0, 1]$. Cantor's result is expressed in terms of notions of countability and uncountability as follows. First of all, the set A is said to be *finite* if it is in one-to-one correspondence with $\{1, 2, \ldots, n\}$ for some positive integer n; and A is *infinite* if it is not finite. A set A is said to be *countable* if A is either finite or has the same cardinal number as the set of positive integers, and *uncountable* if it is infinite but not countable. So what Cantor proved is that the set of points on the line segment

[0, 1] is uncountable. More generally he proved that, for any infinite set A, there is an uncountable set B whose cardinal number is greater than that of A.

What Cantor could *not* prove is the so-called Trichotomy Principle, a seemingly obvious statement according to which, for any two cardinal numbers, either they are the same or one is smaller than the other. In other words, for any two sets A and B, either A has a one-to-one correspondence with B or has one with a subset of B – or vice versa. Cantor thought this was intuitively true by simultaneously "running through" the elements of the sets A and B one by one, matching up elements along the way, until one or the other (or each) of the sets is used up. But this would require a succession of choices of elements from A and B alternately, even where no rule is known by which such choices would be made and even where finitely many or countably many choices are insufficient to exhaust the sets. To explain this idea more precisely, Cantor introduced the notion of a well-ordering relation, by which is meant an ordering relation such that, beyond any proper initial segment of the ordering, there is always a first element. Cantor also asserted the Well-Ordering Hypothesis, according to which – no matter what set is taken – there is always *some* way to impose a well-ordering relation on it. The positive integers with their usual ordering form a well-ordered set, while by contrast, the points on the uncountable line segment [0, 1] are not well-ordered in the usual ordering. For example, there is no first element beyond the initial segment that goes from 0 up to and including 1/2. Nevertheless, the Well-Ordering Hypothesis assures us that there is *some other way* of imposing a well-ordering relation on the points of a line. Although there is no known way of defining such an ordering, Cantor believed this hypothesis to be so intuitively clear that he called it a law of thought; at other times he tried to prove it from more primitive principles, but without success.

The Axiom of Choice

Zermelo offered the sought-for reduction to a more primitive principle when he proved that the Well-Ordering Hypothesis follows from what he called the *Axiom der Auswahl* [Axiom of Choice], a principle he considered to be both intuitively clearer and evidently true. This asserts that if $S = \{A, B, \ldots\}$ is any set of non-empty sets $A, B, \ldots$ that (pairwise) have no elements in common, then there is a set X that "simultaneously chooses" exactly one element from each of the members of S. In other words, for each A in S, the intersection of X with A in S consists of a single element x_A. This is a *purely existential* claim; when S is infinite and the sets in S are not well-ordered in any specified way, there is in general no way to *define* the choice of x_A in A for each A in S. The Axiom of Choice seems to be clearly true from one ("Platonistic") point of view regarding the existence of sets, independently of how they may be defined, but it is rejected by those who think that an existential claim is justified only if one can show how to define or pick out the object claimed to exist. Because of that fundamental difference in basic point of view, the Axiom of Choice was immediately controversial, and even the correctness of Zermelo's proof that it implies the Well-Ordering Hypothesis was challenged. To make clear that there were no errors in his proof of the implication, in a second publication Zermelo made explicit, in the form of axioms, all the basic principles concerning sets together with the Axiom of Choice that he needed to carry out the proof. Thus was born axiomatic set theory. Zermelo's axioms were later extended by Abraham Fraenkel and others in order to further strengthen the system in a natural way, leading to what are now called the Zermelo–Fraenkel axioms for set theory. These constitute the assumptions generally accepted as underlying modern mathematics. In response to the objections raised against the Axiom of Choice, Zermelo had made a number of informal arguments for why all his axioms should be accepted; he also pointed out that the Axiom of Choice is implicitly used in much of standard mathematics. Nevertheless, a number of mathematicians of his day, including the leading French mathematicians Emile Borel and Henri Lebesgue, refused to accept it; they objected on the grounds that existence results had to be supported by explicit definitions, even though their own mathematical work made implicit use of the Axiom of Choice in one form or another.[2]

Tarski's Engagement in Set Theory

For those mathematicians who were attracted by the new world of ideas and theorems concerning finite and infinite sets opened up by Cantor and legitimized to some extent by Zermelo, there was an immense flowering of the new subject of set theory, with many open problems and challenges. Before long there were textbooks; one of the first with a wide readership and impact was Felix Hausdorff's book on the foundations of set theory

(*Grundzüge der Mengenlehre*), published in 1914. Hausdorff accepted the Axiom of Choice unhesitatingly, and one of the first outstanding uses he made of it, in that same year, was in proving a theorem that was the direct precursor to the Banach–Tarski Paradox. In Poland, Wacław Sierpiński was an early expositor of the burgeoning theory of sets and lectured on it regularly at the University of Warsaw from 1916 onward. Seeking to capitalize on the new directions in mathematics, the editors of *Fundamenta Mathematicae* promoted set theory along with logic and topology as one of its main areas of concentration. Even among those who did not accept the Axiom of Choice, its role in various parts of pure and applied set theory was considered to be of special interest. For this reason, in 1918 Sierpiński assembled a survey of statements proved (by him or others) to be implied by or equivalent to the Axiom of Choice. For example, the Trichotomy Principle turned out to be equivalent to it, and the statement that "any countable union of countable sets is countable" could not be demonstrated in any obvious way without some use of the Axiom of Choice. Also, surprisingly, the equivalence between two alternative definitions of what it means for a set to be finite seemed to require an essential use of this proposed axiom.

As a student in Warsaw, Tarski was introduced to the subject of set theory through attendance at Sierpiński's lectures and his reading of Hausdorff's text. (Incidentally, he recommended that classic text to his own students ever after.) Tarski was immediately intrigued by the many interesting results and open problems in the theory of sets. His first article, written in 1921 while he was still a student, was on the axiomatics of well-ordered sets. It was probably this article that marked him as a "comer."

As soon as Tarski completed his dissertation on a purely logical question raised by his Ph.D. supervisor, Stanisław Leśniewski, he returned to set theory as his main area of research; it clearly offered much greater scope for an ambitious and talented young mathematician than Leśniewski's very specialized systems. Three articles that he published on set theory in 1924 quickly boosted him to the forefront of researchers. The first concerned several theorems equivalent to the Axiom of Choice, among them the surprising one that it is equivalent to Cantor's theorem that $m^2 = m$ for all infinite cardinal numbers m. In his second paper Tarski identified five distinct notions of finiteness, increasing in logical strength (including the two previously studied), whose equivalence apparently could not be proved without some use of the Axiom of Choice. The third article in set theory that Tarski published that same year was his remarkable one with

Banach on paradoxical decompositions of sets of points in Euclidean space of various dimensions. Banach, at the University of Lvov in Poland and ten years senior to Tarski, was then already world-famous for his work in mathematical analysis. He and Tarski were both attracted to the problem of measure in set theory and independently arrived at their surprising result; they then decided to collaborate on a joint paper.

The Measure Problem

The origins of this problem lay in geometry, in the computations of areas and volumes of various figures. A traditional way of computing the area of a figure in the plane is to cut it up into a finite number of pieces and reassemble the pieces so that they form a rectangle, whose area is equal to its base times its height. The simplest case is that of a parallelogram, which is shown to have the same area as a rectangle of the same base and height by lopping off a triangle at one end and moving it to the other end. The areas of many polygons such as triangles, quadrilaterals, pentagons, and so on were determined in this way by the Greeks. It was proved in the nineteenth century by a general argument that any polygon can be cut up into a finite number of polygonal pieces that can be rearranged to form a rectangle, and even then rearranged to form a square. But this method does not work for determining the areas of curved figures such as the area bounded by a circle (which is why "squaring the circle" frustrated the Greeks). Still, one might try to cut it up into other kinds of figures and reassemble them into a square. The problem for volumes is correspondingly more complicated, since it turned out that the volume of a polyhedron (many-faceted figure) cannot in general be determined by dissection into polyhedra that are reassembled as a cube.

When mathematicians in the late nineteenth century began to think of arbitrary sets of points beyond traditional geometric ones, they asked to which bounded sets in the plane is it possible to assign numbers as areas by cutting them up and reassembling them as a square, and similarly which bounded sets in 3-dimensional space can be cut up and reassembled as a cube, and so on. This led to the *measure problem* for n-dimensional space, for any $n = 1, 2, 3$ and higher, as follows:

> Is it possible to assign a number as measure to each bounded subset A of n-dimensional space in such a way that (i) if A is cut up into a finite number of disjoint subsets that are reassembled to form a set B then

A and B have the same measure, (ii) the measure of the union of two disjoint sets is the sum of the measures, and (iii) the measure of the n-dimensional unit cube equals 1?

By definition, the n-dimensional unit "cube" consists of all n-tuples of numbers between 0 and 1, so for $n = 1$ it is just the interval $[0, 1]$ of length 1, for $n = 2$ it is the square $[0, 1]^2$ of area 1, and for $n = 3$ it is the cube (in the usual sense) $[0, 1]^3$ of volume 1. Thus, depending on the dimension, a solution of the measure problem would generalize length, area, and volume to arbitrary sets.

Using the Axiom of Choice in an essential way, Hausdorff had shown in 1914 that the measure problem has a *negative* answer in every dimension n greater than or equal to 3; that is, there is no way to satisfy requirements (i)–(iii) for such n. In dimension 3 he established this by a paradoxical-looking decomposition of the surface of a sphere. But Hausdorff's work left open the problem for $n = 1$ and $n = 2$. Then Banach showed in 1923, using the Axiom of Choice, that it *is* possible in both these cases to solve the measure problem positively. Thus the status of the measure problem was completely known for all dimensions: positive for dimensions 1 and 2 and negative for all higher dimensions.

What Banach and Tarski added to this picture in their joint work of 1924 was to transform Hausdorff's Paradox into an even more blatantly paradoxical result in every dimension n greater than or equal to 3. In particular, consider any two *balls* A and B in three dimensions, that is, sets consisting of all points on or interior to a sphere, and suppose they have different volumes. Using the Axiom of Choice in an apparently essential way, Banach and Tarski showed that A can be cut up into a finite number of pairwise disjoint subsets that can be reassembled to form B. This clearly implies the negative solution of the measure problem in dimension 3, for A and B would have to have the same measure, which in that dimension would mean that they have the same volume. Banach and Tarski also obtained similar paradoxical decompositions in every dimension n greater than 3. In the introduction to their paper, they pointed to the role of the Axiom of Choice in their proofs, which – with great understatement – they say "seems to merit attention." Since their result clashes violently with ordinary geometrical and physical intuition, it focused further attention on the role of that principle and on whether its use might not even result in outright contradictions.

Why the Banach–Tarski Paradox Is Not Inconsistent with the Axioms of Set Theory

The logical relationships between the statements dealt with in the Banach–Tarski Paradox were not to be settled until many years later. In the interim, the axiomatic development of the subject at the hands of Fraenkel and others reached a more settled form, and one could tackle questions of consistency and independence of one axiom or another in relation to the remaining ones in a definite way. This constituted a basic shift in viewpoint from a more informal, *mathematical* one to a *metamathematical* one. Kurt Gödel's proof in 1938 of the consistency of the Axiom of Choice with the other axioms of set theory was the first striking result from the new viewpoint.[3] According to his theorem, if the Zermelo–Fraenkel axioms are consistent without the Axiom of Choice then they remain consistent when that axiom is added. This is a *relative* rather than *absolute* proof of consistency, since it makes use of the *assumption* that the Zermelo–Fraenkel system is consistent; no absolute proof of consistency of this system is known.[4] But at least this was of some comfort to those mathematicians who had no problems about accepting the axioms of Zermelo–Fraenkel on intuitive grounds yet had worries about the validity and consequences of the Axiom of Choice. In particular, since the Banach–Tarski "Paradox" is a consequence of the Axiom of Choice and the other axioms, it is consistent to assume that statement even though no way of explicitly defining the seemingly paradoxical decompositions can be provided.

Much later, in 1963, Paul Cohen proved by means of his new method of "forcing" that the Axiom of Choice is independent of the Zermelo–Fraenkel axioms, again assuming the latter system is consistent.[5] An extension by Robert Solovay of Cohen's forcing method allowed him to show that, if the Zermelo–Fraenkel axioms are consistent in a slightly strengthened form, then these axioms by themselves cannot prove the Banach–Tarski theorem.[6] In other words, Solovay's theorem finally showed that the Axiom of Choice is indeed essential in the derivation of the Banach–Tarski Paradox.

Tarski and the Promised Land

Writing of Tarski's work in set theory a few years after his death, Azriel Levy pointed out that Tarski's role in its development was "similar to

[that] of Moses showing his people the way to the Promised Land and leading them along the way, while the actual entry [into] the Promised Land was done by the next generation." For the next forty years after his work with Banach, Tarski made many more signal contributions to set theory. But for the really striking new developments that others began to make in the 1960s, his importance lay in providing, in Levy's words, "a source of energy and inspiration to his pupils and collaborators ... always confronting them with new problems and pushing them to gain new ground."[7]

Despite Tarski's central role in bringing this subject to its present state and the fact that almost all his work in logic and algebra is imbued with the set-theoretical spirit, there was a strange ambivalence in his attitude toward it philosophically because he did not accept the usual "Platonistic" understanding of the axioms of set theory, according to which "arbitrary" sets exist independently of any human definitions or constructions. Although Tarski was chary of expressing his philosophical views in print, in extemporaneous remarks made at the closing of his seventieth birthday symposium he said: "I am a nominalist. This is a very deep conviction of mine. It is so deep, indeed, that even after my third reincarnation, I will still be a nominalist." He went on, "People have asked me, 'How can you, a nominalist, do work in set theory and logic, which are theories about things you do not believe in?'... I believe that there is value even in fairy tales and the study of fairy tales."[8] On another occasion,[9] Tarski described himself as

an extreme anti-Platonist .... However, I represent this very crude, naive kind of anti-Platonism, one thing which I could describe as materialism, or nominalism with some materialistic taint, and it is very difficult for a man to live his whole life with this philosophical attitude, especially if he is a mathematician, especially if for some reason he has a hobby which is called set theory.[10]

In a related vein, Tarski once described himself to a philosophy graduate student as a "tortured nominalist."[11] Perhaps the explanation for his seemingly schizophrenic attitude is that set theory is generally viewed as necessary for both the foundation and the practice of twentieth-century mathematics, and Tarski was, *qua* mathematician, not to be held back by older views of what is admissible to mathematics. Quite the contrary, in that respect he was entirely a man of his time and place.

3
Polot! The Polish Attribute

Work

FROM THE DAY he was born and for all of his student years, Tarski lived with his family in the large apartment on Koszykowa Street. After he was awarded his Ph.D., he continued to live there for five more years, until his marriage in 1929. The inevitable stress and strain that arose from living at home was mitigated by the comfort and convenience. The flat was centrally located, and he could walk or take a tram to the university or to his favorite cafés, cabarets, and restaurants to meet friends and associates. There was a telephone in the apartment; meals were prepared for him by the household help; and no doubt his imperious, fastidious mother, with her textile-magnate heritage, saw to it that he had the right outfit to wear for every occasion. In short, all the small practical matters of life were taken care of, so Tarski never did learn how to cope with the everyday. But even with all the necessities provided for, a young man, especially a bon vivant like Alfred, had expenses, and until he got a job he had to turn to his father for support. This can be uncomfortable under any circumstances, and when Alfred abandoned the name "Teitelbaum" for "Tarski" the tension was exacerbated. Alfred's decision that Polish patriotism was more important than Jewish identity did not sit well with his parents, for whom assimilation did not change the principle that they were Jewish first and Polish second. On at least one occasion when Alfred asked his father for money, Ignacy Teitelbaum was said to have replied, "Money? You need money? Well, why don't you go see your old man Tarski?"

One only has to remember Alfred's characterization of his father as the soft one, the affectionate one, to feel the injury in that response. To salvage his self-respect, Alfred needed an income of his own; but more than that, money was scarce and it was a sore spot in the household. His father's business was not doing well and his mother's complaints about her lowered standard of living were constant.

As soon as he finished his doctorate, Tarski was appointed to the position of docent; at twenty-three, he had the honor of being the youngest docent at Warsaw University at that time. This gave him the right to lecture and hold seminars and, most importantly, to be recognized as a member of the academic community at the university. The problem was that the salary was a pittance and certainly not enough to live on. Had the Teitelbaums been a wealthy family, Tarski might have been content to let his family support him until eventually – one hoped – a position as professor would be offered him. Such a situation was not so unusual for sons of the intellectual elite, but clearly Tarski's family was not sufficiently well-off and, in any case, he was too much his own man to be happy in a dependent position.

Following the pattern of scholars of that era who did not have personal means, Alfred went looking for a position in a high school and found one at the Polish Pedagogical Institute of Warsaw, a school for young women training for the teaching profession. He taught there for two years until one day the headmistress called him into her office and told him she was sorry but she had to dismiss him because there were complaints about the mathematics teacher being Jewish. In later years, Tarski's caustic comment was: "Those were ordinary middle-class girls. If they had been aristocrats, they wouldn't have complained." As it turned out, losing that job was a blessing because he immediately got a better one; but the reason he was fired reflected the anti-Semitism directed toward him personally and toward Jews in general. It was an early example of the fact that, in spite of his "Polish" name and his self-identification as Polish first, everyone in Warsaw thought of Tarski as a Jew and no one forgot that his name had been Teitelbaum. That rankled him no end.

Among his Jewish colleagues, too, many disapproved of his conversion because they considered it a betrayal and a denial of his origins. One person who took a hard view was Samuel Eilenberg, who was to become a distinguished mathematician and professor at Columbia University. In his first year at Warsaw University, Eilenberg had taken a logic course from the docent, Alfred Tarski. He is one of the few who claims to have been "only moderately impressed" by Tarski's lectures. In Eilenberg's frankly hostile opinion, Tarski should neither have changed his name nor – even worse – become Catholic:

He did it so that life would be easier and he did it with great zeal. People poked fun at him because suddenly he was interested in all kinds of things which were strictly not in the Teitelbaum repertoire, if you know what I mean. Things like Catholic liturgy and the rites associated with it. He wasn't very popular and not just because he switched. Others did it; in fact, his colleague Bronisław Knaster did it, and nobody poked fun at *him*. It's just that Tarski acted so pious. People told stories about him and treated him as a comical figure. He, himself, had absolutely no sense of humor.[1]

But Eilenberg conceded that it would have been difficult if not impossible to get a university position if one had a Jewish name and that Tarski would not even have gotten a docent's position had his name remained Teitelbaum. How much Tarski actually knew about the heavy disapproval of other Jews is not clear, but he was teased openly, in a lighter vein, by his friend Knaster, who gave himself license because he too had converted. When Tarski, eager to display his erudition, discussed the details and fine points of the Catholic religion and its hierarchy, Knaster – a physical opposite of Tarski, tall, slim, dark-eyed, and possessed of a flashy wit – would say, "Alfred, I am sure that one of these days you will write a book about all of this." To which Alfred would reply: "Look. Don't be silly. It's just a hobby." Years later when Tarski, by then long in the United States, published his book *Cardinal Algebras,* he received a telegram from Knaster, still in Poland, that read, "Congratulations! You see, I was right, after all."

Despite the general appreciation for his reasons, the stigma attached to the switch was never fully erased. The issue of Tarski's "Jewishness" came up over and over again in diverse ways: privately in expressions of surprise – "What? I never knew Tarski was Jewish!" – or more rarely, publicly, at a conference when someone would ask, "Whatever became of that logician Tajtelbaum with whom you wrote a paper in the 1920s?" Others in the audience could only guess whether the interlocutor was innocent or malicious. The vein in Tarski's forehead became more prominent as he explained. "*I* am Tajtelbaum."

Forced from his job at the Pedagogical Institute, Tarski found a much better situation at a private *gymnasium* named in honor of the great Polish writer Stefan Żeromski (1864–1925), who was known as "the conscience of Polish Literature" because his novels detailed the grim realities of poverty and war.[2] The Żeromski Lycée, on Marszałkowska Street in Tarski's own neighborhood, had high academic standards and a liberal-minded faculty, many of whom were assimilated Jews and socialists. The director, Teofil Wojeński, was the author of a book about Żeromski and an active member of the Polish Socialist Party. Tarski immediately found the atmosphere intellectually, politically, and culturally congenial, and later there were to be further romantic and practical benefits.[3]

The students were good and the teachers serious, none more so than Tarski. A letter from the philosopher Karol Martel, written sixty-five years after his student days, gives early evidence of Tarski's style:

> There was something very special in his role as a gymnasium teacher. Later on I understood that it was the elegance of his argument; the logical consistency of his lecturing; even as a schoolteacher, he was in fact an academic lecturer.
>
> Most impressive is the image I have of Tarski in our classroom at the blackboard, teaching us geometry: first, the statement of the theorem and then proceeding step by step with its proof. He always ended the demonstration with the formal *c.b.d.d.* (the abbreviation in Polish of "which was to be proved") or the Latin *q.e.d.* (quod erat demonstrandum).
>
> This purely logical way of teaching was totally unknown to us pupils – we were probably not mature enough to understand it – but it was in fact our first meeting with the principles of scientific thinking in deductive sciences.[4]

While Tarski was offering his bright lycéens at Żeromski's an elegant introduction to scientific thinking, at Warsaw University he was taking every opportunity to deliver more advanced lectures and to lead and participate in seminars. He also enrolled in courses on physics and astronomy as a postdoctoral student; but above all he was thinking and writing about logic and mathematics, conceiving new ways to look at problems, and proving new theorems. He continued to interact first with scholars in Poland and later, through the Vienna Circle and the Unity of Science movement, with the larger European community of philosophers, mathematicians, physicists, and social scientists.

In the fifteen years following his Ph.D. (1924–1939), Tarski wrote more than fifty articles, mostly in French or German, and in Polish he published his great monograph on the concept of truth in formalized languages as well as the first version of his introductory text on logic and a geometry textbook (with Zbigniew Chwiałkowski and Władysław Schayer) for high-school students. He bitterly resented the burden of working two jobs and constantly complained that he never had enough time to develop all the ideas he had.

How on earth did he accomplish as much as he did and still have time for an active social life, too? He was limitlessly energetic, enthusiastic, organized, aggressive, and competitive, and he had an iron constitution.

Play

Tarski may have been small, but he was sturdy and healthy. Almost until his death he could outtalk, outargue, outdrink, and outlast anyone who tried to stay awake with him into the wee hours. His intellectual fearlessness, the sense that nothing was beyond his grasp, was coupled with physical boldness – surprising because he was neither athletic nor graceful; rather, he had a nervous, jittery way of moving and always seemed to be backing into things. Yet he became a dedicated hiker and even did some mountain climbing.

Like many Poles, Tarski regularly took his summer vacations in the Tatra mountains in the southeastern corner of Poland. These high mountains on the border of Czechoslovakia became the standard against which he judged all others; in his eyes, none ever quite equalled their beauty, although the Trinity Alps in California came close. Beginning in 1923 he kept a journal wherein, with characteristic precision, he systematically chronicled his hikes to the various peaks in the Tatras and other ranges – noting the date, the beginning and the end of the trail, the destination peak, and the other occasions upon which he had made that excursion.

Usually his home base was the charming alpine town of Zakopane, a well-known resort for climbers and vacationers, a center for art and theatre and music, and a meeting place for intellectuals. Attracted by the clear, dry mountain air and the handsome alpine architecture, several of Poland's outstanding musicians, painters, writers, and thinkers had settled in Zakopane in the early 1900s and created an avant-garde community. Leon Chwistek, the painter, philosopher, and mathematician whom Tarski came to know very well, was born there in 1884, and Stanisław Ignacy Witkiewicz – another painter, writer, and amateur philosopher – had come as a child. The father of Witkiewicz, also an artist and an architect, was active in the cultural life of the community. Others of note were the anthropologist Bronisław Malinowski and the composer Karol

The Witkiewicz house in Zakopane.

Szymanowski. These people and their free-wheeling, free-loving, anti-bourgeois modus vivendi made the town doubly attractive to Tarski.

In his memoir *A Year of the Hunter*, the Nobel Prize–winning poet Czesław Miłosz describes Zakopane as having become "the artistic capital" of Poland even before World War I. Adding more spice to the picture, he writes:

> My generation did not experience the orgies that took place in Zakopane in the 1920s and the ... thirties .... Judging by what I have read about artistic-literary London during more or less the same period, about Virginia Woolf for example, erotic freedom was very advanced there. In Zakopane they went in for heavy drinking with the highlanders to the accompaniment of highland music; there were also plenty of love affairs, both homosexual and heterosexual. Apparently for the participants it was a time of Dionysian intoxication with no presentiment of dread.[5]

It's ironic, but not without precedent, that all this "debauchery" flourished in the place that the older generation had, in part, chosen for its clean air and salubrious qualities.

Ignacy Witkiewicz

Most flamboyant, the wildly eccentric, manic-depressive Ignacy "Witkacy" Witkiewicz was the *enfant terrible* of the Polish intellectual avant-garde. As a child prodigy, encouraged and tutored by his painter-architect

Stanisław Ignacy Witkiewicz ("Witkacy"), avant-garde painter, playwright, and "philosopher."

father, he wrote plays, poetry, and autobiographical novellas, designed theatre sets and costumes, and participated in philosophy seminars at Warsaw University and even wrote papers and monographs on philosophy. His magnum opus, *Concepts and Principles Implied by the Concept of Existence*, published in 1935 in an edition of 600 copies, was said by its author to have sold only twelve.[6] He had notoriously tormented love affairs; a fiancée committed suicide; his subsequent "experiment" in marriage to a woman of aristocratic origins, founded on "tolerance and friendship and unconditional freedom," was by agreement not monogamous yet the partners were extremely jealous of each other's open love affairs – especially since they confided the details to each other. All of this became material for his sensational plays and novels.

Witkacy also experimented with a variety of drugs and wrote a monograph entitled *Narcotics: Nicotine, Alcohol, Cocaine, Peyote, Morphine and Ether*, a treatise on addiction in which he described the visions produced by these drugs and the creative spurts that resulted; at the same time he warned the reader about the disastrous effects of all forms of addiction. Fascinated by the role drugs played as a catalyst for art, he admitted that he was unable to shake his habit even while he saw its negative power.[7]

To make a living, Witkacy painted portraits, since his other works produced little real income. Not entirely tongue-in-cheek, he established "The S. I. Witkiewicz Portrait-Painting Firm" with the motto "The customer must be satisfied. Misunderstandings are ruled out," and he published a long list of rules so as "to spare the firm the necessity of repeating the same thing over and over again." Rule #1, for example, spelled out the types of portraits the "firm," meaning Witkacy, would do – from conventional (A), to more emphasis on character "bordering on the caricatural" (B), to type (C) portraits executed "with the aid of C_2H_5OH and narcotics of a superior grade." Witkacy did many type (C) portraits, including one of Tarski, in 1934, and another of Maria, Tarski's wife, in 1938. Rule #2 stated that the customer had the right to reject a portrait but had no right to demand that it be destroyed. Rule #3: "Any sort of criticism by the customer is absolutely ruled out. The customer may not like the portrait but the firm cannot permit even the most discreet comments .... If the firm had allowed itself the luxury of listening to customer's opinions it would have gone mad a long time ago." Rule #4 is a corollary to #3: "Asking the firm for its own opinion of a finished portrait is not permissible nor is any discussion about a work in progress." The list goes on in this vein ending with various statements about how to pay, and the benefits of acting as an agent for the firm by bringing in new clients.[8]

Witkacy's lightly surreal but romantic pastel and charcoal rendering of Alfred Tarski – with sensual lips, clear blue eyes, protruding forehead, and purple feathers brushing his cheeks – seems intended to reveal a troubled intensity, but it somehow misses his full physical fire and intelligence. One feels that even a silent Tarski obeying the "rules of the firm" would have burst through the canvas. By contrast, in the beautiful portrait of Maria – with an aureole of yellow behind her dark hair, dark eyes, pouting mouth, with a heart and snake-like figure on her chest – she is magically present as herself: wise, injured, accepting, and determined to carry on. Both portraits have the characteristic Witkacy signature that includes the date and the code for the drug taken while the work was underway.

Whether or not Tarski participated in the Dionysian revels of Zakopane, it is a safe guess he would have liked to. Moreover, since Leon Chwistek was an accomplished artist as well as logician and since the multifaceted Witkacy did philosophy of sorts, it is very likely that Tarski – who had wide intellectual interests, capabilities, and ambitions – would have wished for artistic talent, too. Since he was not gifted in that way, he could at least become an aficionado and a collector and so he did, with an affinity for drawings, prints, and watercolors, mostly but not exclusively by Polish artists.

Another effect of Tarski's association with the bohemians was his attitude about sexual preference and about sexual freedom in general, which, to put it loosely, seemed to be: the more the merrier. He found nothing to criticize about the sexual morés of the Zakopane crowd and to some extent took them as his own. Same-sex love affairs as well as conventional heterosexual ones in and out of marriage titillated but never scandalized him, which is neither to suggest that he had homosexual encounters himself nor that he might not have had. As far as he was concerned, sexual preferences were irrelevant when it came to friendship or collegiality. His first truly close friend in the United States, as well as two of his Ph.D. students at UC Berkeley, were gay, and he was never uncomfortable about it except for the risks their homosexual encounters presented. This is an easy attitude to adopt since the liberation of attitudes toward the end of the twentieth century, but it was not the norm in Tarski's day, when most homosexuals were in the closet.

It is impossible to say whether the predilection of Witkacy's circle for narcotics, alcohol, and experimentation with mind-expanding drugs influenced Tarski's behavior. The fact that he was a lifelong chain smoker of cigarettes and cigarillos can be attributed to the culture of the time both in Europe and the United States. There was nothing unusual about it; "everyone," especially men, smoked and did so with a general lack of concern about the physical effect, though Witkacy in his tract on narcotics bemoaned the fact that "every smoker is the ruin of what he could have been if he did not smoke." While reiterating, "I am going to stop smoking tomorrow," he said: "I write these words with a lousy 'weed' stuck in my filthy mug, thoroughly poisoned ...."[9]

There is ample evidence that Tarski regularly took stimulants such as Kola Astier (a kola–caffeine mixture he obtained from France) and Benzedrine (an amphetamine) to stay awake and keep a creative edge as he worked late at night and into the early morning hours. It is not clear just how much he experimented with hallucinogenic drugs; in the 1960s he smoked marijuana on occasion but there are no reports that he was a "user." Tarski was more than willing to give himself over to various recreational activities – mountain climbing and sex, for example – but these he saw as truly "re-creative" and good for the mind and the body. He explicitly said he thought having sex was good for doing mathematics and enhanced his ability to think.[10] But taking something like peyote was different. Witkacy's description of what it was like to have peyote visions – a minute-to-minute and hour-to-hour report – is both attractive and repulsive, surreal beyond most dreams. Although surprising untapped associations of sound, color, mood, and image spring forth, so do monsters and snakes and extravagant sexual images. Some of this out-of-body experience may have had its appeal for Tarski, but most of the visions were linked to the visual-emotional rather than the logical-conceptual. This would perhaps be material for a painter or a writer but not something that would feed ideas to a profoundly rigorous thinker; he wouldn't have liked the loss of control – of those images burbling up from who knows where in the subconscious, taking the user on a "trip." For all his interest in experimentation, the risk was probably greater than Tarski was willing to countenance because he was bent on doing something truly great – not just creating a sensation – and he didn't have time to fool around with such stuff.

Yet he liked to be "high" and often was, although only a few people knew the extent to which the bulging eyes and pulsing veins were signs of his having taken "something." In 1963 he wrote to a friend who had been frightened by the way he was looking at her, "I did not feel physically well that day, my mind was not clear and, since I wanted to talk at the seminar, I took Benzedrine and this caused that tense look which scared you."[11] So that "look" in his eyes, which many assumed was a natural physiological condition, seems to have been enhanced pharmacologically; by the 1960s and probably earlier, Tarski was not only taking stimulants at night to stay awake, he was also using "speed" to clear his mind during the day.

In the Warsaw community of intellectuals and artists, everyone was linked together; they all knew one another. The university, the lycée, the cafés, political meetings, and discussions were all part of the social and intellectual fabric. Kotarbiński and Witkacy were good friends, and the Kotarbiński and Leśniewski families were also extremely close and lived in the same apartment building. Witkacy and Żeromski were friends until the latter's death in 1925. Other professors lived nearby in cooperative university flats in the old town.[12] The cloth was tightly knit; when it later unravelled and shredded, the damage was never repaired.

The university did not form a separate enclave (as, for example, in the United States) because so many deeply academic thinkers did not have positions there – either because there were no openings or in the case of Jews because they were considered a liability by the administration. Anyone could attend seminars and lectures at the university, and as a result the atmosphere was enriched by the diversity of interests. Henry Hiż, who would later become Willard Quine's student at Harvard and a professor of linguistics at the University of Pennsylvania, was in high school when Tarski was simultaneously lecturing as a docent at Warsaw University and teaching geometry at Żeromski's Lycée. He recalled timidly asking Tarski if he might attend one of his lectures: "Tarski laughed: 'But of course, come. Even if I wanted to – which I do not – I am not allowed to stop anybody from coming to the class.'" Hiż remembered that lecture as "clever, beautiful, and forceful"; he wished he could continue the course, but his high-school work was so demanding he simply did not have time.[13] But many others did, from eager high-school students to painters and playwrights; the field was open to anyone who wanted to be part of the audience.

Marriage

In the mid-1920s, Tarski became engaged to Irena Grosz, a lively young woman he met in the leftist literary circles of Warsaw. Always interested in political ideas and ideals, he had strong opinions about what was right, just, and practical; and naturally, he had strong feelings about anti-Semitism and social injustice, having felt the sting directly. Also, from early days he was opposed to Zionism and in favor of assimilation, so he gravitated toward socialism of that stripe. He was not an outright committed activist – he would not give up his precious research time for that – but he did attend meetings where he took part in discussions and arguments. Indeed, those conversations were part of everyday interaction at cafés, at the university, and at Żeromski's Lycée, where most of the teachers were of a leftist bent.

In contrast, Irena Grosz – an intense woman who was equally passionate in her political idealism – held fast to her Jewish identity and over the years gravitated from socialism to communism. She was a writer and, during the postwar communist regime in Poland, wrote for an agricultural trade magazine whose audience was working class, not intellectual. She was much more fervently committed to activist politics than Tarski – it was her raison d'être, just as logic was his. This seems to have been at least one of the reasons she broke her engagement to Alfred.

As time went on, their political paths diverged; she went farther left, he went farther right. Each found a more suitable mate, but they remained lifelong friends – first in the circle of Warsaw intelligentsia where they had met and eventually across oceans and continents. In later years, Irena told Tarski's son that, in spite of Alfred's great prominence, she never regretted having rejected him because she had had a very meaningful life on her own terms, which she knew would not have been possible had she married him.[14] Irena and Maria became friends too, which is how the former really knew what it would have been like to be Tarski's wife. Indeed, the couples maintained contact for many years despite their ever-deepening political differences. At a crucial moment immediately after World War II ended, Irena and her husband – who was a general in the army and held a ministerial position in the defense department of the Polish government – would play a role in facilitating Maria Tarski's and her children's escape from Warsaw.

The gods were good to Tarski when they sent him Maria Witkowska as a life partner. In her early twenties, she had come to Warsaw from Minsk, where there had been a sizeable Polish-Catholic colony, most of whom (including her family) were ardent Polish nationalists allied with Piłsudski in his fight for independence. Maria, a year younger than Alfred, was small, dark, pretty, socially adept, and above all warm, understanding, and loyal. Many of the Tarskis' later acquaintances in the United States had the impression that Maria's family was gentry and descended from nobility and even that she was a second cousin to Marshal Piłsudski – rumors that are a testimony to her genuinely refined character. However, the facts according to the Tarski children are that Maria's parents owned a butcher shop where they both worked long hours; they may also have had some small land holdings that qualified them as "minor gentry," but they were definitely not related to Piłsudski and certainly did not think of themselves as nobility. According to Ina Tarski, "They made a decent living, they always had enough to eat and, my mother told me, they always had some soup ready to give to a poor person who might come to the door asking for food. But they weren't rich. My mother's mother worked!"

When she moved to Warsaw, Maria lived with her sister Józefa, who was already established there. Józefa, several years older, had been like a mother to Maria (their mother had died young and their father had remarried). There was another bond between them: both Józefa and Maria had been couriers in Piłsudski's army during the last battles over territory between Poland and the Soviet Union, battles that continued after

Maria Witkowska-Tarski, c. 1929.

World War I had ended. At great peril, they had crossed enemy lines to deliver letters and gather information; when Piłsudski became head of state, they were decorated for their valor. Oddly, it was Alfred rather than Maria who made capital of this. While he told of her courage under fire, she would smile modestly. Then, in a tone that almost conveyed congratulations to himself for the admission, he would add, "I don't think I would have been so brave." Indeed she was brave and, more to the point, self-effacing. She was never in competition with Alfred and had no need for the limelight; in short, she was the perfect mate for him.

They met at Żeromski's school, where Maria had a job teaching the young children in the elementary wing while he taught the advanced mathematics courses in the *gymnasium*. He was attracted by the enthusiasm and warmth that complemented her seriousness of purpose, and naturally he appreciated her dark good looks; he probably was not aware of her qualities of patience and indulgence until much later. She, on the other hand, was dazzled by his brilliance, his wide-ranging interests, and his stature in the academic world. She knew he was "special" – a genius – and was willing to defer to him and do what she could to further his scientific vision and career. He courted her gallantly, for he was a romantic through and through and had *polot* – flair, inspiration, and extravagance of style – a highly valued attribute among Poles. In later years, Celina Whitfield, one of Maria's closest friends in Berkeley, would say: "Her only fault was that she was too damned good. She was a saint. She put up with anything Alfred did."[15] Almost true, but even the saintly Maria had her limits.

Alfred and Maria married in 1929. Because they were teachers and members of the teachers' union, they were able to get a low-cost loan and buy a flat on Sułkowski Street (number 8) in one of the avant-garde cooperative housing estates that were springing up all over Zoliborz in the northern section of the city. Inspired by the Bauhaus "form is function" principles of design, this area of stylish new apartments, flats, and houses surrounded by greenbelts was considered a choice place to live, just a little bit out of the hustle and bustle of the center of town yet still part of the city, only a ten-minute trolley ride to the university. With two incomes, these were relatively good times for the newly married couple; the atmosphere at Żeromski's was congenial, and they were surrounded by friends and like-minded people both at home and at work. Although his teaching load, divided between the university and the lycée, was heavy, at least Alfred's position was secure. His biggest problem was finding enough time to do research and write papers; his solution, when the pressure became too great and he absolutely needed uninterrupted time to bring a piece of work to conclusion, was to resort to the age-old tactic of calling in sick.

The Lvov Affair

Lucky in love and marriage, Tarski was not a winner in his search for a professorship. However, he came so close to being appointed to a university chair that the reasons for why he was not chosen were discussed in the academic community for years afterward. The competition began in 1928 when, in accordance with the growing interest in the field of logic in Poland, the faculty at Lvov University decided to create a new professorship in the faculty of mathematics and natural sciences. The candidates for the position were Leon Chwistek and Alfred Tarski.[16]

Chwistek was a painter as well as a logician; born in 1884 in Zakopane, the Tatra mountain town, he grew up in the same arty bohemian milieu as Ignacy "Witkacy" Witkiewicz. Like his intimate friend Witkacy, Chwistek threw himself into the role of "the artist living an artistic life," a style that made him well known in society as an extravagantly eccentric character, with *polot* to spare. However, unlike Witkacy, Chwistek was a very serious scholar who, while attending the Academy of Fine Arts in Cracow, also studied mathematics and philosophy at the Jagiellonian University, where he was eventually awarded a Ph.D. He then did further studies abroad, attending Hilbert's lectures in Göttingen and Poincaré's lectures in Paris. Inspired by Bertrand Russell's work, Chwistek went on to do significant research in Russell's theory of types. Allegedly, in the early years (according to Russell himself), he was one of only six individuals throughout the world – three of whom were Poles – who had read and understood *all* of *Principia Mathematica*.[17]

Hugo Steinhaus and Stefan Banach, two of Poland's greatest mathematicians, supported Chwistek's candidacy for the Lvov position. When the choice was made, it did not go unnoticed that Chwistek was Steinhaus's brother-in-law. According to gossips, Steinhaus had said, "Don't worry, dear sister, your husband will get the job." Banach, for his part, was in a touchy situation because of his connection with Tarski as coauthor of their famous joint paper.

Tarski's support came from the grand old man of the philosophy department in Lvov, Kazimierz Twardowski, and his former student Kazimierz Ajdukiewicz. From Warsaw, Łukasiewicz, Leśniewski, and Kotarbiński weighed in in favor of Tarski, while the Cracow faculty supported Chwistek. Outside opinions were solicited as well; of these, Bertrand Russell's letter of 29 December 1929 was probably decisive:

I much regret that owing to my absence in America your letter on the 31st of October has remained hitherto unanswered. I know the work of Dr. Chwistek and think very highly of it. The work of Mr. Tarski I do not at the moment remember and do not have access to at present. In these circumstances, I can only say that in choosing Dr. Chwistek you will be choosing a man who will do you credit, but I am not in a position to compare his merits with those of Mr. Tarski.[18]

The decision was made in 1930: the job went to Chwistek. Twardowski was much offended, as were the other philosophers, because, in their view, Tarski was far and away the superior candidate. The residue of personal rancor lasted, and as a result logic did not bring the mathematicians and philosophers of Lvov together into a unified circle as it had in Warsaw. Nevertheless, relations between Chwistek and Tarski remained good.

As always in Poland, the question of the role of anti-Semitism reared its head. Was that why Tarski didn't get the job while the non-Jewish Chwistek did, or was it because Russell (who should have remembered Tarski's work because he referred to two papers by Tajtelbaum-Tarski in the 1925 edition of *Principia Mathematica*)[19] was enthusiastic about Chwistek? The former possibility is not easily dismissed. Mark Kac, a probability theorist of Polish origin, later wrote:

To the best of my knowledge, Steinhaus was the only professor of Jewish origin in Poland who had not converted to Catholicism. His international reputation combined with his own and his family's ties to the Polish patriotic movement were probably responsible for his being able to break through the antisemitic barrier. A few years later he might not have made it.[20]

Steinhaus was appointed in 1920. By 1930, as Kac pointed out, the picture had changed and even Catholic converts like Tarski were not getting jobs – but neither were many Catholic Poles, because openings were so few and far between.

Also, it must be remembered that Tarski was twenty-eight when the competition was opened and had only had his Ph.D. for four years, whereas Chwistek was forty-four. No doubt the feeling was that the brilliant Tarski would soon have other opportunities. As it turned out, that was not what happened – far from it. On the other hand, had Tarski won the job in Lvov and been secure in his position there, he might not have left Poland in the summer of 1939. In that case he might very well have shared the fate of his family and a large number of his Polish-Jewish colleagues who died in the ravages of World War II.

INTERLUDE II

The Completeness and Decidability of Algebra and Geometry

TARSKI PROBABLY FELL in love with the subject of geometry in his school days, and it began to figure in his own work as soon as he started publishing research papers in 1924. Euclidean geometry, then 2300 years old, had long been considered to be the model of deductive reasoning and was settled in its results, but in the nineteenth century mathematicians started to pick serious holes in Euclid's arguments. For one thing, the proofs turned out to make use of tacit assumptions that seemed evident on the basis of diagrams yet did not follow from Euclid's postulates. Another reason for carefully examining the deductive basis of Euclidean geometry was the discovery – by Bolyai and Lobachevsky in the early part of the nineteenth century – of the logical possibility of developing non-Euclidean geometries. Euclid's Fifth Postulate, the so-called Parallel Postulate, was shown to be independent of the other postulates because it is consistent with them to assume its contrary. In a famous book published in 1899 on the foundations of geometry, David Hilbert undertook a careful systematic study of geometry from an axiomatic point of view, including questions of the independence of each of the axioms from the others. However, by the time Tarski started his work, even Hilbert's treatment did not meet the logical standards and sensitivity to logical distinctions that had then come to be the norm, which was especially due to the great influence of Whitehead and Russell's *Principia Mathematica*. In particular, it was recognized that Hilbert's axiom system assumes a certain portion of set theory. Thus it was that Tarski undertook to develop a system of elementary geometry, where the distinction "elementary" signals that it is formulated without the use of set-theoretical notions, *not* that it is elementary in the sense of Sherlock Holmes.

Completeness and Decidability vs. Incompleteness and Undecidability

The significance of this distinction between Tarski's and Hilbert's form of axiomatization concerned the following natural questions, which Tarski was to address. Are the axioms *complete* – that is, do they serve to determine the truth or falsity of every statement in the language of geometry? Is geometry *decidable* – that is, do we have a systematic step-by-step method to determine the truth or falsity of any such statement? The answers to these questions turn out to be very sensitive to the question of *which statements* are counted as belonging to the language of geometry. That was not fully appreciated until 1931, when the Austrian logician Kurt Gödel proved his famous incompleteness theorem according to which *no sufficiently strong consistent system is complete*. Six years later, the American logician J. Barkley Rosser proved that *no such system is decidable*. Rosser built on the work of his teacher, Alonzo Church, and on related work of the young English mathematician, Alan Turing, who gave a definitive general characterization of decidable and undecidable problems.

The negative results due to Gödel and Rosser require a precise explanation in logical terms of what it means for an axiom system S to be "sufficiently strong." The idea, roughly speaking, is that (a) the set N of natural numbers 0, 1, 2, ..., together with the operations of addition and multiplication of natural numbers, can be defined in the language of the system S, and (b) under these definitions, the basic axioms about N can be derived from the axioms of S. It turns out that if we allow – as Hilbert did – reference to arbitrary sets of points, then the axiom system S for geometry is sufficiently strong in that sense. For, given any two distinct points p and q, we can define in S the smallest set A of points $p, q, r, s, \ldots$ spaced equidistantly on a straight line emanating from p through q – that is, a line on which the distance from p to q is the same as that from q to r and as that from r to s, and so on:

The set A of points thus looks exactly like the set of natural numbers 0, 1, 2, 3, ..., and we can define suitable operations on the points of A so as to correspond to the operations of addition and multiplication of elements of N. Thus Hilbert's language and axiom system for geometry allowing reference to arbitrary sets of points is "sufficiently strong" in the sense of Gödel's incompleteness theorem and Rosser's undecidability theorem.

First-Order Logic vs. Higher-Order Logics

None of this was known explicitly at the time that Tarski began his work on geometry. But there was already an awareness, coming from several sources, that some kinds of restrictions of language would probably need to be made if one were to obtain positive results of completeness and decidability for a system of axioms. First of all, in the theory of types developed by Whitehead and Russell in their grand opus *Principia Mathematica* (1910–1913), the general language of logic was stratified into *types*, of which the following is a simplified version. At the lowest type one has variables ranging over a domain of individuals, the nature of which is left unspecified; statements at this level – built using such *propositional connectives* as "and", "or", and "not" and the *quantifiers* "for every individual" and "for some individual" – are said to be *first-order*, and the system of rules of reasoning with such statements is called *first-order logic*. For example, given a relation xRy between individuals x and y, the statement

for every individual x there is some individual y for which xRy holds

is expressed in first-order logic. If the domain of individuals is taken to be the set of all human beings and if the relation xRy holds when x is a child of y, then this statement expresses that every human being has at least one parent.

At the next higher type of language one allows, in addition to the propositional connectives and the quantifiers over the domain of individuals, also the *second-order quantifiers* of the form "all sets of individuals" and "some set of individuals"; the resulting logic is called *second-order logic*. For example, given a relation R between individuals as before, one can write out in this second-order language the following statement:

For every individual z there is some set X of individuals which contains z as a member and which has the property that, for every individual x in X, there is some individual y in X for which xRy holds.

In the case of the child–parent relation R in the set of individual human beings, the truth of this would imply that every human being ("z") has infinitely many ancestors.

In the theory of types one also has variables of higher orders for sets of sets, for sets of sets of sets, and so on. All the complexities of the system of *Principia Mathematica* emerged at its higher orders, and thus it was reasonable to begin by isolating first-order logic in order to study its properties. This was begun in 1915 by the German logician Leopold Löwenheim and continued in the 1920s by the Norwegian logician and number theorist Thoralf Skolem. The decisive result for first-order logic (to which Skolem came very close) was finally established by Gödel in his 1929 Vienna dissertation, in which he showed that every first-order statement that is true no matter *how* we interpret the domain of individuals and the basic predicates and relations in it can be derived using the axioms and rules of inference of *Principia Mathematica*. Again, this was after Tarski began his work on geometry in 1926, but at least he knew of the work of Löwenheim and Skolem that led up to it.

The Method of Eliminating Quantifiers

Tarski was even more influenced in the direction his work took on elementary geometry by known examples of systems of axioms in first-order logic for which completeness and decidability had been established. This was by a method of "eliminating quantifiers" that had been initiated by Löwenheim and then put to work by Skolem and an American logician, C. H. Langford. For any statement built from basic statements by the propositional operations "not", "and", and "or", there is a simple finite decision procedure for determining its truth or falsity by using what are called the truth tables for those operations. On the other hand, if a first-order statement contains the quantifiers "for every individual x" or "for some individual x" and if the associated variables are taken to range over an infinite set of individuals, then in general one can't decide its truth or falsity in a finite number of steps. However, for certain systems of axioms one can show that every statement is provably equivalent to one without quantifiers and hence *can* be decided. Löwenheim had dealt in this way with a system of axioms in which equality of individuals is the only basic relation but in which any number of monadic (one-placed) predicates are allowed. (Actually, owing to the poverty of the language involved, Skolem was able to extend this procedure, as a rare case, to monadic second-order logic.) And Langford had applied the method to the first-order theory of dense order. This is a system of axioms about a linear ordering relation

R with the property that, for every x and y, if xRy then for some z we have xRz and zRy, that is, z is between x and y. This system of axioms is not complete, because it does not determine whether there is a least individual in the relation R or whether there is a largest one. But Langford could show by the method of eliminating quantifiers that every statement in this system is provably equivalent to a propositional combination of those two particular statements. Thus, if one adds to the axioms for a dense ordering relation R an axiom stating that there is no least individual and no largest individual, then the truth or falsity of every statement is completely determined *and* the enlarged system of axioms is complete and decidable. Langford also used the method of eliminating quantifiers to deal with the theory of discrete ordering relations with a least element but no largest element, of which the ordering of the set N of natural numbers is a familiar example.

Tarski's Seminar on the Elimination of Quantifiers

In the years 1927–1929, Tarski conducted the "exercise sessions" for the seminar at the University of Warsaw led by Jan Łukasiewicz, and he used the opportunity to pursue a systematic development of the method of eliminating quantifiers. After going over the earlier applications of the method, he obtained some new results with its use; for example, he was able to apply it to the theory of arbitrary discrete ordering relations. For more general applications of the method one must sometimes augment the basic predicates and relations of an axiom system by suitably defined new ones. As an "exercise," Tarski suggested to one of the students, Mojżesz Presburger, that he find an elimination-of-quantifiers procedure for the additive theory of the natural numbers N in which the basic relation is that given by the operation of addition, $x + y = z$.

Presburger succeeded in arriving at this in the spring of 1928 by a suitable augmentation of the language, and thus he established the completeness and decidability of the additive theory of N. This result served as Presburger's thesis for a master's degree; published two years later, it was his sole paper in logic. The importance of this paper, a mere nine pages in length, was only recognized in later years – among other reasons because it served to provide one dividing line between positive and negative results about decidability. For, as mentioned previously, within the decade Rosser proved that the theory of the natural numbers with the operations of both addition and multiplication is undecidable. In retrospect, some think that Presburger deserved a Ph.D. for this work, slim as it was. Be that as it may, he left the university soon after to work in the Polish insurance industry throughout the 1930s. A sad coda to this story is that Presburger, a Jew, perished in the Holocaust around 1943.[1]

From Geometry to Algebra

In comparison with Hilbert's work on the foundations of geometry, Tarski's system is distinguished not only by its restriction to first-order logic but also by its simplicity and all-round economy.[2] Where Hilbert has different sorts of variables corresponding to points and lines in plane geometry, Tarski has variables for points only. This economization depends simply on the fact that a line is determined by two distinct points. One then needs only the relation of *betweenness* in order to say that a point is on a line as specified by two such points. The only other basic relation in Tarski's system is that of *equidistance,* expressing that the distance between one pair of points p and q is the same as that between another pair r and s. These two relations also serve as basic notions for geometry in all higher dimensions. By contrast, Hilbert made use of further basic notions, both in plane and higher-dimensional geometry, while Tarski was able to show that those can all be defined in terms of betweenness and equidistance.

The restriction to first-order logic is required for one axiom only: that of *continuity,* according to which there are no gaps in any straight line. Formulated in set-theoretical terms, this says that if X and Y are any two disjoint, complementary, non-empty sets of points on a line such that every point of X is to the left of every point of Y in a specified direction of the line, then either X has a largest point or Y has a smallest point. This requires second-order logic for its statement; within first-order logic one can only state the axiom as (what is called) a *continuity scheme,* consisting of a separate statement for each pair X and Y of sets defined by first-order formulas. Tarski called the system restricted to its first-order language *elementary geometry*; the second-order system is sometimes called *full geometry.* By the introduction of a Cartesian coordinate system, every model of full geometry of the plane, for example, can be represented as the set of ordered pairs (x, y), where x and y are any real numbers; then the relations of betweenness and equidistance are algebraically definable in the ordered field R of real numbers – that is, the structure with R as its set of individuals, addition and multiplication as its basic operations, and the ordering relation as its basic relation. But the same model serves to reduce elementary geometry to *elementary algebra* – that is, to the axioms for the field of real numbers – with the usual axioms for addition, multiplication, and order together with the continuity scheme.

The Completeness and Decidability of Elementary Algebra and Geometry

Tarski's own reason for passing from elementary geometry to elementary algebra was that the latter is more amenable to the method of elimination of quantifiers. Moreover, there was a long history of work in algebra on special cases of the elimination problem that served as background to Tarski's goals. In particular, he centered attention on an algorithm (or step-by-step procedure) developed by Charles-François Sturm, a nineteenth-century Swiss-French mathematician. Sturm's algorithm allows one to compute, given any interval in the real numbers, how many roots a given polynomial has in that interval; the determination is made in terms of the truth or falsity of certain algebraic relations between the coefficients of the polynomial. Because the algorithm determines in particular whether there is *some* root of the polynomial in any given interval, in effect it is a special elimination of quantifiers. Tarski's major effort went into generalizing Sturm's algorithm to determine, for any finite system of polynomial equations and inequalities, how many real numbers satisfy all its equations and inequalities. His struggle to establish such a procedure did not succeed until 1930. Once achieved, though, the completeness and decidability of the axiom system for elementary algebra followed readily and, by the foregoing interpretation of geometry into algebra, the same followed for the system of elementary geometry.

In later life, Tarski (as did many others) considered the decision procedure for algebra and geometry to be one of the two most important research contributions in his entire career, the other being his theory of truth. Yet, though the procedure had been discovered around 1930, no publication of its actual details was made until 1948. This is surprising since Tarski, as a rule, was quick to publish his results soon after they were obtained. The full story of the vicissitudes surrounding the eventual exposition of this work may be found in Interlude IV, which follows Chapter 7.

4
A Wider Sphere of Influence

A STRIKING ABSENCE of self-doubt set Tarski apart from most of humanity, especially in intellectual matters. The assurance that he would achieve whatever he set out to do was embodied in the way he carried himself, in his forward-thrusting walk. If he encountered obstacles along his path, placed there by prejudice, economic forces, or unfavorable academic situations, he simply persevered with a stunning single-mindedness and a modicum of cunning. Those who were unenthusiastic about his projects thought he was "cocky" rather than graciously self-assured, but he was essentially impervious to criticism. This is not to say he always got what he was after, but he was exceedingly tenacious and rarely gave up on anything. Losing the Lvov competition was a big disappointment, and he may or may not have brooded and raged for weeks or months; but he continued to work with the conviction that, after all, what mattered most were the results of his research and the influence that his work was beginning to exert in a wider arena.

At about the same time, he had a series of providential encounters that would have important consequences for his future. The first of these was with the brilliant young Viennese mathematician Karl Menger, who was invited to lecture in Warsaw in the autumn of 1929. Menger, a year younger than Tarski, worked in topology, one of the areas that had been staked out for special concentration in the *Fundamenta Mathematicae* journal. Particularly in that subject, the policy of publishing only in international languages (such as French and German) had succeeded admirably in putting Poland on the map. Menger began reading the journal in 1923 and corresponded with several of the leading Polish contributors whose work interested him, especially Kuratowski and Knaster. Menger, like Tarski, received his Ph.D. in 1924, but then he spent two years as a docent at the University of Amsterdam working with the idiosyncratic mathematician Luitzen Egbertus Jan Brouwer, famous for his fundamental work in topology and his iconoclastic "intuitionistic" approach to the foundations of mathematics. On his return to Vienna in 1927, Menger was appointed professor of geometry at the unusually young age of twenty-five. There he immediately became a member of the Vienna Circle, a discussion group for scientific philosophy and logic that had been inspired by the revolutionary developments in physics, especially Einstein's theory of relativity, as well as by the *Tractatus Logico-philosophicus* of Wittgenstein and the *Principia Mathematica* of Whitehead and Russell. When Menger came to talk to the Warsaw mathematicians, he was also primed to make contact with its logicians and philosophers.

He first had a brief encounter with the eccentric Professor Leśniewski, who had been described to him as "a Polish Wittgenstein" and who lived up to his reputation, for upon meeting Menger he took a good look at him and exclaimed, "Are youngsters made professors in Vienna?" Menger recalled, "I felt greatly amused and flattered since I seemed to have successfully camouflaged my receding hairline, but my mathematician hosts were very angry, apologized in front of Leśniewski for his remark which despite my protest they considered as an insult to the University of Vienna and me, and dragged me away.... I never saw Leśniewski again."[1] While undoubtedly envying Menger's rapid ascent, Tarski might also have taken it as a positive omen for his own application in Lvov, which in 1929 was still pending. And, he was probably doubly chagrined when later he was rejected for that position.

In Warsaw, talking at close range with the Polish logicians, Menger was much impressed by the precision of their work, so much so that he immediately invited Tarski to Vienna. As he later wrote:

They were interested in philosophical problems similar to some discussed in the Vienna Circle, but they attacked them in connection with ... their exact logical studies. They always confined themselves to concrete questions and completely eschewed those vague generalities which seemed to me to becloud some of the Vienna discussions in the late 1920s. So I decided to familiarize the Vienna Circle [and] my Mathematical Colloquium with the logico-philosophical work of the Warsaw school and invited Tarski to deliver three lectures before the Colloquium, to two of which I planned to invite also the entire Circle.[2]

Menger chose Tarski because he was the most dynamic member of the logic group and an engagingly clear lecturer; moreover, personally they were immediately in tune with one another, both being quick-witted young men with broad cultural enthusiasms and interests outside of mathematics.

Tarski leapt at the invitation. Vienna's attractions as a vibrant crossroads were well known; although it was no longer the capital of an empire, the city had recovered to a great extent from the effects of its World War I losses and was again flourishing culturally and intellectually. It was a key place to make his mark; the university was first rate, and outside the academic community an unusually large number of professional and business people took an intense interest in scholarly activity. The city was replete with "movements" and groups with special interests, and its citizens belonged to various circles (*Kreise*) that met regularly in cafés or peoples' homes to discuss issues of particular interest to them. There were political groups left, right, and center; a "paneuropean" group – way ahead of the times – urging the political and economic union of Europe; literary and linguistic groups; psychoanalytic groups, of course; a respected (at least by some) parapsychology group; and philosophical study groups. The intellectual community was hospitable and gracious to foreigners, and café life was in full swing. What with the traditional ambience of *Freundlichkeit* and *Gemütlichkeit*, not to mention the heavenly tortes, who could resist?

The leaders of the Vienna Circle, whom Tarski met during his visit in February 1930, were a fascinating collection of individuals brought together by common interests in science as the touchstone for philosophy. They traced their ideas back to the great physicist Ernst Mach, who held the chair in the philosophy of the inductive sciences at the University of Vienna from 1895 until his retirement in 1901. Mach believed in the unity of science and was deeply suspicious of anything metaphysical. He argued that science should deal only with what could be directly experienced, and he espoused a radical form of empiricism called positivism; in particular he disbelieved in the atomic theory of matter, despite the considerable indirect evidence already available. In the years prior to World War I, a proto–Vienna Circle met on Thursday evenings to discuss philosopical issues raised by the natural sciences; its most prominent members were the mathematician Hans Hahn, the physicist Philipp Frank, and Hahn's brother-in-law, the sociologist-economist Otto Neurath. Though strongly sympathetic to Mach's anti-metaphysical positivism, they recognized that his program to reduce *all* scientific concepts to those of immediate experience faced serious problems. In particular, the language of mathematics – in which the laws of physics and other sciences are expressed – embodies high-level abstractions that cannot be reduced to concrete experience in any obvious way, and its truths seem to constitute a priori knowledge.

Meetings of this circle were suspended during World War I; when philosophical discussions were resumed in the early 1920s, the chair that Mach and his successors had occupied was long vacant. At that point Hahn determined to have it filled again, and he succeeded in bringing the German-born philosopher Moritz Schlick to Vienna in 1922. Schlick had done his doctoral work in Berlin under the atomic physicist Max Planck, a leading proponent of relativity theory. He was the perfect choice, having written a book on space and time in modern physics and another book on epistemology; moreover, he was noted for his broad erudition, his lucidity of thought in science and philosophy, and his calm disposition and elegant manner. Two years after his arrival in Vienna, Hahn encouraged Schlick to revive and lead the Thursday evening meetings. Once more, Hahn, Frank, and Neurath were centrally involved, but now they were joined by younger colleagues and students.[3]

In its examination of the role of mathematics in the sciences, the Schlick Circle (as it was first called) was particularly influenced by *Principia Mathematica*. The aim of the *Principia* had been to reduce all of mathematics to logic, a program called *logicism* that had been initiated in the latter part of the nineteenth century by the mathematical philosopher Gottlob Frege. Russell had found a fundamental inconsistency in Frege's system; to avoid contradictions and repair the program, in *Principia Mathematica* he introduced complicated restrictions on the form of its basic principles. But mathematics could no longer be reduced to logic when restricted in that way and so, to make up for the loss, Russell added some assumptions that were not clearly logical. This step then raised the question of whether Russell's repair of Frege's system could still be counted as fulfilling the logicist program; Russell, optimistically, believed that it did. If the logicist thesis in one form or another was accepted then it would seem to solve, at least in principle, the problem raised by the failure of positivism to account for the role of mathematics in science. The idea was that all admissible knowledge could be seen as generated by logical deduction from statements validated by direct experience. This was also the thesis

Ina and Rudolf Carnap, Prague 1933.

of Wittgenstein's *Tractatus Logico-philosophicus*, at least as interpreted by the Schlick Circle; they studied it closely alongside *Principia Mathematica*. By weaving these ideas together, the philosophical program of the Vienna Circle came to be called *logical positivism*.

The first systematic attempt at spelling out the logical positivist program was made by Rudolf Carnap, who had been Frege's student at the University of Jena in Germany. When Schlick and Hahn learned that he had a substantial work in progress along these lines, they invited him to become a docent at the University of Vienna and to join the Circle. Carnap's opus appeared in 1928 under the title *Der Logische Aufbau der Welt* [*The Logical Structure of the World*]. In it, Carnap aimed to apply the logic of *Principia Mathematica* to Mach's program of reducing all scientific knowledge to that of direct experience, while also attempting to do justice (as Mach had not done) to the role of mathematical knowledge in science. Eventually, Carnap became recognized as the most important philosopher in the Vienna Circle. Because of Carnap's known expertise in logic, the prospect of meeting him was an added attraction to Tarski. Also, there was a hidden attraction: the logician Kurt Gödel, who had completed his Ph.D. under Hahn's direction just the year before, was a regular attendee at the meetings of the Circle and of Menger's Mathematical Colloquium. Although years later Gödel said that he disagreed with the basic philosophical tenets of the Vienna Circle, at the time he did not voice his contrary views;[4] by character very private and reserved, he typically held his own counsel.

The Visit to Vienna: Gödel and Carnap

At his first lecture for the Mathematical Colloquium, Tarski chose to report on some results in set theory; for the remaining two lectures he asked Menger to make a selection from several topics in logic. In his memoir, Menger wrote that he chose two that not only illustrated the work done in Warsaw but also seemed to him to fill the needs of the Circle. He announced the lectures to its members prior to Tarski's visit, adding a description of his impressions in Warsaw.[5] Despite the encouragement, of the Circle's members only Hahn and Carnap came to the second of Tarski's lectures, which dealt with some fundamental concepts of the methodology of deductive sciences, framed in terms of the most general properties of the consequence relation. Menger had picked this topic among those offered by Tarski because he thought it showed the possibility of obtaining valuable "if not necessarily earth-shaking results" developed within a metalanguage for logical systems. Though Carnap "soon recognized the importance of the topic for philosophy and expressed his appreciation," Menger was greatly disappointed at the poor turnout of the Circle's members and took it upon himself personally to invite each and every one of them to be there for Tarski's third and final lecture. He did it, he said, "with great – perhaps too great – urgency" but "it was for their own good," and on the last evening the Colloquium was joined by the full membership of the Circle. For that meeting, Tarski presented various results concerning the sentential calculus that were the distinctive product of researches in Warsaw, including work that he had done on Łukasiewicz's three-valued logic.

In the end, the lecture series was significant not for the size of the audience but for its quality, and in particular for the important connections that it opened up with Kurt Gödel and Rudolf Carnap, both of whom were deeply impressed by Tarski's presentations. Gödel, socially shy but completely self-confident intellectually, asked Menger to arrange a private meeting with Tarski so that he could tell him in detail about his recently finished doctoral thesis on the completeness of first-order logic, the reasoning system at the bottom of *Principia Mathematica*. What a meeting

Tarski and Gödel in Vienna.

it was, and what a study in contrasts! Tarski, the ebullient, scrappy, experienced man of twenty-eight, sitting, cigarette in hand, at a table at the Café Reichsrat and following with riveted attention as the frail, ascetic Gödel, five years his junior, carefully explained how he arrived at his decisive result. Years later, Gödel's completeness theorem and the closely related compactness theorem were to play a fundamental role in the field of model theory, one of Tarski's principal areas of research. Although the importance of the completeness theorem in and of itself was immediately evident on general grounds, at the time there were only glimmers of appreciation of its future ramifications.[6]

Rudolf Carnap interacted with Tarski on a different level. Here again the contrasts were dramatic, although their taste for logical precision was similar. Carnap, ten years Tarski's senior, was a tall, generous, courteous, warm-hearted man who expressed himself in measured and thoughtful speech.[7] Following Tarski's second lecture, they had a long discussion on the metamathematical approach and its potential value to philosophy. Of this, Carnap later wrote:

Of special interest to me was his emphasis that certain concepts used in logical investigations, e.g., the consistency of axioms, the provability of a theorem in a deductive system, and the like, are to be expressed not in the language of the axioms (later called the object language), but in the metamathematical language (later called the metalanguage).... My talks with Tarski were fruitful for my further studies of the problem of speaking about language, a problem which I had often discussed, especially with Gödel. Out of these problems and talks grew my theory of logical syntax.[8]

In the course of their discussion, Tarski invited Carnap to Warsaw. He accepted, and that acceptance not only forged an important new personal connection but also launched a new decade of scientific interaction.

At the train station, at the moment of his departure from Vienna, Tarski thanked Menger effusively for his hospitality and for giving him the opportunity to present his work. Menger demurred, saying all he had wanted to do was further the common interest of the two scientific-philosophical groups – by which he meant the Lvov–Warsaw school and the Vienna Circle – while Tarski insisted that he had done much more. Bursting with warmth and gratitude, he embraced Menger and said, "I will never forget what you have done for me."[9]

If nothing else, the excitement of Tarski's sojourn in Vienna mitigated the sting of losing the competition in Lvov. Moreover, when Carnap came to Warsaw in November to continue his conversation with Tarski and to meet the other Polish logicians, Tarski's influence as an entrepreneur became manifest. Carnap, like Tarski, gave three lectures on his work: one on the reduction of psychological concepts to physical ones, another on the elimination of metaphysics, and the third on the nature of logical inference. He was impressed with the extent of interest in these kinds of problems among the attending faculty and students – and in how well-versed they were in modern logic. While there, Carnap had extensive discussions with Tarski, Leśniewski, and Kotarbiński, whose philosophical work in Polish had largely been inaccessible to most non-Polish philosophers; these discussions prepared the ground for the eventual enlargement of his ideas about logical syntax by semantics, the theory of meaning and truth.

Buoyed by these expanding encounters with the intellectual world beyond Poland, Tarski worked with ever-increasing intensity. He held his seminars at the university and taught his classes at Żeromski's. Then, one fine day in January of 1931, he received a letter from Gödel saying that he had arrived at some new results in metamathematics that he thought "should interest Tarski." In one of the understatements of the century, Gödel signaled his stunning *in*completeness theorems that were to change the face of logic. These theorems showed that, in various axiomatic systems for the foundations of mathematics (including that of *Principia Mathematica*), there are arithmetical propositions whose truth cannot be settled by the axioms, one way or the other, assuming that the system in question is consistent.[10] To establish the incompleteness of such systems, Gödel made use of a novel way of coding metamathematical notions by arithmetical ones, called the *method of arithmetization,* in order to construct arithmetical statements that indirectly refer to themselves so as to deny their own provability. While the world at large was still struggling to comprehend these arguments and results, Tarski immediately grasped their significance and was the first to present them outside of Vienna, at the April meeting of the Philosophical Society in Warsaw. All this was splendid for Gödel. But for Tarski it was a blow, since he felt that he had been close to discovering the incompleteness theorems himself.[11] In an article on the fundamental concepts of metamathematics just a year before, he had written:

The concept of absolute completeness is of great importance for ... "poor", elementary disciplines of an uncomplicated logical nature .... On the other hand this concept has not yet played an important part in investigations on "rich", logically more complicated disciplines (e.g. the system of *Principia Mathematica*). The cause of this is perhaps to be sought in the widespread, perhaps intuitively plausible, but not always strictly well-founded, belief in the incompleteness of all systems developed with these disciplines and known to the present day.[12]

At least Tarski was able to apply Gödel's precise method of arithmetization to cap his own theory of truth, which he was in the process of developing. Prior to this, Tarski had shown how to define the concept of truth for a formal language within a metalanguage. Using Gödel's new method to construct arithmetical statements that indirectly refer to themselves in order to deny their own truth, he was then able to show that the concept of truth for a formal language cannot be defined within the language itself. Thus, such a definition always requires the additional resources of a metalanguage. Later, in the historical notes to the 1935 German translation of his Polish monograph on the concept of truth for formalized languages, Tarski went to great lengths to emphasize that this was the *only* thing that he had garnered from Gödel's work:

I may say quite generally that all my methods and results, with the exception of those where I have expressly emphasized this ... were obtained by me quite independently.... In the one place in which my work is connected with the ideas of Gödel – in the negative solution of the problem of the definition of truth for the case where the metalanguage is not richer than the language investigated – I have naturally expressly emphasized this fact ...; it may be mentioned that the result so reached, which very much completed my work, was the only one subsequently added to the otherwise already finished investigation.[13]

What rankled Tarski even more was that he had been close to developing the method of arithmetization himself, though he acknowledged in the same notes that Gödel had done it "far more completely." Gödel's triumph was a major disappointment and the beginning of a lifelong unspoken competition.[14]

Years later, Tarski spoke resentfully of having had to devote so much energy in his early creative years to making money.[15] While his wealthier colleagues had the luxury to spend all their time in research, he had to hold two jobs to support his wife and children and help his parents because of his father's financial difficulties.[16] Implicit is the suggestion that, had he not had such obligations, he would have been first. Nevertheless, disappointment never stopped him. He soon launched what would be some of his most significant and richest work in the conceptual analysis of logical notions, work that would serve to redefine the subject.

Quine

Could anyone have possibly guessed, when the tall young American, Willard Van Orman Quine, came to Warsaw in May 1933 to meet with the logicians of the Polish School, that he, of all people, would have the most important effect upon Tarski's future? Quine, a midwesterner, was in his own words "an unknown neo-doctor of twenty-four"; he had just completed his Ph.D. in philosophy at Harvard and was spending a *Wanderjahr* of study and travel abroad with his adventurous wife, Naomi. Advised by friends at Harvard, he had arrived in Vienna the previous fall hoping to meet Wittgenstein and Carnap – neither of whom, as it turned out, was there: Wittgenstein was in Cambridge and Carnap had moved to a professorship in Prague. Instead, Quine attended Moritz Schlick's philosophy

Willard Quine and his first wife, Naomi, 1932.

lectures and was invited to participate in and even present his own work at the gatherings of the Vienna Circle.

In due time Carnap invited Quine to visit him in Prague and so, at the end of February 1933, he and his wife moved on. In his autobiography, Quine wrote:

We were overwhelmed by the kindness of the Carnaps. He had written me twice .... I attended his lecture the day after our arrival, and he invited us to their house for dinner. Meanwhile his Viennese wife, Ina, hearing of our lodging problems, tramped the streets with us for three hours, talking broken Czechish [sic] with the landladies.[17]

This description of the Carnaps and particularly of Rudolf is echoed time and time again by those who were to meet him in the United States in later years as students or colleagues. An inspiration to young thinkers, he was universally loved for his humanity as well as his philosophy. At the memorial service for Carnap at UCLA in 1970, Tarski, in his eulogy, spoke of all those endearing qualities in a way that gave the audience the sense that Tarski envied Carnap's sweetness and goodness. Although Tarski, too, was admired and respected, and though his friends and students had great affection for him, they frequently feared his anger and disapproval. Never an easy man to deal with, feelings about him were always mixed, whereas about Carnap people would simply say, unequivocally and with no modifying phrases, "I loved him."[18]

For Quine, becoming acquainted with Carnap in Prague was a marvelously heady experience. As a result of his stay in Vienna, his German had become just good enough for him to engage in abstract philosophical discussion with Carnap. "It was my first sustained intellectual engagement with anyone of an older generation, let alone a great man," he wrote, "my most notable experience of being intellectually fired by a living teacher rather than by a book."[19] A further beneficial side effect was that by the time he and Naomi moved on to Warsaw – after a typical Quinean whirlwind tour of Italy and North Africa – his German was even better.

In a letter written on the occasion of Tarski's eightieth birthday, Quine paints a vivid picture of the beginning of their friendship in the context of the intellectual climate he encountered in Poland in May 1933:

Dear Alfred,
When I came to Warsaw you were 32. Logic in America, as in England, had been at a standstill since *Principia Mathematica*. On the continent of Europe it was thriving. Poland was in the forefront, and you, a struggling young instructor, were already the leading logician of the Poles.
I came to Warsaw from Prague. It had been through Carnap in Prague that I began to catch up with the latter-day Continental logic and it was you that opened up the whole bright scene.
... you asked your seminar students to use German or French for my sake, instead of Polish. It was an impressive seminar, a research center. You were already a great teacher and trainer of research logicians, as you have been now for fifty years.
... six weeks of your seminar and our own conversations and your published papers, and I came away a happier and wiser man.[20]

Quine was welcomed as warmly in Warsaw as he had been in Prague. He and Naomi were a novelty – the first Americans to visit the Polish logicians – and Łukasiewicz, Leśniewski, and Tarski were all unstinting with their hospitality and time. The Łukasiewiczs and Leśniewskis invited them to dinner at their homes in the charming seventeenth-century old town, and the Tarskis invited them to their modern Bauhaus flat in Zoliborz. Many hours were lavished in discussion with "Van" after seminars, and conversations continued in the cafés. He read their papers and in turn presented a version of the talk he had given to the Vienna Circle; the response in Warsaw was friendly but critical. The impact upon Quine was immediate: "My ideas were getting healthily dislodged," he reported. "I was catching up with latter-day logic ... new thoughts flooded in."[21] By the time he left Warsaw in June he was happily outlining drastic revisions to the book that contained his thesis work.

Just as Hitler was taking over in Germany, Quine returned to the United States to become one of the first junior members of the newly endowed Harvard Society of Fellows, with a three-year fellowship that allowed him to use his time as he pleased. To that group and others he brought information about the great happenings in logic in Europe and especially the ideas of the Vienna Circle and logical positivism.[22] The crucial importance of Quine's role in Tarski's life would not become evident until six years later.

Maria Kokoszyńska-Lutmanowa

Every summer Tarski went on a hiking vacation to the spectacular Tatra mountains in the southeast of Poland on the border with Czechoslovakia. As was his habit, he chronicled the particulars of his hikes in the journal he had begun in 1923 when he was still a student, naming the passes, the peaks, and the previous occasions he had traversed the same routes. However, in 1933 he changed his manner of notation and cited his companions, putting their names after the date but before the mountaineering data. For July 11, his wife Maria and her sister Józefa appear as the first names in the journal in this fashion: "I. 11. VII. M. Tarska, J. Zahorska. Zakopane," followed by the name of the dale, "Dol. Kościeliska," which was their destination.

What is the reason for this sudden change? It is not that he had never had company before that time nor that he was newly married – he had already been married four years. But there *is* the coincident fact that, in the very next entry of the journal, "M. Kokoszyńska-Lutmanowa" appears next to "M. Tarska" and, for the next five years until the end of the summer of

Alfred and Maria Tarski, out for a walk in Zakopane, 1933.

1938, her name is written over and over again – in fact, more frequently than any other name in the journal. Maria Kokoszyńska-Lutmanowa, four years younger than Tarski, had been a student of Twardowski and Ajdukiewicz in Lvov; she was married to R. Lutman, whose name is also occasionally noted in the journal.[23] She is there in entry after entry: for day hikes, for week-long hikes, for hikes with a small group of friends, for hikes when her husband was part of the crowd, and on hikes where she and Tarski are apparently alone for many days. Other companions come and go, but M. Kokoszyńska-Lutmanowa is a constant. She would go with him to Vienna in the winter and spring of 1935; it was she who took the famous photograph of Tarski and Gödel standing hatted and coated in the wintry streets, clutching their briefcases, Gödel wearing white spats over his shoes. She was with him in France in August and September of that same year when the first Unity of Science congress met in Paris. A snapshot, perhaps caught by a street photographer, shows the happy couple in front of the Hotel Herbert, sharing an amusing thought, Maria's round pretty face in full smile, and Alfred, debonair, looking for all the

Tarski and Gödel in Vienna, 1935.

world like a Parisian – his ubiquitous briefcase in one hand, an umbrella hanging over his other arm, and of course a cigarette between his fingers.

Leaving from Paris prior to the meeting scheduled for the following month, Tarski and Kokoszyńska-Lutmanowa went to Biarritz and Pau and spent a week hiking in the Pyrenees. In September, again from Paris they went to the French Alps, using Chamonix as a base for their climbs. For several days they had a guide, which indicates that some serious mountaineering was going on, but obviously something else was happening. Alfred and Maria were having a full-blown, completely open "affair" that everyone surely knew about, including her husband and his wife; apparently all was friendly, and occasionally they all spent days together.

Kokoszyńska was a committed philosopher of logic and science who eventually became a professor in Wroclaw; in her work on semantics she was strongly influenced by Tarski. As was the case in almost all of Tarski's many affairs, there was an intellectual and professional basis for the relationship, with Alfred in the role of father/mentor/lover. Thirty years

A Wider Sphere of Influence

Tarski and Maria Kokoszyńska-Lutmanowa, in Paris, 1935.

later she wrote him poignant letters recalling their time together, indirectly alluding to their love affair and quite directly saying she did not realize how much he had taught her. She said that those times had sustained her during the war and that, after the war, she took up climbing again and taught it to others.[24]

Tarski had applied for and was awarded a Rockefeller grant for research in Vienna and Paris and obtained a leave of absence beginning in January 1935 from both of the institutions at which he worked. His wife, Maria, stayed home in Zoliborz with their infant son Janusz, who had been born in December 1934. Four years later they had a second child, a daughter Krystyna (Ina), born in March 1938. There is no evidence that Maria Tarska accompanied Alfred on any of his travels in the mid-1930s except for excursions to the mountains when visitors came from abroad.

If Maria's demeanor in later life is an indication, her way of dealing with her husband's absences, flirtations, and more serious affairs was to accept them as graciously as she could. The ethos of the culture at large was that husbands would have their mistresses and wives would look the other way; moreover, in their circle of friends, monogamy was seen as an unnatural bourgeois convention that scarcely anyone adhered to. After Alfred's death, a visitor to the Tarski home in Berkeley recalled Maria showing him a family photograph album with pictures of Alfred and his friends and colleagues; with a little smile she pointed out: "Now here is one of Alfred's girl friends ... and here is another." Perhaps it is not too much to say that she took some pride in the fact that other women found him attractive; perhaps Maria had admirers and lovers too, but about that there is no hint either way. What is clear, though, is that up to a point she was willing to share her "genius husband," and felt that it was part of the bargain of marriage to an exceptional man.

Vienna, Paris, and the Unity of Science

The Unity of Science movement was the brainchild of Otto Neurath, one of the most active promoters and propagandists of the ideas of the Vienna Circle. Physically powerful, energetic and enterprising, he was a talented organizer, very witty, and prone to heavy use of sarcasm in intellectual discussions and controversies.[25] He was contemptuous of the so-called parapsychological investigations that were followed by a number of Viennese, including his brother-in-law Hans Hahn. Neurath said these were for "uncritical, run-down aristocrats and a few supercritical intellectuals such as Hahn.... Listen to the gibberish the mediums pass off as the words of Goethe."[26]

The first vehicle for popularizing anti-metaphysical and empiricist views of the Circle was the *Verein Ernst Mach* [Ernst Mach Society], which sponsored lectures of public interest. Then, in 1929, Carnap, Hahn, and Neurath co-authored a manifesto entitled "The Scientific World Conception: The Vienna Circle," with a print run of 5000 copies, which laid out the division of knowledge into empirical and logical statements and declared all other statements outside of those categories to be meaningless. The long-term goal of the "scientific world conception" was to be "unified science," to be obtained "by applying logical analysis to the empirical material."[27]

International conferences devoted to these goals began in 1929, with a meeting in Prague organized by the Ernst Mach Society in collaboration with the Society for Empirical Philosophy begun in Berlin by the German philosopher of science, Hans Reichenbach. A second conference was held in Königsberg the following year; it was there, incidentally, that Kurt Gödel informally announced his world-shaking incompleteness theorem. Except for the brilliant and quick-thinking mathematician John von Neumann, the audience did not appear to grasp its significance; only after the publication of his results did people begin to realize its importance.[28]

Although there was much activity – especially in the form of publications – during the next few years, the Unity of Science movement was not officially launched until 1934, when a preparatory conference and planning session for the 1935 Paris conference was held in Prague. This planning session brought together scholars from Austria, Czechoslovakia, France, Germany, Poland, Scandinavia, and even the United States. Naturally, Carnap, Frank, Neurath, Schlick, and Reichenbach were there; among the Poles who attended were Ajdukiewicz, Tarski, and the philosopher of science Janina Hosiasson-Lindenbaum, the wife of Tarski's student and collaborator, Adolf Lindenbaum. France was represented by Louis Rougier, who was to lead the organization of the Paris conference in 1935. The Americans were Charles Morris of Chicago and Ernest Nagel of New York. One of the later consequences of that meeting was that Morris was instrumental in bringing Carnap to a position at the University of Chicago in 1936, and Nagel would be one of those to help Tarski in 1939.[29]

While in Prague, Tarski met the philosopher of science, Karl Popper. In many respects their situations were similar: they were close in age, both of them were high-school teachers, and both were brilliant and ambitious and, later, famous and influential. Though Viennese, Popper had kept himself at arm's length from the Vienna Circle and produced a thesis of *falsifiability* that he considered to be the distinguishing characteristic of scientific hypotheses, which was contrary to the Circle's view of how science develops. According to this thesis, which he explained to Tarski, science advances not through successive empirical verification of hypotheses but rather via a recurrent cycle of problems, tentative solutions, and bold conjectures that overreach and are eventually refuted and corrected. When Popper first met Tarski (he later recalled),

I had written in my book, *Logik der Forschung* [*The Logic of Scientific Discovery*], whose page proofs I had with me in Prague and showed to Tarski ...: "the striving for knowledge and the search for truth are ... the strongest motives of scientific discovery." Yet I was uneasy about the notion of truth; and there is a whole section in that book in which I tried to defend the notion of truth as commonsensical and harmless by saying that, if we want, we can avoid its use in the methodology of science, by speaking of deducibility and similar logical notions.[30]

When they met again in 1935 during Tarski's extended visit to Vienna, Popper asked Tarski to explain his theory of truth to him,

and he did so in a lecture of perhaps twenty minutes on a bench (an unforgotten bench) in the *Volksgarten* in Vienna. He also allowed me to see the sequence of proof sheets of the German translation of his great paper on the concept of truth, which were then just being sent to him.... No words can describe how much I learned from all this, and no words can express my gratitude for it. Although Tarski was only a little older than I, and although we were, in those days, on terms of considerable intimacy, I looked upon him as the one man whom I could truly regard as my teacher in philosophy. I have never learned so much from anybody else.[31]

≈≈≈

Apart from the numerous weekend skiing and hiking excursions outside the city with Kokoszyńska, Tarski spent at least four months in Vienna – talking about logic, mathematics, philosophy, and other subjects dear to his heart with Gödel, Menger, and Popper. While Menger's Mathematical Colloquium was still meeting regularly, the Vienna Circle itself was falling apart. Hans Hahn died the year before; Carnap and Frank had long since left for Prague, where Neurath joined them briefly before reestablishing himself in the Netherlands following the political turmoil of 1934. The Nazis' rise to power in Germany had drastically affected the situation in Austria. According to Menger,

there were periods when life in Vienna was almost intolerable.... Groups of young people, many wearing swastikas, marched along the sidewalks singing Nazi songs. Now and then, members of one of the rival paramilitary groups paraded through the wider avenues. I found it almost impossible to concentrate and rushed out hourly to buy the latest extra.[32]

The Social Democrats, whose stronghold was "red" Vienna, were steadily losing power and, following a brief civil war in 1934, were smashed by the rightist Dollfuss regime. The Ernst Mach Society, accused of spreading Social Democratic propaganda, was dissolved. Neurath's unique Museum of Economy and Society was closed and his papers confiscated.[33] Schlick's position at the university was precarious. Though a liberal and not politically active, he was tainted by his association with Neurath (a known radical) and by his association with the Ernst Mach Society, and he insisted on keeping Friedrich Waismann, a Jew, as his assistant.[34] The university was closed for long periods because of the continuing unrest, but the shrunken Schlick Circle tried to go on. Then, when Dollfuss – who was anti-Nazi as well as anti–Social Democrat – was assassinated by the Nazis in the summer of 1934, the political situation became even worse.

One of the most decisive discussions Tarski had during his 1935 visit to Vienna was with Carnap, who made a brief sojourn there. Carnap had by this time become a convert to Tarski's theory of truth and was to play a major role in how its presentation would unfold at the forthcoming Unity of Science meeting in Paris. Relating the circumstances that led him to accept and promote the theory published in Tarski's "great treatise on the concept of truth," he wrote:

When Tarski told me for the first time that he had constructed a definition of truth, I assumed that he had in mind a syntactical definition of logical truth or provability. I was surprised when he said that he meant truth in the customary sense, including contingent factual truth .... In his treatise Tarski developed a general method for constructing exact definitions of truth for deductive language systems, that is, for stating rules which determine for every sentence of such a system a necessary and sufficient condition of its truth. In order to formulate these rules it is necessary to use a metalanguage which contains the sentences of the object language or translations of them .... In this respect, the semantical metalanguage goes beyond the limits of the syntactical metalanguage. This new metalanguage evoked my strongest interest. I recognized that it provided for the first time the means for precisely explicating many concepts used in our philosophical discussions.[35]

Carnap urged Tarski to report on the concept of truth at the forthcoming congress in Paris. "I told him that all those interested in scientific philosophy and the analysis of language would welcome this new instrument with enthusiasm, and would be eager to apply it in their own philosophical work." Tarski was very skeptical. "He thought that most philosophers, even those working in modern logic, would be not only indifferent, but hostile to the explication of his semantical theory." Carnap convinced him to present it nevertheless, saying that he would emphasize the importance of semantics in his own paper. In any event, Tarski's presentation in Paris was to cause a sensation, and it led to controversial discussion.

Although the Vienna Circle was dying out as a discussion group, it had a spectacular rebirth and transformation in the Unity of Science movement. The first congress, at the Sorbonne from the 16th to the 21st of September 1935, was attended by 170 participants from more than twenty countries. Bertrand Russell, the most famous philosopher of the century, gave the opening address. A year later, in the publication of his lecture, he wrote:

The congress of Scientific Philosophy in Paris in September 1935, was a remarkable occasion, and for lovers of rationality, a very encouraging one. My first impression, on seeing the opening session, was one of surprise: surprise that there should be in the world so many men who think that opinions should be based on evidence. My second impression, on hearing the papers and discussions, was one of further surprise, to find that the opinions advocated actually conformed to this rule: I did not discover any of the signs of unfounded and merely passionate belief which, hitherto, has been as common among philosophers as among other men.[36]

In tune with the rational-empiricist theme of the congress, which he traced back to Leibniz, Russell continued:

In science, this combination existed since the time of Galileo; but in philosophy, until our time, those who were influenced by mathematical method, were anti-empirical, and the empiricists had little knowledge of mathematics. Modern science arose from the marriage of mathematics and empiricism; three centuries later, the same union is giving birth to a second child, scientific philosophy, which is perhaps destined to as great a career. For it alone can provide the intellectual temper in which it is possible to find a cure for the diseases of the modern world.[37]

Given the irrational social, economic, and political "diseases" that were already in considerable evidence and that would, before long, engulf the world, this statement was remarkable for its naive faith in the curative power of rationality that was so characteristic among the proponents of scientific philosophy.

The meetings of the Paris congress were organized in eight sections under the following titles, with some ninety presentations in all (in French, German, and English). It is interesting to note the prominence of logic and related fields among the topics:

   I. Scientific philosophy and logical empiricism
  II. Unity of science
 III. Language and pseudo-problems
  IV. Induction and probability
   V. Logic and experience
  VI. Philosophy of mathematics
 VII. Logic
VIII. History of logic and scientific philosophy

Russell's opening lecture was followed in the first section with contributions by Frank, Reichenbach, Ajdukiewicz, Morris, Carnap, Neurath, Kotarbiński, and Chwistek. The Poles were well represented in most of the other sections: Tarski, Kokoszyńska, Zawirski, Hosiasson-Lindenbaum, Jaśkowski, Lindenbaum, and (again) Ajdukiewicz and Chwistek. Also among the speakers were scholars whom Tarski met for the first time and who would play a role in his life in the next few years: the philosophers of science, Carl Hempel and Olaf Helmer (both students of Reichenbach), philosopher of physics Jean-Louis Destouches, logician Heinrich Scholz, and the biologist Joseph Henry Woodger (fondly referred to as "Socrates").[38]

Tarski gave two talks at the Paris congress, both in German. The first was entitled "Grundlegung der Wissenschaftlichen Semantik" [Foundations of Scientific Semantics], in which he presented his theory of truth; the second, "Über den Begriff der Logischen Folgerung" [On the Concept of Logical Consequence]. Of all his work, these two subjects have had the most impact in philosophy and constitute Tarski's most important conceptual contributions to logic. But Tarski had been right to worry about the negative response his theory of truth might have on the audience in Paris, despite Carnap's assurances in Vienna. Some of the reactions to his and Carnap's papers were decidedly negative. The bone of contention was whether the semantical concepts could be reconciled with a strictly empiricist and anti-metaphysical point of view. To Carnap's surprise, there was vehement opposition even on the side of our philosophical friends. Therefore we arranged an additional session for the discussion of this controversy outside the offical program of the Congress. There we had long and heated debates between Tarski, Mrs. Lutman-Kokoszyńska, and myself on one side, and our opponents Neurath, Arne Naess, and others on the other.[39]

Though Tarski's theory of truth did not immediately win the day, the sensation caused by his presentation increased his growing international reputation. In the following years, Carnap continued to champion the semantical approach in his own writings and helped win philosophers over to the acceptance of Tarski's treatment as a paradigm that would be at the core of considerable future work by him and others.

Piłsudski's Death

The Poland that Tarski returned to in September 1935 was different from the one he had left in January in one very visible and important respect: the hero of the people, Józef Piłsudski, had died of cancer in May 1935. Immediately following Polish independence in 1918 and the establishment of the second republic, Piłsudski had bowed out of politics, but the subsequent years of political and economic turmoil brought him back as leader of a coup against the Sejm (parliament) in 1926. From then until his death, in one position or another – sometimes as prime minister, sometimes as minister of war – he was the de facto leader of the government. After he died, his body lay in state for two days at the Belvedere Palace and was then moved to the main cathedral in an elaborate procession up the Royal Way, lined with hundreds of thousands of weeping Poles. His body was then transported by special train to Cracow, where it was finally laid to rest in a crypt of the cathedral of Wawel Castle across from the tomb of a great Polish warrior, King Jan III Sobieski. Despite his many failings, Piłsudski had commanded supreme respect and admiration. His death left an enormous void in Polish politics, and in the following years "the government of colonels" was led by a succession of army officers.[40] Anti-Semitism became increasingly virulent, and the universities were particularly troubled. Already in 1931, as described in the article "From 'numerus clausus' to 'numerus nullus'" by the historian Szymon Rudnicki, incidents started at the Jagiellonian University [in Cracow], beginning with the affair of the corpses: Jews were not admitted to anatomy lectures or practical experiments. Because of these incidents, lectures were suspended for a week.... At Warsaw University trouble broke out in the faculty of law, which was a bastion of MW [an anti-Semitic student organization] and where the greatest number of Jews in relative terms were studying. The result of these incidents, as MW admitted, was to leave several dozen Jews severely beaten.[41]

There was a series of anti-Semitic street demonstrations following Piłsudski's death in 1935, and in the universities, serious proposals were being made for the segregation of Jewish students on a "ghetto bench" just as Tarski returned to Poland. Though individual professors spoke out against what was happening, "they did not affect the perpetrators of the violence, especially as many professors tolerated the thugs or even supported them."[42]

If it were somehow possible, by magic potion or selective amnesia, to be ignorant of the horrendous political and economic events in Poland and Europe – to be oblivious of the facts that Hitler was in power in Germany, that fascism was in ascendance and the Great Depression in full sway – one might think, looking at Tarski's career and even his social life, that the mid-1930s were excellent years for him, particularly after the Paris Unity of Science conference. Naturally, as his stature rose his discontentment grew and, never one to be silent, he let it be known. But it was a fact of life that there were no positions for logicians in Warsaw or anywhere else in Poland and Tarski, at least, had his job at the Lycée – no small matter in a period of widespread unemployment. At the university he was promoted to adjunct professor and assistant to Łukasiewicz. After helping to prepare the 1935 German translation of his earlier work on the concept of truth in formalized languages, "Der Wahrheitsbegriff in den Formalisierten Sprachen,"[43] he wrote an elementary logic text in 1936. Ultimately, this would be his best-selling book, never out of print and translated from Polish into well over a dozen languages during Tarski's lifetime. In keeping with his love of words, the more exotic the language, the happier he was.

Tarski criss-crossed Europe many times between 1935 and 1938 – returning to Prague and Paris, traveling to the Netherlands, England, Belgium, and Germany – mostly to speak at conferences but also to visit colleagues. In England he stayed at the home of his biologist friend J. H. Woodger; at the 1935 Paris congress, Woodger had given a short exposition of his efforts to axiomatize biology in the language of *Principia Mathematica*. Tarski then and there offered to help by reading and criticizing his further work on that project. In the following years they had considerable correspondence and Woodger visited him in Poland three times, with many enjoyable discussions and hiking excursions.[44] The fruits of that interaction were evident in Woodger's book published in 1937, *The Axiomatic Method in Biology,* to which Tarski contributed an appendix. It was on the occasion of Tarski's visit to Woodger at his estate in Epsom Downs in that year that, fortuitously, he met Olaf Helmer, who had also been at the Paris congress. Helmer, a graduate student of German origin, was studying at the London School of Economics while living at Woodger's home and tutoring his children. Tarski approached Helmer with the idea of doing an English translation from the German edition of the *Introduction to Logic*. Since he was a native speaker of German who was fluent in English and interested in logic, Helmer accepted the task. Serendipitously, Helmer and his friend Carl Peter Hempel would be the first people to greet Tarski upon his arrival in the United States in 1939.

ಎ ಎ ಎ

Given the bleak economic picture in Poland in the 1930s, there was nothing especially unusual about Tarski's not having a professorship, even though it was recognized by everyone who mattered that he deserved one. There simply were no openings; Warsaw already had its two eminent senior logicians. Nevertheless, a letter from Tarski's professor Leśniewski to his own teacher Kazimierz Twardowski, the grand old man in Lvov, sheds a harsh and disturbing light on the situation and reveals with stunning honesty Leśniewski's feelings about Tarski and about Jews in general. Written from Zakopane on 8 September 1935, the long letter begins, in Leśniewski's own words, with "a flood of lyricism." He describes his feelings of affection and gratitude and his need to tell his mentor how much he means to him and how he misses him. The tone shifts, though, as Leśniewski responds to a point in Twardowski's previous letter which intimated that, since Tarski was recently appointed adjunct professor, it was no longer urgent to be concerned about a position for him. Surprisingly, Leśniewski disagreed:

1) I am inclined to think that it could be immensely useful to create a chair in our university for Tarski, whose specialty differs significantly from those of Łukasiewicz and myself, since it would enable Tarski to conduct his scientific and pedagogical operations in a significantly broader and more independent field than the subsidiary and, at any rate, subordinate position of an adjunct professor permits.

2) When I suggested this past year creating a chair of metamathematics ... I was acting in favor of the scientific interests of the University of Warsaw, and not in favor of the scientific needs, though very much justified, of Tarski.

The letter continues with a shocking turn in its next point:

3) In connection with a series of facts in recent years which I could tell you sometime if you were interested, *I feel a sincere antipathy towards Tarski* and though I intend, for the reasons of which I have spoken above, to do everything in my power so that he can get a chair in Warsaw; however, I admit ... that I would be extraordinarily pleased if some day I were to read in the newspapers that he was being offered a full professorship, *for example in Jerusalem*, from where he could send us offprints of his valuable works to our great profit.[45] [emphasis added]

Finally, Leśniewski turns to a brief summary of his own recent work and, after describing the various time pressures he faces, writes:

I am trying to keep as secret as possible not just the details of this work but also the very fact of its development, so that certain Jew-boys or their foreign friends do not play some filthy trick on me again, as they have already done.[46]

Who are the "Jew-boys" and the "foreign friends" he alludes to? Tarski and his co-worker Lindenbaum; or Tarski and the Vienna Circle, a number of whom were Jewish; or Tarski and Carnap, who was "foreign" but not Jewish? Perhaps all of the above are suspects. Leśniewski had great concern about his students and others cribbing his ideas and not giving him credit for his work on semantic categories and the language–metalanguage distinction;[47] indeed, his attitude may be the origin of Tarski's own obsession with attribution. But more to the point is Leśniewski's fear of plagiarism and of "filthy tricks" being played upon him – and of betrayal by "Jew-boys." This, coupled with his wish for Tarski to end up in far-away Jerusalem, does not give much credence to the statement that he'll do everything he can to see his former student appointed in Warsaw.

Relations between teacher and student had been strained for years, for more reasons than Leśniewski's anti-Semitism. Nothing pinpoints their feelings better than the old cliché "a love–hate relationship," an early intellectual romance gone sour; yet through all the years of later disaffection, Tarski and Leśniewski continued to meet privately, once a week, at Leśniewski's home for discussions – as they had done since Alfred's student days – a privilege the teacher accorded no other colleague or student.[48] But seldom censoring his thoughts and feelings, Leśniewski gave loud voice to his disappointment with Tarski for not following his original program – for, after writing a brilliant doctoral dissertation that to Leśniewski's great satisfaction had rounded out his system of logic in a most beautiful way, Tarski never worked on those systems again. Instead, once launched, he had sailed into new and uncharted waters, aggressively promoting a new program of research interests – his own – without, as Leśniewski saw it, giving sufficient credit and attribution to him. Tarski had surpassed his teacher and become the most prominent and the most ambitious figure in Polish logic; only someone like the saintly Carnap would not have had jealous and resentful feelings. So the friendship was replaced by personal and professional bitterness that was compounded by Leśniewski's overt anti-Semitism. Years later, Tarski complained that neither Leśniewski nor Lukasiewisz would welcome him to their table at their café for fear of being seen with a Jew in public. If Tarski was aware that there might have been other reasons for his teachers' hostility, he never acknowledged them.

Another Job Lost

At last, in 1937, a job opening did occur – not in Jerusalem, as Leśniewski had hoped, but at the University of Poznan – when Zygmunt Zawirski, a logician and philosopher of science, moved to Cracow. The Ministry of Education asked all the relevant professors in Poland to suggest a candidate to fill the vacancy, and Tarski was unanimously recommended. However, Poznan, always a stronghold of right-wing conservatism and dominated by the Catholic church, had, since Piłsudski's death in 1935, moved even farther to the right and become outright fascistic and anti-Semitic.

Unanimous recommendations notwithstanding, Poznan University did not appoint Tarski, and since there would have been no way to appoint anyone else without making the reasons for denying him the professorship patently clear, the position was eliminated.[49] This was quite different from the situation in Lvov in 1929, where at least a good academic case could be made for choosing Chwistek. Now the door really was slammed shut, and everyone agreed anti-Semitism had provided the force.

Tarski was furious; on the other hand, if they had appointed him, he might not have liked the idea of living in Poznan. In any case, while vociferously protesting what had been denied him, he just kept on doing what he had been doing in Warsaw.

In September of 1938, returning from a meeting in Amersfoort, Holland, Tarski stopped in Berlin, where the mathematician and philosopher Kurt Grelling was his host. Grelling, fifteen years Tarski's senior, had obtained his Ph.D. under Hilbert in Göttingen in 1910. He then became a school teacher in Berlin and there joined Reichenbach's society for scientific philosophy. Part Jewish, in 1933 Grelling was forced to give up his job, and his circumstances thenceforth were increasingly difficult. Tarski recalled the 1938 visit to Berlin forty years later, in an interview with the logician Herbert Enderton:

In the morning, directly after I arrived in Berlin, Grelling took me for a walk. We went to the Chancellery where Hitler was giving his violent speech after his return from Munich. I was among the crowd of people who listened to this talk – I was present at this historic event. After Hitler's "heroic" speech, Grelling invited the remainder of Berlin logicians to his house. There was [Leopold] Löwenheim and then there was this German aristocrat whose name I can't recall. I know that this man published something in Boolean algebra and he was surprised that I was one of the few people he met who ever noticed this article. Anyway we spent there many hours and talked about logic. Löwenheim told me I was the first logician with academic status that he had met or talked to in his life.[50]

In retrospect, it is amazing that Tarski would risk stopping in Germany that late in the game, actually hear Hitler raving to the crowd after signing the infamous Munich pact, and then go to Grelling's house and discuss logic for hours (which for Tarski would mean until three or four o'clock in the morning) with the three logicians who had not fled the Nazi terror – and that he would then return home without the sense that something terrible was about to happen in Poland, too. But Tarski was not alone in thinking that somehow Poland would be protected by England and France. This is not to say that he was unaware of the dynamics of what was occurring in Europe and in his own country. Quite the opposite. He was well informed politically and had strong opinions about social justice, Polish patriotism, assimilation, and most other issues. He belonged to the teachers' union and to a socialist club and had leftist sympathies, although he was strongly anti-communist as well as anti-Zionist. But his actual involvement in political action was minimal; his own projects and those of his students consumed him, and he had those two full-time jobs.

*** *** ***

One of Tarski's first students at the doctoral level was Andrzej Mostowski, who had also taken courses from many of Tarski's former professors and had gone abroad to attend Gödel's lectures on set theory in Vienna in the spring of 1937. From there he went to Zürich to study actuarial mathematics, thinking he would not be able to obtain a position in Poland in pure mathematics, but he gave that up out of boredom.[51] Mostowski returned to Warsaw and was awarded a Ph.D. in 1938 for his work on the independence of definitions of finiteness in axiomatic systems of set theory. Tarski was universally acknowledged as his mentor and dissertation advisor; nevertheless, Tarski, as *adjunct* professor, could not sign the dissertation because he was not a *full* professor. Instead, Kazimierz Kuratowski's name appeared on the document as the official supervisor.[52] For the proud Alfred, this was a bitter pill.

Tarski's working relationship with the students in his seminar was intense: inspiring, but demanding. Every aspect of a paper or an oral presentation was subjected to close scrutiny, all done in the spirit of assuring that the student achieve the best possible results. In the process, he and his students often became personally close. Andrzej Mostowski, a courtly and subtly witty man, became a lifelong friend.

At about the time Mostowski was finishing his graduate studies, Wanda Szmielew, another gifted new student on the scene, was just beginning hers. Tarski was immediately interested in her and, when he went to Zakopane in the summer of 1938, she and her husband were invited to join him on a weeklong climbing circuit in the Tatra mountains; both of their names are written in his journal for the dates 10–20 August, with a precise notation for exactly which part of the journey each was

Wanda Szmielew, 1938.

present. Three days later, Wanda gave Tarski a small but striking photograph of herself, inscribed on the back: "To Dr. Alfred Tarski, with thanks for the most beautiful of trips." Significant by her absence, Maria Kokoszyńska-Lutmanowa is not mentioned in 1938 or ever again in Tarski's journal, although they did correspond and see one another after the end of World War II. While Wanda may not have usurped Kokoszyńska's place in Tarski's heart as early as 1938, she had already established herself in his affections, and in varying degrees they would be close colleagues and intimate friends for life.

After Mostowski's doctorate, his first job was with the Polish Meteorological Institute; later, when the university went underground during the war, he was a teacher in that clandestine enterprise.[53] When all of Warsaw was put to the torch by the retreating Germans, he and his mother were ordered out of their house, and he was forced to make a choice between taking his "big, wonderful notebook" with many of his accumulated mathematical discoveries and taking some bread. "I decided to take the bread, so all of my notes were burnt." Fortunately, he was later able to reconstruct from memory most of what he had lost.[54] Mostowski was one of the few logicians to survive those terrible years and to remain in Poland. After the war, as the leading mathematical logician in Warsaw, Mostowski succeeded, even with limited means, to pick up the pieces and re-create a major center in logic in the spirit of what had existed during the interwar period. During these difficult years of reconstruction, Tarski provided him with important moral and professional support from the United States.

To Go or Not To Go?

In the spring of 1939, Tarski was presented with a major dilemma. He had received an invitation from Quine and others to speak at the Fifth International Unity of Science Congress to be held at Harvard University in early September. Normally he would have considered this an extremely attractive opportunity. He would have wanted to visit America, to take part in an important international conference, and to meet again with his colleagues from Vienna and Prague – Menger, Carnap, and others whom he had not seen since their emigration – and, not least, to satisfy his endless curiosity about new places and people. Instead he wavered and delayed his response. Why?

Obviously, 1939 was not a normal year, yet the cause of Tarski's hesitation is not so obvious. The crisis in Europe or anxiety about leaving his family in uncertain times were not the reasons, or at least not the major ones. Odd though it may seem after the fact, most Poles had somehow convinced themselves that their situation was not precarious; many were taking their August vacations in the countryside as usual even though talk of war, sooner or later, was in the air.[55] Instead something else was holding Tarski back. In May, Leśniewski had died suddenly of thyroid cancer a few days after surgery. As recalled by Henry Hiż, then a student at the university,

> the operation was conducted without anaesthetic, because the anaesthesiology of the time had no methods which did not constrict the blood vessels around the thyroid. He was permitted to smoke during the operation! When Kotarbiński [his dear friend and colleague] went to the hospital the next day, I gave him a packet of "Plaski" ["Flat"] brand cigarettes, which Leśniewski smoked at that time.... One day before his death, Leśniewski said, "I am still below zero."[56]

Hiż's account may be inaccurate since it contradicts known surgical procedure; but in any case, the story provides another example of Leśniewski's eccentricity.

By dying, the man who had first anointed Tarski, his only student, as a "genius" with his much-repeated statement "All my students are geniuses!" had done what he claimed he wanted to do: he had created an

Jan and Ina Tarski with their grandmother, Rosa, c. 1938.

opening for a new logic professor at Warsaw University. Everyone in the academic world, not least Tarski, knew who was, indisputably, the best candidate to fill the vacancy. The big question was: Could (or would) it happen in Warsaw in 1939?

If politics had made Poznan problematic in 1937, the situation in 1939 Warsaw was even worse. Yet the suggestion is that Tarski, with his supreme absence of self-doubt, thought that because he was clearly the pre-eminent logician in Poland and because, in his own mind, he had disidentified himself as Jewish, it was logical and correct that he be appointed to the vacant chair. In the abstract the case was clear, so what could be the grounds for a negative decision? He was not, however, so completely naive as to think his appointment a certainty, and he wanted to be physically present in Warsaw when the deliberations were taking place. Even if his being there did not enhance his candidacy, he would at least be privy to information about how the matter was to be decided.

That was the real reason he delayed in responding to Quine's invitation to speak at Harvard: he wanted the job that was his due. Quine knew less about Tarski's personal situation and more about the danger in Europe, and was baffled at his hesitation. By 1939, most scholars were eager to seek refuge in the United States and, as already indicated, a good number of the Vienna Circle and its associates had done so. In all likelihood, Quine did not understand just how strong Tarski's feelings of entitlement were and that – for whatever reason – the Poles felt less threatened than other Eastern Europeans. He wrote again, proposing that Tarski consider the possibility of a position in America, adding that he would arrange a few lecture engagements for him after the meeting at Harvard. Those words had an effect. After waiting until it was almost too late, Tarski decided to let matters take their course in Warsaw in his absence; and he began to entertain seriously the idea of looking for a position in the United States. At the very last minute, he accepted Quine's invitation. As a proud Polish patriot rooted in his country's ways, it was hard to think of leaving, but at this point he thought he was only testing the waters. He would lecture at universities in the United States and would see what opportunities America offered; if he moved, it would only be temporary. His visitor's visa was granted on 7 August 1939, and his ship departed four days later. Janina Hosiasson-Lindenbaum, who was to attend the same conference, applied for passage on the next ship to North America but her visa was denied. In those few days, the bar had come down and there was no legal way to leave Poland. That rejection cost her her life, as she and her husband Adolf Lindenbaum perished, along with countless others, in the hellish descent to come – a fate Tarski escaped by the skin of his teeth.

INTERLUDE III

Truth and Definability

THE QUESTION OF WHAT IS TRUE comes up constantly in all avenues of life: the fibs of a child, stories in the newspapers, disputes in the courts of law, and tests of scientific theories. It may be very difficult to determine what is true in each situation and it is recognized that it may not even be a black-or-white matter, yet the idea of what is true or false is not ordinarily considered to require explanation or analysis. But as soon as one inquires into what, in general, the grounds are for one's beliefs or claims, the perennial philosophical question of "What is truth?" comes into play. Many theories have been advanced to answer that question. To mention only a few: the *correspondence* theory identifies what is true as that which corresponds with the facts, the *pragmatist* theory as that which proves to be useful in the long run, and the *verificationist* theory as that which can be demonstrated. Each of these has been found unacceptable for one reason or another; even the one that appears most unobjectionable – the correspondence theory – is criticized on the grounds that it fails to supply satisfactory explanations of "corresponds" and "the facts."[1]

Tarski formulated a very precise version of what he took to be the correspondence theory of truth. Although it applies only to certain relatively narrow and strictly regimented languages, this theory has been recognized as one of the most important examples of conceptual analysis in twentieth-century logic. First published in full in a long article in Polish in 1933, it reached a wider audience only when translated into German in 1935 under the title "Der Wahrheitsbegriff in den Formalisierten Sprachen" [The Concept of Truth in Formalized Languages], often referred to simply as the *Wahrheitsbegriff*.[2] That article spread Tarski's fame and influence far beyond the sphere of specialists in the field of logic, although the reasons for that are by no means straightforward and have led to some controversy over the nature of his accomplishment and hence of its significance. A source of confusion is that there are two ways in which he formulated his theory, tied together by a common technical procedure called the recursive definition of the satisfaction relation. The first formulation, in the *Wahrheitsbegriff,* was directed primarily toward a philosophical audience; this concerns truth in some sort of absolute sense. The second formulation was directed primarily toward a logical audience and was not published until the late 1950s, although Tarski had already arrived at it by 1930; it concerns truth in the relative sense of what is true in a mathematical structure such as a model of geometry or algebra. To reach a mathematical audience, in 1931 Tarski modified the technical apparatus for his theory of truth in order to explain the notion of definability in a structure in a way that does not require the recursive definition of satisfaction, even though it was directly motivated by it.[3]

The Problem of Truth in Ordinary Language

A classical expression of the correspondence theory, due to Aristotle, runs:

> To say of what is that it is not, or of what is not that it is, is false, while to say of what is that it is, or of what is not that it is not, is true.

This leads to a puzzling aspect of truth – namely, that it seems to be a redundant notion, since to say of an assertion that it is true seems to say no more than the assertion itself. A frequently cited example due to Tarski is:

(∗) "Snow is white" is true if and only if snow is white.

A variant of (∗) starts by singling out the sentence

(1) Snow is white.

Then another way of expressing (∗) is:

(∗∗) (1) is true if and only if snow is white.

Still another way is to point to the sentence (1) in a book or on a blackboard and then say:

(∗∗∗) *That* is true if and only if snow is white.

Tarski formulated these sorts of consequences of the correspondence theory as a general scheme, sometimes called the *T-scheme*:

(T) S is true if and only if P,

where the letter 'P' is to be replaced by a statement (or proposition) of one's language and the letter 'S' is to be replaced by a *name* of that statement. There are various ways of naming statements; as just illustrated, one way is simply to quote it, while another is to give it a number or other kind of mark or even to point at it; still another is to enumerate the words in the statement, letter by letter.

The T-scheme by itself has no meaning, it simply describes the *form* of a group of statements that *ought* to be accepted under the usual notion of truth. It would seem that each instance of the scheme is so platitudinous as to be hardly worth stating. But even that minimal requirement on what it means to be true turns out to make the notion of truth in ordinary language seriously problematic, since it leads to contradictions through one version or another of the Liar Paradox: the statement "I am lying", if true, is false, and if false, is true. A variant is the statement, "This statement is not true", where the word 'This' points to the very statement with which it begins. To see how that works in conjunction with the T-scheme, let (2) be the statement "(2) is not true":

(2)   (2) is not true.

According to the T-scheme, if in it we replace the letter 'S' by '(2)' and 'P' by the statement it names (i.e., what is displayed to the right of it), then we ought to accept the instance:

(3)   (2) is true if and only if (2) is not true,

which is contradictory on its face. According to Tarski, the problem of such contradictions lies with the *universality of ordinary language,* which allows one to express in it not only statements about extra-linguistic matters but also about matters of meaning and truth in the language itself. It is not that all uses of language to talk about language are fraught with difficulty; for example, there is no problem with saying that "This statement is in English" is true or that it has the subject–predicate form, and no problem in saying that "This statement is in French" is false as well as that "This in English statement" is not grammatical in English. Roughly speaking, there is no problem in using ordinary language to talk about its syntax; it is only when it is used to talk about its semantics that we can run into problems.

At first sight it might appear that such problems can be avoided simply by declaring peculiar self-referential sentences like (2) to be illegitimate in one way or another – as appearing to make statements when in fact they don't – and then legislating them out of the language. However, the contemporary philosopher Saul Kripke has pointed out that there are paradoxes hidden in very ordinary talk that would not be affected by such gerrymandering.[4] As an example, he asks us to suppose that the following assertions have been made by Jones and Nixon, respectively:

(4) Most (i.e., a majority) of Nixon's assertions about Watergate are false.
(5) Everything Jones says about Watergate is true.

Suppose it happens that (4) is the only thing that Jones has said about Watergate and that, other than (5), Nixon's statements are evenly balanced between those that are true and those that are false. Then it can be seen that (5) is true if and only if it is false, and the same holds for (4). The paradox in this case hinges on the (hypothetical) empirical facts, not on any peculiarity as to the form of (4) and (5). This reinforces Tarski's view in the *Wahrheitsbegriff* paper that

> in [everyday] language it seems to be impossible to define the notion of truth or even to use this notion in a consistent manner and in agreement with the laws of logic.

Instead, he says that he will "consider exclusively the scientifically constructed languages known at the present day, i.e. the formalized languages of the deductive sciences."[5] Except for change in notation and some terminology, the following is a step-by-step exposition of how Tarski proceeded in the *Wahrheitsbegriff* to carry out that restricted program.

Formalized Languages

Examples of the deductive sciences that Tarski had in mind were the axiom systems for geometry and algebra or parts of the theory of types. In particular, his definition of truth in the *Wahrheitsbegriff* is illustrated by *the language of classes of individuals*; this is part of the theory of types, where the sole basic relation between two classes x and y is that of inclusion or being a subclass, symbolized mathematically as $x \subseteq y$ and formally as xIy.[6] That relation becomes a sentence only if we substitute, for the variables x and y, names of specific classes. For example if 'h' names the class of human beings and '*m*' the class of all mammals, then *hIm* is a true sentence while *mIh* is a false sentence. Since every class is regarded as a subclass of itself (though not a proper subclass), also *hIh* and *mIm* are true sentences. The sequences of symbols *xIy* are examples of expressions that contain variables and that become sentences when we substitute designations for specific objects (in the case of this language, names of specific classes) for the variables. They are thus called *sentential functions* and act just like mathematical functions such as $x + y$, which have a specific value only when we substitute for the variables names of specific numbers like '2' for x and '5' for y to yield $2 + 5 = 7$.

There are other ways than substitution for the variables to lead to sentences from sentential functions: we can form the *negation* of a sentential function F by preceding it by the symbol '¬' to form $¬F$ (read "not *F*"), and we can form the *disjunction* of two sentential functions F and G by joining them by the symbol '∨' to form $F \vee G$ (read "*F* or *G*"). Finally, we can form the result of *universal quantification* of a sentential function F with respect to a variable x that may occur in F, by preceding F by '$\forall x$' to form $(\forall x)F$ (read "for all x, F").[7]

The following are examples of sentential functions constructed by these three means:

(6) $¬xIx$, the negation of *xIx*, which expresses for any given class x that x is not included in x;
(7) $xIy \vee yIx$, the disjunction of *xIy* and *yIx*, which expresses for any given classes x and y that x is included in y or y is included in x;
(8) $(\forall x)xIx$, the universal quantification with respect to x of *xIx*, which expresses that for *all* classes x, x is included in x.

Of these three, only (8) expresses a completed sentence, which in this case is true. In general, we can construct sentences from sentential functions by applying the universal quantifier to the variables that occur in them. Thus, for example:

(9) $(\forall x)(\forall y)(xIy \vee yIx)$ expresses that, for all classes x and y, either x is included in y or y is included in x.

That sentence is false, since there are classes neither of which is included in the other (e.g., the class of living humans and the class of humans who ever visited France). If we precede the sentence in (9) by the symbol '¬' we thus obtain a true sentence. To form more interesting sentences it is convenient to make abbreviations. The symbol '∧' for *conjunction* is used to form $F \wedge G$ (read "F and G"); this is regarded as an abbreviation for $\neg(\neg F \vee \neg G)$, which expresses that neither F nor G is false. The symbol '→' for *implication* is used to form $F \to G$ (read "F implies G"), which is regarded as an abbreviation for $(\neg F) \vee G$; this, when true, is such that if F is true then G must be true, and it is false only if F is true and G is false. The symbol '↔' for *equivalence* is used to form $F \leftrightarrow G$ (read "F is equivalent to G"); it abbreviates $(F \to G) \wedge (G \to F)$. Finally, the symbol '∃$x$' denotes *existential quantification* with respect to a variable x; here $(\exists x)F$ is read as "there exists an x such that F" and is regarded as an abbreviation for $\neg(\forall x)\neg F$, which expresses that it is not the case that for all x, F is false.

Using the preceding abbreviations, we can form such sentences as:

(10) $(\forall x)(\forall y)(\forall z)((xIy \wedge yIz) \to xIz)$, which expresses that, for any three classes x, y, and z, if x is included in y and y is included in z then x is included in z.

(11) $(\exists x)(\forall y)xIy$, which expresses that there is a class x which is included in every class y.

Both (10) and (11) are true; in the case of (11), the required x is simply the empty class.

The Metalanguage

So far we have considered only specific examples of sentential functions and sentences in the formalized language L of classes, together with their informal explanations. But to provide a general definition of what it means to be a *sentential function* (or *formula* in more modern terminology), and then what it means to be a *sentence*, and finally what it means to be a *true sentence*, Tarski argued that one must move out of the formal language L into a richer language L*, called a *metalanguage* for L, which is then referred to as the *object language*. A number of syntactic and mathematical notions are required in L*, and it is much more complicated than the language L used by Tarski to illustrate his theory of truth. In the *Wahrheitsbegriff* he sketched how L* could be formalized and then have its own metalanguage L**; that is not done here. In L* we can talk about sets of formal expressions in L, their syntactic structure, operations on expressions, and so on. Besides the symbols I, ¬, ∧, ∀, and left and right parentheses, one must provide for an unlimited supply of variables $x, y, z, \ldots$ to be able to formulate increasingly complicated statements.

The *basic sentential functions* (or *atomic formulas*, in more modern terminology) are defined to be all those of the form xIy, where x and y are any variables (possibly the same). The following is a recursive definition of what it means for F to be a sentential function in general:

> Defn. 1: F is a *sentential function* if and only if either (i) F is a basic sentential function, or (ii) F is of the form $\neg F'$ for some sentential function F', or (iii) F is of the form $(F' \vee F'')$ for some sentential functions F' and F'', or (iv) F is of the form $(\forall x) F'$ for some variable x and some sentential function F'.

This is not an explicit definition of what it means to be a sentential function, since the required notion is evidently defined in terms of itself. What the definition does do, however, is explain what it means for a given F to be a sentential function: either outright in (i); or in terms of less complicated expressions from which F is built by one of the operations of negation, disjunction, or universal quantification and which themselves are already recognized to be sentential functions. It is a standard matter, as Tarski pointed out, to turn a recursive definition into an explicit definition by use of the language of set theory; in this case the set-theoretical definition takes the following form.

> Defn. 1*: F is a sentential function if and only if F belongs to every set X of expressions such that (i) every basic sentential function belongs to X; (ii) whenever F' belongs to X then $\neg F'$ is in X; (iii) whenever F' and F'' belong to X then $(F' \vee F'')$ is in X; and (iv) whenever F' belongs to X and x is any variable then $(\forall x)F'$ is in X.

All the recursive definitions in this interlude can by similar devices be eliminated in favor of explicit set-theoretical definitions.

To single out the set of sentences from the set of sentential functions, one uses the standard notion of the *set of free variables of a sentential function*, again defined recursively:

> Defn. 2: (i) The set of free variables of a basic sentential function F consists of the variables that occur in it; (ii) the set of free variables of $\neg F'$ is the same as the set of free variables of F'; (iii) the set of free variables of $(F' \vee F'')$ is the union of the set of free variables of F' with

that of F'''; and (iv) the set of free variables of $(\forall x)F'$ is the set of free variables of F' which are distinct from the variable x.

Clearly the set of free variables of a sentential function is finite. A sentential function with at most one free variable x is written $F(x)$. Variables of a sentential function that are not free are called *bound*. In order to produce a *sentence* (or *closed formula*), eventually all variables must be bound by quantification.

> Defn. 3: A *sentence* is a sentential function whose set of free variables is empty.

In the following, the letter S, with or without primes, is used to range over sentences.

Satisfaction and Truth

The truth of a sentence depends in a regular way on the truth of its parts; for example, a sentence S of the form $\neg S'$ is true if and only if S' is not true, and a sentence S of the form $(S' \vee S'')$ is true if and only if S' is true or S'' is true. But these recursive conditions for truth don't work for sentences of the form $(\forall x)F(x)$, where F is a sentential function that contains x as a free variable, for then $F(x)$ is neither true nor false; it may be true for some values of the variable x and false for others. What we need is a generalization of the notion of truth that applies to sentential functions as well as sentences. This is provided by Tarski's notion of *satisfaction*. To take a simple example, consider the sentential function xIy and the classes h of humans and m of mammals. This is satisfied by the assignment of h to x and m to y but not by the assignment of m to x and h to y; it is also satisfied by the assignment of h to both x and y. By an *assignment a* is meant an association with each variable x of a class a_x; the same class might be assigned to distinct variables.

> Defn. 4: For a sentential function F and an assignment a, the relation *a satisfies F* is defined recursively as follows: (i) if F is of the form xIy then the assignment a satisfies F if and only if $a_x \subseteq a_y$; (ii) if F is of the form $\neg F'$, then a satisfies F if and only if a does not satisfy F'; (iii) if F is of the form $(F' \vee F'')$ then a satisfies F if and only if a satisfies F' or a satisfies F''; (iv) if F is of the form $(\forall x)F'$, then a

satisfies F if and only if every assignment b that agrees with a on the variables other than x also satisfies F'.

The idea of part (iv) is that b_x can be chosen arbitrarily, but b_v must be the same as a_v for all variables v distinct from x.

Assignments attach values to all variables, not only those free in a given sentential function, but that is not essential. For, if F is a sentential function and a and b are assignments such that $a_v = b_v$ for each variable v that is free in F, then a satisfies F if and only if b satisfies F. This fact is readily proved by induction on the complexity of F. It follows, in particular, that if F is a sentence and *some* assignment satisfies F then *all* assignments satisfy F, since F has no free variables. Thus, Tarski defined truth for sentences of L as follows:

Defn. 5: A sentence S is *true* if and only if every assignment satisfies S.

Having defined truth, the final task that remains for Tarski's program is to verify each instance of the T-scheme, in the form

(T) S is true if and only if P,

where S is any sentence of L and P is a translation of S into the metalanguage L*. For example, one must verify

(12) $(\forall x) xIx$ is true if and only if, for all classes c, $c \subseteq c$

as well as

(13) $(\forall x)(\forall y)(xIy \vee yIx)$ is true if and only if, for all classes c and d, $c \subseteq d$ or $d \subseteq c$.

By (8) and (9), respectively, the sentence on the left-hand side of (12) is true and that on the left-hand side of (13) is false. But both sentences (12) and (13) are easily seen to be true sentences of the metalanguage L*. It is beyond the scope of this interlude to explain how every instance of the T-scheme for L can be verified in L*.

Even though the universality of everyday language makes it unavoidably inconsistent, we can still (according to Tarski) readily formalize fragments of that language, and the method of defining satisfaction and truth can be applied to the resulting formal language L within a part L* of everyday language functioning as the metalanguage. For example, if the domain of interpretation consists of all human beings (past, present, and future), then as the basic predicates of L we might take: xDy, interpreted as "x is a descendant of y"; xEy, interpreted as "x is the same as (i.e., is equal to) y"; and Wx, interpreted as "x is a woman". Then in L the sentential function xCy, interpreted as "x is a child of y", can be defined by saying that x is an immediate descendant of y, which in turn is expressed by the sentential function $xDy \land \neg(\exists z)(xDz \land zDy)$. The sentential function xMy, interpreted as "y is the mother of x", is then expressed by the sentential function $xCy \land Wy$. After satisfaction and truth are defined for L – as was done previously for the language of classes, but now with variables restricted to range over human beings – one thus obtains the following sentences as instances of the T-scheme:

(14) $(\forall x)(\exists y)(xDy \land Wy \land \neg(\exists z)(xDz \land zDy))$ is true if and only if everyone has a mother;

and

(17) $(\exists y)(\exists z)(Wy \land \neg Wz \land (\forall x)(\neg xEy \land \neg xEz \rightarrow xDy \land xDz))$ is true if and only if there is a woman and a man from whom everyone else is descended.

This last sentence is thus true if and only if there were some "ur-parents" such as Adam and Eve.

Formal Languages Cannot Define Their Own Truth

As a complement to his definition of truth for a formal language L within a metalanguage L*, Tarski showed by adaptation of the Liar Paradox that we cannot define the notion of truth for L within L itself. First of all, the expressions of L can be effectively enumerated by simply listing all finite sequences of basic symbols as if they were words in an alphabet (assuming we have imposed some alphabetic order on the basic symbols). Among the totality of expressions of L are the sentences, whose enumeration can be obtained by dropping out all the expressions that are *not* sentences as one goes along. If the sentences are thus put in an order as $S_0, S_1, S_2, \ldots, S_n, \ldots$, where n is any natural number, and if L has a method for naming each natural number by an expression, call it $\#n$, then every sentence of L can be named in L itself via its associated number. Tarski used this to show that, for L containing the language and axioms of elementary number theory, the notion of truth for L cannot be defined in L unless L is inconsistent. That is, there is no sentential function $T(x)$ – with one free variable x and expressing that the sentence with number x is true – for which the T-scheme can be proved for each sentence S_n in L in its following formal version $(T)_L$:

$(T)_L \quad T(\#n) \leftrightarrow S_n$.

For, using the technique of self-reference introduced by Gödel in 1931 to prove his incompleteness theorem,[8] Tarski showed that there exists an n such that

$$S_n \leftrightarrow \neg T(\#n)$$

is provable; that is, there exists a sentence which – via its number – expresses of itself that it is not true, assuming $T(x)$ indeed expresses that x is the number of a true sentence. Combining this with the presumptive formal T-scheme $(T)_L$ immediately leads to a contradiction. Tarski viewed this result as corroboration of his argument for the strict necessity of limiting attention to formal languages whose truth predicate can be defined only in a metalanguage.

Truth and Definability in a Structure

It was not until 1957 that Tarski, together with his student Robert Vaught, published definitions of the notions of *satisfaction and truth in a structure*[9] for applications to logic and mathematics, though there is considerable evidence that Tarski already had these clearly in mind by 1930, well before embarking on the *Wahrheitsbegriff*. The framework for their definitions is as follows. Suppose L is a formal language with some basic relation symbols R of various numbers of arguments; a structure (or model) M for L is given by (i) a non-empty set of objects, called the *domain* of M, and (ii) an n-ary relation R_M between elements of the domain of M for each relation symbol R of L with n argument places. (When $n = 1$, R_M is simply a subset of the domain of M.) The basic sentential functions of L are the relation symbols followed by a variable in each argument place; and the sentential functions are generated from these by negation, disjunction, and universal quantification, just as explained in Defn. 1. Then the definition of the set of free variables of a sentential function of L is given just as in Defn. 2, with the sentences of L being those without free variables as in Defn. 3. It is only in the Defn. 4 of satisfaction that some minor modifications need to be made. First, by an *assignment a in M* is meant an association with each variable x of an element a_x of the domain of the structure M. Then, in place of Defn. 4 we must explain the notion *a satisfies F in M*, where F is a sentential function of L and a is an assignment in M. The recursive clauses for the modified definition are obtained from those in Defn. 4 simply by adding "in M" at each appropriate point. In particular, the modified clause (iv) reads: if a is an assignment in M and F is a sentential function of the form $(\forall x)F'$, then a satisfies F in M if and only if every assignment b in M that agrees with a except possibly at the variable x, also satisfies F' in M.

Having thus obtained the definition of satisfaction in the structure M, Tarski and Vaught defined "S is true in M", for S a sentence of L, to hold (as in Defn. 5) just in case every assignment (in M) satisfies S in M.

For Tarski, an important example of the notion of truth in a structure was given by the structure R basic to elementary algebra, whose domain is the set of all real numbers and which has four basic relation symbols corresponding to the relations of equality ($x = y$), inequality ($x < y$), addition ($x + y = z$), and multiplication ($x \cdot y = z$). Tarski called a set D of real numbers *definable in R* if there is a sentential function $F(x)$ with one free variable x such that D is the set of all real numbers d such that the assignment of d to x satisfies $F(x)$ in R; the notion of an n-ary relation being definable in R is explained similarly. Using his elimination-of-quantifiers procedure for the elementary theory of real numbers, Tarski was able to show that: (i) if the sentential function s does not contain the relation of multiplication, then D is a finite union of intervals with rational endpoints; and (ii) when multiplication is also included, D is a finite union of intervals with algebraic endpoints. Tarski hoped these results would attract the attention of mathematicians, but he thought that they would be put off by the metamathematical employment of formal languages. In a paper published in 1931 on definable sets of real numbers,[10] he devised a method that circumvented this problem by directly generating the definable relations from certain basic relations by closing them under the set-theoretical operations of complementation, union, and projection – which correspond to the logical operations of negation, disjunction, and existential quantification.[11] His stated reasons for using this approach in the 1931 article were as follows:

> Mathematicians, in general, do not like to deal with the notion of definability; their attitude toward this notion is one of distrust and reserve. The

reasons for this aversion are quite clear and understandable. To begin with, the meaning of the term 'definable' is not unambiguous: whether a given notion is definable depends on the deductive system in which it is studied.... It is thus possible to use the notion of definability only in a relative sense. This fact has often been neglected in mathematical considerations and has been the source of numerous contradictions .... The distrust of mathematicians towards the notion in question is reinforced by the current opinion that this notion is outside the proper limits of mathematics altogether. The problems of making its meaning more precise, of removing the confusions and misunderstandings connected with it, and of establishing its fundamental properties belong to another branch of science – metamathematics.[12]

Tarski went on to say that "without doubt the notion of definability as usually conceived is of a metamathematical origin" and that he has "found a general method which allows us to construct a rigorous metamathematical definition of this notion." But then he wrote that

by analyzing the definition thus obtained it proves to be possible ... to replace it by [one] formulated exclusively in mathematical terms. Under this new definition the notion of definability does not differ from other mathematical notions and need not arouse either fears or doubts; it can be discussed entirely within the domain of normal mathematical reasoning.[13]

Despite Tarski's efforts to attract the interest of mathematicians in this way, here and in later articles, there is no evidence that they saw a significant difference between the metamathematical notions of definability in a structure and their mathematical surrogates in the sense just described.

The Impact of Tarski's Theory of Truth

During the period of the 1920s in which Tarski established his research career in Warsaw, there were close relations between logicians, mathematicians, and philosophers. Thus, though he identified himself first and foremost as a mathematical logician, it was natural for him to try to reach out to all three audiences with appropriate versions of his theory of truth – not only in Poland but also on the international scene. At first the reception was mixed. Logicians such as Löwenheim, Skolem, Langford, and Gödel had been using the informal notions of satisfaction and truth in a structure in a sure-footed way for some years, and Tarski himself seems to have used them in his own work on the elimination of quantifiers for various axiomatic theories. Among logicians, Tarski's precise definition did not become the norm until much later – in fact, not until the 1950s. Outside of logic, those relatively few mathematicians who were interested in applying logic to such areas as algebra, geometry, and number theory were undisturbed by the supposed differences between metamathematical and mathematical definitions of satisfaction and truth in a structure; like the logicians, they were generally satisfied with an informal explanation.

In the mid-1930s, the first philosophers to whom Tarski communicated the ideas of his theory of truth were Karl Popper and Rudolf Carnap, and both took to it quickly. Carnap urged Tarski to communicate it to the Unity of Science meeting in Paris in 1935; at that meeting, however, other members of the Vienna Circle (most prominently, Otto Neurath) were openly critical. The bone of contention was whether the semantical concepts could be reconciled with the strictly empiricist and anti-metaphysical point of view of the Circle. In the write-up of his talk "The Establishment of Scientific Semantics" for the Paris conference,[14] Tarski tried to make the views compatible, but he still found it necessary to respond to critics as late as 1944 in an expository article, "The Semantic Conception of Truth and the Foundations of Semantics."[15] It is fair to say that, since then, at least Tarski's T-scheme has been central to many philosophical discussions of the nature of truth, though the philosophical significance of his *definition* of truth – or whether truth in the everyday sense is even definable – continues to be a matter of considerable dispute. The T-scheme has also been important to extensive efforts by workers in philosophical logic toward loosening the strict separation of language from metalanguage so as to allow a language L to contain its own truth predicate while still respecting the scheme to some extent or other.[16]

Tarski's theory of truth was to have its greatest impact in technical developments of logic and its applications in the field of model theory. The central notion of this subject is that of a structure M being a model of a set A of sentences, considered as the axioms of a theory; M is a model of A if each sentence S of A is true in M. At the hands of Tarski and others, model theory underwent an extraordinary development starting in the 1950s, to begin with for languages of first-order logic and later for logics with more general sorts of propositional operations and quantifiers. In each case, Tarski's definitions provided the paradigms for the requisite notions of satisfaction and truth needed to carry out such extensions of logic in an exact and systematic way. (These developments are elaborated in Interlude V, which follows Chapter 11.)

Beyond logic, Tarski's approach even reached out to the semantics of everyday ("natural") language despite his insistence that this was not suitable for scientific study. One of Tarski's own students from the 1950s, Richard Montague, showed how significant parts of natural language could be regimented formally and how Tarskian semantics could then be applied to them; this approach came to be known as Montague grammar, though it involved an extension of both the semantics and the syntax of classical logic.[17] Since then, many questions of the semantics of natural language have been treated informatively in precise, Tarskian, style – for example, concerning such quantifiers as "few", "many", "more ... than", "almost all", and so on.[18] More recently, applications have appeared in computer science, where the "denotational semantics" of programs is used to talk in a rigorous way about the meaning of statements that make up a program and about the correctness of programs for their intended applications.[19]

Aside from the value of all these specific adaptations, Tarski's approach from 1930 on,to logic in general and to the theory of truth in particular, has been most influential in eventually creating a substantial shift away from the entirely syntactic ways of doing things in metamathematics – as first promoted in the late 1920s by David Hilbert with his theory of proofs – and toward set-theoretical, semantic ways of doing things. Unlike Hilbert, who was concerned with proving the consistency of formal theories for a foundation of mathematical practice by the most restricted means possible, Tarski thought there should be no restriction on using the methods of twentieth-century mathematics when applied to what he called the methodology of deductive sciences. In that respect, mathematical logic has in large part followed his lead in all its subfields, including parts of proof theory.

5

How the "Unity of Science" Saved Tarski's Life

Crossing

THE GDYNIA-AMERICA LINES M/S *Pilsudski* bound for New York sailed out of the Baltic Seaport of Gdynia, Poland, on 11 August 1939 with Alfred Tarski on board. With his steamship ticket, Polish passport, and a temporary visa for the United States in hand, he had kissed his family good-bye in Warsaw and boarded the train for the port, carrying one light suitcase with summer clothing; no need for winter clothes because he just planned to stay for about a month. In his journal he noted only the date: 22 August 1939, the city: New York, and the words: "arrived on the Piłsudski" to record what turned out to be a momentous voyage.

The crossing of the Atlantic had begun auspiciously. The very first day, Tarski discovered that the mathematician Stanisław Ulam, from Lvov, and his younger brother Adam were fellow passengers. By happy coincidence, Ulam, a protégé of John von Neumann of the Institute for Advanced Study at Princeton, was also on his way to Cambridge, Massachusetts. He was not headed for the Unity of Science meeting like Tarski but rather for a lectureship at Harvard, where he had already spent three years in the prestigious Society of Fellows. Ulam recalled:

I suddenly noticed Alfred Tarski in the dining room. I had no idea he was on the boat. Tarski, famous logician and lecturer in Warsaw, told us he was on his way to a Congress on the Unity of Philosophy and Science.... It was his first trip to America.... We ate at the same table and spent a good deal of time together. I still have an old shipboard photograph which shows Adam, Tarski, and me dressed in dinner jackets ready for the gay American social life.[1]

Stan Ulam was the perfect shipboard companion for Tarski: lively, urbane, and a seasoned traveler to the United States. Already sensitive to

Tarski with Stanisław Ulam (left) and Adam Ulam (right) aboard the M/S *Piłsudski*, August 1939.

the ways a Pole might misunderstand America and Americans, he could alert him to the mistakes a foreigner might make. Although younger than Tarski, Ulam had made a splash with some outstanding early work at the University in Lvov, where he had studied with Kuratowski, who had a high opinion of Tarski. In the small, intense world of Polish mathematics, Alfred and Stan had met many times; at Harvard in the Society of Fellows, Ulam had met Tarski's advocate Quine, who talked about Tarski with great enthusiasm and respect.

Aboard the *Piłsudski*, Tarski and Ulam talked mathematics, gossiped about colleagues, and went over the simple practical details Tarski would need to know once the ship docked. Inevitably, they also discussed world affairs and worried together about the menacing political atmosphere in Poland and Europe. Ulam was Jewish and from a wealthy family of bankers and lawyers whose position in Polish society was the highest Jews could attain. He did not have Tarski's financial worries and, unlike Tarski, he had not changed his name or religion. Nor had he sought a university position after receiving his Ph.D. in 1933, for, brilliant though he was, he knew that it was an impossibility for a Jew.[2]

By the time the ship arrived in New York, all thoughts of "the gay American social life" they jokingly had anticipated early in the crossing were submerged by rumors of the alliance between Hitler and Stalin.

Following the Munich Pact in 1938, the Germans had occupied all of Czechoslovakia; this led the British and French to abandon their policy of appeasement and form an anti-aggression front with Poland and other smaller threatened European countries. But the Poles refused to let the Russian armies cross its territory in case of war. The Germans and the Russians, whom all had assumed to be deadly enemies, then suddenly transformed themselves into even deadlier allies. First they made a trade agreement and then, within a few days, signed the non-aggression pact that stunned the world. Although this pact was not actually signed until after the *Pilsudski*'s arrival, the ship's radio had broadcast the startling rumor that it was about to happen; indeed, two days later the rumor became fact.

Adam Ulam – not quite seventeen – was about to enter Brown University as a freshman that fall. Intelligent and sophisticated beyond his years, he would later become a distinguished professor of history and political science at Harvard University and one of the world's foremost authorities on Russia and the Soviet Union. Of that August 1939 he said:

In hindsight we realized we must have been blind, but when we left Poland we really didn't think war was *immediately* imminent. Sure, everyone talked about the possibility of war, of the German takeover in Austria and Czechoslovakia, but there was also a lot of talk and bravado about how France and England would defend Poland. In August, when we left, all the Poles were going off on their traditional August vacations, as if everything was normal, as if nothing was about to happen. My father, for example, after seeing us off on the ship, went to the same Baltic Sea resort that he always went to. And my brother who had been traveling back and forth between the United States for the three years previous didn't think this year was different from any other.[3]

Waiting to greet the Poles as they disembarked were John von Neumann for the Ulams and Carl Hempel for Tarski. Von Neumann and Hempel, themselves European exiles, had already lived in the United States for several years, both having seized earlier opportunities to leave Europe. They were much more distressed than the newly arrived passengers. Ulam noted, "Johnny appeared very agitated. People in the United States had a much clearer and realistic view of events than we had had in Poland." Even the officer in charge of Ulam's army reserve unit in Poland had, without batting an eyelash, given the necessary permission for him to travel abroad.[4]

Unquestionably, it was a stroke of luck for Tarski to have landed in New York at this moment, out of range of the bombs that would devastate Warsaw and safe from the Nazi roundup and murder of Jews, intellectuals, artists, and "deviants" of any stripe that was to follow. At the same time, his wife, his two small children, his mother, father, brother and sister-in-law, aunts, uncles, and cousins were in mortal danger. The next few weeks were agonizing. Even if there had been a way for him to return home, to do so would have been suicidal. So there he was, safe but stranded, with the wrong kind of visa, no job, no money to speak of, a mountain of anxiety and guilt, and that one suitcase with summer clothing.

Carl Hempel, a kind man who knew from his own experience what Tarski would be facing as a stranger in New York, took him on a quick tour of the city and its environs the day after he arrived. Using two full lines in his journal – more space than he accorded for the day he set foot in the United States – Tarski listed the places they visited: New Rochelle and Orchard Beach, and his companions: E & C. G. Hempel; E. & O. Helmer, L. & S. Broadwyn. Helmer, like Hempel, had come to America sponsored by Paul Oppenheim, a wealthy philanthropist interested in the philosophy of science, and both had grants from the Rockefeller Foundation to work as Carnap's research assistants. The Broadwyns, like the Oppenheims, were wealthy intellectuals and patrons of a sort, happy to meet distinguished scientists and to extend hospitality to the refugee scholars who were flooding into the country. At their home in New Rochelle, in the company of his European friends, Tarski ate his first American family meal.

Tarski did not linger in New York. On the 24th of August – the day after the German-Soviet non-aggression pact was signed – he took the train from New York to Boston a week before the Unity of Science meeting was to begin. This gave him ample time to prepare his talk as well as to worry about what was happening at home. And there was plenty to worry about. Stan Ulam recalled, "I spent much of my time with the other Poles who had found their way to Cambridge – Tarski, Stefan Bergman and Alexander Wundheiler. They were all terribly unhappy.... We would sit in front of my little radio which I left on all day long and listen to the war news."[5]

In the three short weeks since Tarski left Poland, his world had collapsed. On September 1, the Germans bombed Warsaw and invaded from the West. The *New York Times* headline of September 6 announced: "GERMANS SHELL WARSAW, RESIDENTS IN FLIGHT". A few weeks later, the Russians entered from the East; the treaty between Hitler and Stalin had secret clauses about how the country was to be partitioned between them, and once again Poland was occupied and torn apart. The collective feeling of stunned surprise that this had all happened so unexpectedly and so quickly gave way to terrible anxiety as the reality of war sank in. It would be weeks before Tarski knew what had happened to his wife and children, and it was much longer still before he knew the fate of the rest of his family.

Three months after Tarski's arrival in the United States, the Polish ship that had carried him across the Atlantic was torpedoed and sunk by a German submarine.[6]

The Unity of Science Conference

The Unity of Science Conference opened the evening of September 3 with a "smoker" at Eliot House at which Harvard President James B. Conant gave the opening address. The sessions began the following day with a program listing sixty-five speakers and twenty sessions, many going on concurrently, with topics ranging from issues in the unity of science and the philosophy of science to biology, sociology, psychology, language and semantics, probability, and (of course) logic. Always able to work and concentrate under any conditions, Tarski seems to have been fully involved in the matters of the conference; a photo taken outside Harvard's Emerson Hall shows him in a typical stance, clutching his briefcase and gazing directly into Otto Neurath's eyes. Was he as detached as Rudolf Carnap claims *he* was when he wrote, "In spite of the exciting world events, we found it possible to devote ourselves to the theoretical discussions of the Congress"?[7] The *New York Times* heading for an article on the conference was: "Setback In Science Seen. War Effect Weighed on Eve of World Parley at Harvard"; the tone of the piece was pessimistic, for obvious reasons.

Tarski presented his paper on September 9, the last day of the meeting. On the program, substantiating his original vacillation about coming, there is a question mark after his name instead of the title of his talk.[8] In spite of his capacity for separating work from emotion, he was bound to have been very worried by the absence of his colleagues Leon Chwistek of Lvov and Kurt Grelling of Berlin, who had been scheduled to speak before him that very day, and even more anxious about Janina

At the Unity of Science Conference, Harvard, September 1939; (from left to right) Otto Neurath, J. H. Woodger, Alfred Tarski, E. C. Berkeley, and Willard Quine (hidden) between Haskell Curry and Stephen Kleene, on the steps of Emerson Hall.

Hosiasson-Lindenbaum of Warsaw, who was to have sailed from Gdynia one week after him. It was not until the war was over that he learned of their terrible end.

   ❧   ❧   ❧

From a professional point of view, being present at the conference was good. Quine jokingly characterized the meeting as "the Vienna Circle, with some accretions, in international exile"[9] and Tarski, besides renewing old acquaintances, benefitted from the accretions by becoming friends with some of the leading American logicians, such as Haskell Curry of Pennsylvania State University and Alonzo Church of Princeton, as well as Church's students Stephen Kleene and J. Barkley Rosser. Of equal value, the assembled scholars had the chance to discover that Tarski "live" lived up to Tarski in print – and then some! In short, he impressed them. Along with Quine, they became his closest comrades-in-arms in logic for many years, but more importantly at that time, they took him and his dilemma to heart and did what they could to make his life bearable. They invited him to their homes, took him on excursions and outings, invited him to lecture, wrote glowing recommendations for jobs, and tried – without success – to help him bring his wife and children to the United States.

At the end of the conference, a group of logicians and their wives went to Mount Monadnock, a favorite hiking spot in southwestern New Hampshire. Although the lone mountain was a far cry from Tarski's beloved Tatras, he was pleased to be out walking. Unfortunately, his pleasure was short-lived. At the summit, he slipped, fell, and cut his scalp on the sharp rocks. Quine in his autobiography reports: "Two of us made a cradle with our fists and wrists and carried him down the mountain. A doctor in Jaffrey sewed him up."[10] By contrast, Tarski in *his* journal says not a word about being injured or requiring medical attention, although he does note Mount Monadnock and Jaffrey for that day. Only on the rarest of occasions does a glimmer of comment about what happened on any day appear; for example, on his first visit to Haskell Curry at Pennsylvania State University he noted: *okoliczne góry* [mountains in the vicinity] but nothing more. One may only guess at his first impressions of his new friends: the methodical, solid Alonzo Church; the tall, raw-boned New Englander Stephen Kleene; the hearty, effusive, entrepreneurial Barkley Rosser; and the proper, cosmopolitan Willard Quine, whom he had already met in Warsaw. No doubt he registered the differences between them as Americans and himself as an Eastern European, but only later did he reveal what he thought of them as individuals.

Although an experienced climber, Tarski was a jittery person and prone to physical clumsiness, in contrast to his intellectual adroitness. Even in the best of circumstances he might have slipped and fallen; but surely the turmoil, anguish, and uncertainty of the autumn and early winter of 1939 made him unusually distracted. While desperately trying to get his family out of Poland, he was also looking for a job. To succeed in these endeavors he first had to become a permanent resident of the United States but, as it turned out, to do *that* he would be required to leave the country and re-enter with a permanent visa rather than the temporary one he possessed; and to get that precious permanent visa, he had to convince the State Department that he would be an asset rather than a liability to the United States. In his case, it was not as bad as it might have been. It was relatively easy at that time to leave the country and go to Havana. Once there, by submitting letters in support of his permanent residency, he could apply to the U.S. Consul for readmission.

What letters they were! Rallying to his aid, no fewer than seventeen of Tarski's academic friends – including Marshall Stone, Rudolf Carnap, Willard Van Orman Quine, Eric Temple Bell, Ernest Nagel, Haskell Curry, and Bertrand Russell – wrote to extol his virtues. Quine said, "without reservation I regard Professor Tarski as the greatest logician of our time."[11] Although Quine did not yet have the stature of his later years, he was at Harvard and his ranking Tarski as "the greatest" carried weight. Equally important was the letter from Professor Marshall Stone, also of Harvard, which stated that the mathematics and philosophy departments had recommended that Professor Tarski be given a position as research associate for that year and possibly longer.[12] As it happened, Marshall Stone's father was a sitting U.S. Supreme Court justice; surely this did not go unnoticed.

Others from Harvard, Columbia, Chicago, Yale, and the California Institute of Technology chorused Tarski's greatness, putting him in company with Aristotle, Frege, and Russell (which some of his colleagues continued to do ever after), and proclaimed that any country in the world would be enriched by his presence. The U.S. Consul was duly impressed and the permanent visa was granted in Havana on December 29, thus enabling Tarski to re-enter the United States. He spent New Year's Day, 1940, in Miami. This was an optimistic moment, and one might easily imagine that he was not prepared for the difficulties he was to face in the next few years.

Job Hunting

Anyone who knows of Tarski as "the great logician" may find it difficult to believe that between 1939 and 1942 he was not snapped up and given a position at a major university – but the fact is, he was not. The United States, like the rest of the world, was barely emerging from a decade-long economic depression. In the academic world as elsewhere, jobs were scarce and competition was fierce. To make matters worse, there was an influx of brilliant intellectual refugees from Europe also desperate for positions; and, finally, Tarski had an additional problem: his area of expertise, mathematical logic, was not a mainstream field at any university. To tide him over until he found something, funds were cobbled together at Harvard to appoint him as a research associate, and lodging was found for him at 340 Harvard Street.[13]

To make his presence more widely known, a number of lectures were arranged at various universities in the Northeast. The first of these was at Columbia, where Tarski was a guest lecturer in Professor Ernest Nagel's undergraduate philosophy class and where a very young Leon Henkin was in the audience. Many years later, after doing graduate work at Princeton with Alonzo Church, Henkin would become Tarski's right-hand man in Berkeley, but in 1939 he had never heard of him. His account of that early talk gives a taste of Tarski's charisma:

I had no idea who he was before Professor Nagel told us that a famous logician from Poland was coming to talk to us. He spoke on decision methods and on Gödel's incompleteness proofs; although he had a very heavy Polish accent, his English was understandable and I was terribly impressed. When he finished and it was time for questions, I raised my hand. I had a "real" question to ask but it was also clear to me that my purpose in asking it was not merely to gain information; I wanted to establish contact with him; I wanted him to notice me; I wanted to be more than just a passive member of the audience.[14]

In the search for regular teaching possibilities, Richard Courant at New York University was approached about a position for Tarski. Courant had no interest in logic; his response was downright nasty and reportedly included the gratuitous suggestion "If he needs a job, why doesn't he go join the Free Polish Army?"[15] Courant, a German Jew, had fought on the German side in World War I. He was not alone in implying that Tarski was a coward not to be actively fighting in the war, and whether or not Tarski ever heard such words directly, he must have been sensitive to the notion. However, romantic and passionately Polish as he was, he was also a realist and in the circumstances the best he might possibly do was to try to extricate his family from Poland; in this respect he did everything he could. Nevertheless, a few malicious tongues wagged and continued to do so for years afterward.

New York University's rejection was countered by an offer of a visiting professorship from the City College of New York for the following spring (1940); it was then that Tarski gave his first full-semester university course in the United States. Although it was only a temporary appointment, the announcement made the *New York Times* of 21 November 1939, with Tarski's photo at the top of the column. This certainly served him well when he appeared before the immigration authorities in Havana in late December of that year, for it provided proof that he had a job that was of sufficient importance for the *Times* to consider newsworthy.

How the "Unity of Science" Saved Tarski's Life

News of Tarski's appointment to CCNY;
New York Times, 21 November 1939.

Serendipity played its part at CCNY. One of the students in that class was young Kenneth Arrow, a future Nobel Prize winner in economics. Arrow's description of Tarski's command of English is at odds with Henkin's, but his appraisal of Tarski's genius is not:

The first lecture, nobody knew what he was talking about. We wondered if he had learned any English at all. Then we began to analyze it and realized that all the stresses were off so we learned to interpret his way of speaking – to decode him – and everything was fine. Of course *he* improved too.

It was a great course, Calculus of Relations. His organization was beautiful – I could tell that immediately – and he was thorough. In fact what I learned from him played a role in my own later work – not so much the particular theorems but the language of relations was immediately applicable to economics. I could express my problems in those terms.[16]

For his part, Tarski quickly recognized Arrow's gifts not only for mathematics but also for language, and at the end of the term he asked his bright student if he would be willing to proofread Helmer's translation of *Introduction to Logic* that was about to be published by Oxford University Press; Tarski wanted a native speaker to check the text. Arrow thought it was a great honor to be asked and readily agreed.

So I spent a good part of that summer reading proof at zero pay. I noticed that Tarski had a nose for what was idiomatically correct even though his own English was not necessarily so. He would catch things and ask me, "Is this really good English?" And he was right most of the time. We had a number of interesting conversations and he kept asking me about words.[17]

Arrow did his job well and Tarski graciously gave him a separate last line of thanks in the preface, after he had thanked Helmer and the other colleagues who helped him.

In capital letters, on a separate page he dedicated the book "TO MY WIFE". He had not seen her for more than a year when he wrote those words in September 1940, and she would not see the book until January 1946.

The Chair of Indecency

While Tarski was teaching at CCNY, a remarkable episode occurred in the annals of academic freedom versus political and ecclesiastical pressure. In February 1940, Bertrand Russell was appointed Professor of Philosophy at CCNY for the following year (February 1941 to June 1942) and slated to teach logic and its relation to science, mathematics, and philosophy.[18] When the news appeared in the press, a bishop of the Episcopal Church wrote a letter to the editor of the *New York Post* denouncing the appointment and demanding that it be rescinded on the grounds that Russell was "a recognized propagandist against religion and morality who specifically defends adultery." This stimulated a barrage of hostile letters

Bertrand Russell in 1938.

from religious and political quarters, and a taxpayer's suit was filed in the New York Supreme Court asking that the appointment be vacated on the grounds that Russell was an alien and an advocate of sexual immorality. The suit was initiated in the name of a Mrs. Jean Kay, who stated that she was afraid of what might happen if her daughter were to enroll at CCNY – despite the facts that, at the time, her daughter was too young to attend college and CCNY did not admit women. In his brief to the court, Mrs. Kay's lawyer described Russell as "lecherous, salacious, libidinous, lustful, erotomaniac ... and bereft of moral fiber."[19]

It took little more than a month for a judge, backed by the infamous Tammany Hall political machine, to rule that the Russell appointment must be voided because it "adversely affects public health, safety, and morals" and because it "encouraged conduct tending to a violation of the penal law." The judge stated that the Board of Education had in effect established a "chair of indecency," and he later said that he had had to take a bath after reading one of Russell's books. The press, naturally, had a field day; a highlight is the *New York Post*'s cartoon showing a pipe-smoking Russell seated in a chair perched on top of *Principia Mathematica* and seven more of his philosophical works, with the trial judge pointing at the "chair of indecency."[20] In newspapers and magazines, Russell's colleagues cried out their indignation over the threat to academic freedom, but none of this helped.

Bertrand Russell's "Chair of Indecency"; *New York Post* cartoon, 2 April 1940.

What Tarski thought about the affair as it swirled around him on the campus and in New York is easy to guess: his sympathy would lie with Russell, but he made no public statements. His silence and desire to keep a low profile is understandable, since he himself was not yet a citizen and was just in the process of obtaining his first naturalization papers.[21] His concern over what was happening to his family in Warsaw under the Nazis was paramount and consumed a huge amount of mental and emotional energy. He wrote to many people he thought might be able to help. One was Father Józef Bocheński, a Polish logician who was in Rome and had heard of Tarski's plight through the grapevine. In understandably obsessive detail he told of his position in the United States as a permanent resident with a nonquota visa and his so-far unsuccessful attempts to get his family out of Poland – although, he said,

by current law, my wife and children have the automatic right to obtain similar visas.... For many months now I have been making every effort to bring my family here. Unfortunately I have been met with unsurmountable hindrances. My plan was to get my family to Copenhagen and from there to the U.S.A. This plan was almost carried into effect. Regrettably due to recent developments [Germany had invaded Denmark two weeks earlier], the plan collapsed and I have to seek new ways.[22]

Tarski gave a complete account to Bocheński of which documents he had sent to the various relevant consulates in Copenhagen, Warsaw, Washington, and Berlin; of two possible avenues that still remained open for his family to escape; of his contact with the United States Steamship Lines; and of the difficulty of communicating directly with his wife even though "we have been sending each other an infinite number of letters. They all disappear somewhere on the way. As far as I know, my wife has received only one letter."

The picture of a nightmare of bureaucracy, in the face of the increasing threat that all of Europe would soon be at war, is painfully clear: "Time presses," Tarski continued, "I am not sure that by the time you hold this letter in your hands, it will not already be too late," and he concluded by saying he will be grateful for anything Bocheński can do even if it is only to convey the information in the letter to his wife.

Nine months later, nothing had changed, but Tarski nevertheless was still hoping that Maria and the children would be able to escape via Berlin. In a letter to Heinrich Scholz (his colleague in Münster) written 7 January 1941, eleven months before the United States entered the war, Tarski sounds almost optimistic:

You can easily imagine how exhausting my life is now. Preparing lectures in a language that one doesn't master well takes up terribly much time. Also the English translation of my introduction to mathematical logic has kept me very busy. I read the final corrections draft just today; the book will appear on February 1st.

More importantly, I make new efforts ... to make the trip of my wife and children to the United States possible. Maybe my wife wrote you already that in the last weeks the state of affairs seems to be more favorable: she hopes that she will receive permission in the near future to leave Warsaw, after which she will go to Berlin and try to obtain the necessary visas. Finally she will take the big journey – unfortunately not through Spain and Lisbon (since it will presumably be hard to obtain transit visas) but through Siberia. I know very well that if you will be able to make her stay in Berlin in any way easier you will do it without my request – and you also know that I thank you from all my heart for your willingness to help.[23]

A final attempt to get an exit visa for Maria Tarski and the children was orchestrated by Anders Wedberg, a young Swedish philosopher who had studied at Harvard and Princeton and was teaching at Cornell when Tarski came to the United States. The sympathetic Wedberg wrote to his father, a Supreme Court justice in Sweden, hoping to enlist his help. In his letter, after describing Tarski's distress and outlining the dangerous plan by which Maria would attempt to get a visa to America from the consulate in Berlin, the younger Wedberg asks,

> Dad, do you think that Mrs. Tarski and her children could get a visa to Sweden for a short period of time ...? Tarski could support them there. He has recently got a $2500 Guggenheim Fellowship and he also has a fellowship at Harvard. I would be tremendously glad if you could inquire into this case and even put in a good word for them at the Swedish authorities deciding such matters.[24]

Although Wedberg's father did indeed petition the Swedish Royal Foreign Office and travel visas were authorized, a letter of 15 July 1941 from the foreign office to Supreme Court Justice B. Wedberg states that, since there is no longer any possibility of traveling from Sweden to the United States (presumably because German U-boat warfare had intensified), a visa cannot be granted. Sadly, Tarski's assessment of "too late" in his letter to Bocheński was correct. Notwithstanding the attempts by powerful friends in the United States and Europe, Tarski's family was to remain in Poland for the entire war period. From time to time he received news that his wife and children were alive in Warsaw, and there is some indication that through Scholz, who had contacts in the German foreign ministry, they may have had some kind of protection, at least before the uprising of 1944. It was not until after the war was over that he would hear the full story and learn what happened to the other members of his family.

Living on Benzedrine

Meanwhile, in New York, Tarski continued his teaching, his research, his publications, his speaking engagements, and his search for a permanent position. "I lived on Benzedrine and made my living going from one mathematics department to another, giving lectures," is what he told his students in the 1950s about his first years in the United States. As usual, he granted himself a little license for dramatic effect; following the semester job at CCNY he did, after all, still have his Harvard research appointment, then two lecture series of twelve sessions each at the Young Men's Hebrew Association in New York, and finally a Guggenheim Fellowship for the twelve months beginning April 1941 and renewed for another six months to October 1942.[25] Nevertheless, it was a grueling schedule and he was constantly in motion.

Intellectually, the fall semester at Harvard in 1940 was especially stimulating. Russell was in residence to give the William James Lectures, and Carnap was visiting for the year. A group was formed to discuss logical problems, with Carnap, Russell, Tarski, and Quine its most active members. Carnap gave several talks on defining logical truth in semantical terms whose basis lay in Tarski's work on the concept of logical consequence. As Carnap relates,

Even though my thinking on semantics had started from Tarski's ideas, a clear discrepancy existed between my position and that of Tarski and Quine, who rejected the sharp distinction I wished to make between logical and factual truth.... In other problems we came to a closer agreement. I had many private conversations with Tarski and Quine, most of them on the construction of a language of science on a finitistic basis.[26]

Tarski's booking at the Young Men's Hebrew Association in New York, popularly known as "The 92nd Street Y," was most unusual because his lectures were delivered not in a traditional academic setting but rather in what is now the oldest Jewish Community Center in the United States.[27] The Y was famous for its educational activities and cultural events, which included lectures, concerts, poetry readings, and dance performances. Tarski was engaged as a lecturer in the fall of 1940/41 and again in the fall of 1941/42 to give twelve two-hour evening lectures on logic and the methodology of the deductive sciences. He was paid fifteen dollars per session with the proviso that there be at least ten students enrolled. The students' fees were fifteen dollars each for the entire semester. Other professors teaching at the Y in those years included Professor Sidney Hook, who gave a course on famous heresy trials beginning with Socrates and ending with the recent Russell affair, and Dr. Edgar Zilsel on pioneers and iconoclasts in the history of thought running from Plato to Einstein.[28] Tarski's course in logic was probably as demanding as any ever given there, although in his mind it was designed for the educated layman. There were also courses on psychology and literature and a concert series both years that featured the renowned Budapest String Quartet. The Y's individual lecture series that began in the 1930s and continues to this day included a long list of distinguished speakers in a wide spectrum of fields, ranging from serious academicians to film stars and radio, television, and theater personalities, politicians, presidents, and cabinet ministers. To name just a few of the more recent ones: Henry Kissinger, John Kenneth Galbraith, Dustin Hoffman, Bill Gates, and Dr. Ruth (Westheimer), the popular expert on sexual mores.

From New York, where Tarski had an apartment on West 82nd Street in Manhattan, and from Cambridge, where his home base continued to be at 340 Harvard Street, he commuted to universities all over the East Coast and the Midwest by bus and train, staying a night or two or occasionally a week. Between 1940 and 1941 he made a continual round of visits to Yale, Brown, Dartmouth, Princeton, Cornell, Penn State, the University of Pennsylvania, Chicago, Michigan, Notre Dame, and Illinois – giving talk after talk about his current work on undecidable statements in enlarged systems of logic and on the calculus of relations.

As was his custom, he slept little and worked most of the night. It sounds harrowing, but if his earlier and later life can be taken as example, his enormous energy and enthusiasm for his subject and his general appetite for life kept depression at bay. He was long accustomed to a huge workload. In Warsaw he had taught thirty hours a week in high school, had given regular lectures and seminars at the university, and had written what some would say were his most important papers. Wherever he went, he loved meeting people and exchanging ideas, getting high on talk, food, and drink. In spite of his great anxiety about what was happening at home, his innate exuberance permitted him to have a good time at least some of the time.

For a brief period he had a romantic relationship with Olaf Helmer's estranged wife, Eileen Holding, a painter then living in New York whom he had met the day after he arrived in the United States. After their liaison ended, Eileen confided to Olaf and others: "It was an intense and torrid love affair."[29] Understandably, Alfred looked for solace where he could find it; the company of women had always been and would always be extremely important to him, and chances are good that if other opportunities presented themselves during those early years, he was not shy. In any case, he was always looking.

On the weekends and during holidays, friends from mathematics and philosophy included him in their leisure-time activities and invited him to their homes and to visit scenic spots on the East Coast. He went to the Kleenes's ancestral farm in Maine, to Nelson Goodman's vacation home in Rockport, to Haskell Curry's house in Pennsylvania, and on countless trips to Peekskill and Mohegan Lake in New York with J. C. C. McKinsey. He visited Alonzo Church's family in Princeton where Mary Ann, Church's little daughter, was urged by her parents to be very nice to that Polish man because he was far, far away from his family and needed to be cheered up. Years later, after she was married to John Addison, one of Tarski's closest colleagues in Berkeley, Mary Ann related how she had obligingly toddled over to Tarski to sit on his lap.[30]

"Chen"

In New York, J. C. C. (John Charles Chenoweth) McKinsey, known familiarly as "Chen," became Alfred's closest friend. A slender man known for his keen intelligence and generosity, he was teaching at New York University during Tarski's stay on the East Coast but was in a precarious position because of his open homosexuality in an era when practically no one "came out." He did not have tenure, and the fear of scandal at nervous administrations obliged him to move from position to position. In the first English edition of his *Introduction to Logic*, Tarski singled out McKinsey for his "unsparing advice and assistance." Anyone who ever had contact with McKinsey echoed the sentiment. The philosopher and logician Ruth Barcan Marcus, who was his student at NYU in 1940, recalled:

He [McKinsey] took me under his wing and invited me to do a tutorial in logic. We met two or three times a week in Bickford's Cafeteria near Washington Square and we reviewed my work on the exercises he had given me from his own translation of Hilbert and Bernays. He urged me to go to graduate school but not to Harvard where, he said, Quine would clip my wings. I subsequently realized how much I took his mentorship and generosity for granted; I thought that was what all professors did.

He told me about Tarski, how much he admired him and how concerned he was about him, and that he spent a lot of time and effort trying to find a suitable job for him. He thought it was appalling that no one was seizing the opportunity to hire him.[31]

J. C. C. McKinsey, 1941. Photo for his
Guggenheim Fellowship application.

As further testimony to McKinsey's great regard for Tarski, and perhaps to help him get a position, a year later he wrote Benjamin A. Bernstein, his former Ph.D. advisor at UC Berkeley:

I have been very fortunate, the last two years, to be able to study with Alfred Tarski, who has been spending most of his time here in New York. I attended a seminar he gave here, and have also spent a good bit of time with him in private conversation, and I really feel that I have profited more by this connection alone than by anything else since I left California.... I'm sure that my paper on relations ... would never have been written without his suggestions and advice. At present he is writing a book on completeness of elementary algebra and geometry, and I have been spending a little time helping him express things in good style – but I am afraid he teaches me far more mathematics than I teach him grammar.[32]

Even more than friendship, open admiration was the key to an easy relationship with Tarski at every point in his life. No matter how celebrated he became in later years, a bald declaration of appreciation of his work or his person was the direct way to his heart, or at the very least predisposed him to look with great favor on the declarer. Not everyone understood how deep his need was for approval and glory, but those who did, like McKinsey, benefitted in one way or another. This is not to say that Tarski ever tailored his views on anything to garner approval; on the contrary, his honesty could be brutal.

That Chen McKinsey was homosexual did not unsettle Alfred at all; it may even have made him feel more at home. In Warsaw he had enjoyed the frisson of the risqué; he had been part of the cosmopolitan, anti-bourgeois crowd. So what good luck it was to find a friend like McKinsey to talk to about anything – from advice on how to formulate a phrase or concept precisely, to what his favorite drink was, to exotic sexual practices that he would certainly not discuss with his more straight-laced colleagues. They met often in those early days, spending many evenings and weekends together, and (until McKinsey's untimely death in 1953) continued to do so, especially after they both moved to the West Coast. Much later, Tarski told his younger colleague and friend Steven Givant that, in the United States, the only real friend he had had in the European sense was McKinsey.[33]

The period between 1939 and 1942 is described rather loosely in Tarski's curriculum vitae, despite his usual insistence on precision in all things. One must suspect that in this case, because he did not like the reality, for the practical purpose of finding a position he (understandably) embroidered it a bit. For example, his c.v. shows him as "Visiting Professor at CCNY for 1940–1941" when he taught there only for the spring semester of 1940; and for 1941–1942 he is listed as "Member, Institute for Advanced Study, Princeton" even though his journal shows him to be in New York for all of 1941; his stay in Princeton under the Guggenheim Fellowship did not begin until January 1942.

Tarski and Gödel

In Princeton Tarski lived on Chambers Street, where Kurt Gödel and his wife Adele had stayed for a brief period the year before. The Gödels found their apartment unsatisfactory and moved, but Tarski was not as fussy because his time in Princeton was short and the location convenient. From Chambers Street he could easily walk to the Institute, which was then housed in Fine Hall, the mathematics building at the university. Also, he went to New York frequently, for a weekend or longer, to see his friends and colleagues there. Presumably social life in Princeton for Tarski was dull compared to the high life and excitement of New York City with its crowded sidewalks, its restaurants, cafés, bars, movies, and theatre. On the other hand, in Princeton there was no lack of intellectual stimulation and social contact with the Gödels, the Churches, and the Oppenheims – Peter Hempel's friends and benefactors, who had settled in Princeton. The Oppenheims entertained frequently, "salon style," and at their lavish parties Tarski would meet Einstein, von Neumann, and Oskar Morgenstern, the Viennese economist later famous for his work with von Neumann on the theory of games.

The Gödels had not left Austria until the end of 1939. Their tardy departure forced them to travel eastward at considerable risk on the trans-Siberian railway with German exit visas and Russian transit visas to Manchuria and Japan and then across the Pacific. They arrived in Princeton in March of 1940, half a year later than Tarski's arrival in New York.[34] This may have been the reason for Tarski's hopes that, with help, his family could follow the same route to freedom.

Although he visited the Gödels soon after their arrival in 1940, it was not until he became a visiting fellow at the institute that Alfred and Kurt had prolonged close contact. In spite of their profound differences in character, outlook, and attitude toward life, they had a world of understanding in common – beginning with their condition of exile, their European heritage and culture, and above all the interconnectedness of their work and their commitment to it. The tension and excitement of the unspoken competitiveness between them in its own way heightened their closeness. In their later correspondence, Tarski was one of the very few people Gödel addressed by first name and with the familiar *du*; Tarski, of course, reciprocated but for him it was not so unusual.

With Adele Gödel, Tarski took special care to be charming, recognizing and sympathizing with how out of place and uprooted she felt in the Waspish, recherché Princeton atmosphere. Alfred and Adele had not been acquainted when Tarski was in Vienna in 1935, but he knew the city well enough for them to have common ground for conversation about the good life in cafés, the delicious pastries, the lively music halls, and the passing of the unique Viennese atmosphere. Said by some to have been a dancer in a cabaret (she claimed to have been a ballet dancer),[35] Adele was out of her element in the academic world; a homey, down-to-earth person, a sympathetic woman but not an intellectual and certainly not a standard-brand academic wife. For her it was *gemütlich* to be able to

Alonzo Church at the Tarski Symposium, 1971.

speak German with Tarski (he and Gödel also spoke and corresponded in German during those years, although later they switched to English). Unlike Oskar Morgenstern, who thought her "garrulous, uncultured, and strong-willed" and noted in his diary that he found it almost impossible to talk to Gödel when Adele was present,[36] Tarski felt comfortable with both of them. A year later, in a postcard to Adele, Tarski wrote of the good food she served him and how it "affects my heart at least as strongly as my sense of taste. Please do not forget me," he added, "and write from time to time."

Alonzo Church and Paul Erdös were others with whom Tarski had extensive contact in that period. Church had a long string of brilliant students, including J. Barkley Rosser, Stephen Kleene, and Alan Turing, and later Leon Henkin, John Kemeny, Martin Davis, Michael Rabin, Dana Scott, and Simon Kochen – to name only the first half of the list. As one of the founding fathers on the American logic scene, Church had helped launch the Association for Symbolic Logic and its journal in 1936 and continued to be one of the driving forces of that publication for almost half a century. Tarski was immediately drawn into his sphere of influence and within a few years would himself become president of the Association for Symbolic Logic. Church was a solid citizen and helpful to Tarski but he was just as stubborn as his new colleague, and when the latter had what

Paul Erdös at the Tarski Symposium, 1971.

he thought were excellent ideas for innovation – for example, in the title and contents of the *Journal of Symbolic Logic* – the determined Church was a force that Tarski had to reckon with.

By contrast, Paul Erdös, the itinerant Hungarian mathematician, was a sprite who seemed to live and breathe on mathematical collaboration. Erdös had known of Tarski's work since the early 1930s but had never met him until his arrival in New York. They would write two papers together about large transfinite cardinal numbers, one in the 1940s and one in the 1950s. Between 1941 and 1942 – in New York, Princeton, and Rockport (Massachussetts) – they saw one another often and became good friends. Recalling those days, Erdös wrote, "I never had trouble with Tarski; he very politely criticized the careless way I corrected the proof sheets of our joint paper. He was right, of course."[37]

On another topic Erdös remarked, "I think Tarski was very fond of bosses [as Erdös called women]. This might explain why he had many 'girl' students." While on the subject of the differences between them, Erdös added, "We often discussed politics and the war. He was more anti-Russian than I, and I did not like Joe [Stalin] at all." Erdös also offered an anecdote connected with Tarski's name change:

I do not guarantee that this really happened but it is a nice story. Once, in a talk, Tarski presented a theorem and somebody in the audience said, "You have been anticipated by Teitelbaum." Tarski, somewhat annoyedly said, "*I am Teitelbaum!*"[38]

There are dozens of versions of this story, always told as mildly malicious gossip. However many times Tarski was confronted in this way, he was always discomfited – especially when it was brought up in public. The vein in his forehead pulsed and his face reddened; the remark from the audience was taken as a deliberate provocation, not an innocent observation or question.

Berkeley Beckons

In 1942, after three years of constant uncertainty, Tarski's professional fortune finally began to look almost good. An offer was extended from the University of California at Berkeley, where Griffith Evans, the chairman of Mathematics, was on a mission to build a world-class department. Widely recommended by respected mathematicians and logicians, Tarski had been under consideration since late 1940, but the administrative wheels ground slowly. When the United States entered the war at the end of 1941, appointment questions were laid aside as the country mobilized its forces. Not until the spring of 1942 did funds become available to actually make the appointment. Part of Tarski's salary would be paid by a Rockefeller Foundation grant for "displaced scholars" and part would come from a university emergency fund.[39] He would be appointed lecturer with a salary of $3000 for one year, subject to review and to sympathetic consideration – the implication being that UC would keep Tarski employed for at least the duration of the war. If this was not the best or the most secure of all possible jobs, it was nevertheless a toe in the door, the best offer he had received, and he accepted it.

On a sadder note, before leaving Princeton, Tarski sent a letter to the liquidating agent of the Hamburg-American Lines asking that the $525 he paid for passport fees and exit permits for his wife and children be returned to him, because:

My family never received the exit permits although I had been promised they would ... in consideration of this payment.

I should, of course prefer above everything else to be able to bring my family from Warsaw to the United States, but since the Hamburg-America Lines was unable to secure the exit permits which they promised, I should like to file my claim for the refund of the amount paid.[40]

Did Tarski get his money back? Not likely. But much more importantly, the letter brings home the news that, after the Japanese attack on Pearl Harbor in December 1941 and America's full engagement in the war with Japan and Germany, Tarski had very little hope of being reunited with his family by normal means.

In July of 1942, Tarski left Princeton, returned to Cambridge for a few weeks, and then rented a room in Rockport to spend his final month on the East Coast in the company of his many colleagues, particularly those from the Boston area who summered in that arty resort community. He enjoyed various excursions at neighboring beaches and, among others, he mentions both S. Eilenberg and P. Erdös in his journal for that period, which means he was undoubtedly talking mathematics while enjoying the sea breezes. Eilenberg's disapproval of Tarski's "conversion" did not cause a rupture in their collegial relationship, and on Tarski's eightieth birthday "Sammy" would send Alfred an amusing letter of congratulations.

Just as some stability or at least a promise of a professional future in Berkeley was filtering into his life, the Princeton draft board sent Tarski a letter ordering him to appear for a physical examination at Princeton Hospital at 7 P.M. on Friday, August 7. As a now permanent resident of the United States he had, as required, registered with his local draft board. This was a startling turn of events, for at that time it was customary to call men up for service within two weeks if they passed the physical exam. The letter, forwarded from Princeton to Rockport, did not reach him until August 4. Stunned and dismayed, he immediately wrote a long, polite, but panicky letter to Dr. Frank Aydelotte, Director of the Institute for Advanced Study, saying in effect, "Quick! Help! What do I do now?" Tarski wrote: "This is an urgent matter for me if only in view of the California appointment. It would, of course be quite unreasonable for me to undertake the long and expensive trip to California only to be inducted a short time afterwards."[41]

Because Aydelotte was away for a few days, the letter was passed on to the mathematician Marston Morse, a permanent member of the institute and president of the American Mathematical Society. Morse took it upon himself to visit the Princeton draft board on Tarski's behalf and to write Griffith Evans at Berkeley urging him to fill out the necessary deferment forms certifying Tarski as "a necessary man in a critical occupation, namely the teaching of mathematics." At the same time, Morse also arranged to have the army physical exam transferred to Gloucester, a city near Rockport. A flurry of letters between Ayedelotte, Morse, Tarski, the institute secretary and the draft board reveals the help Tarski got from his friends in Princeton. No doubt he was also given suggestions for what to do by his various friends in Rockport as well. His diary records several trips to Gloucester by "autobus" in the company of G. (Gabrielle) Oppenheim, Paul Oppenheim's wife. It is not clear whether Tarski actually took the physical or how close he came to being drafted, but in the end he was granted the deferment on the grounds that his occupation was essential to the war effort.

Breathing a huge sigh of relief that the matter was settled, he left Rockport for Boston in early September, went from Boston to New York, and on the 11th of September began his *podróż* to California. *Podróż*, which means "journey," is a word that never appears before or after in Tarski's journal; he made this rare distinction when he used it to describe his trip across the continent.

6

Berkeley Is So Far from Princeton

Podróż to Berkeley

THREE THOUSAND MILES from the East to the West Coast of the United States was a vast distance compared to the European travel from one small country to another to which Tarski was accustomed. A trip from Warsaw to Vienna, Prague, Paris, Amsterdam, or even London was nothing like this. As if to emphasize its length and importance, he noted Boston rather than New York as the starting point of his *podróż* and gave the dates: September 8–17, making it a ten-day journey, although he stopped in New York for two days before actually leaving for California.

Tarski's train took him to Chicago and then to Denver, Colorado, where he slowed down to experience the spectacular Colorado landscape and indulge his passion for high mountains. By train, bus, and on foot, he visited Pueblo, Cañon City, Royal Gorge, Montrose, Delta, and Grand Junction; he even had the good fortune to be given a short tour by auto by a "Dr. and Mrs. ?." He had forgotten their names by the time he made his journal entry – but not the ride from Montrose to Delta via the spectacular Black Canyon of the Gunnison River. One can almost hear him giving voice to the beauty of the scene surrounding him and making comparisons with the Tatra mountains in Poland that he knew so well. Two days later, he arrived in Berkeley and real life began again.

Jerzy Neyman, a fellow Pole and already a *force majeure* at the university, was among the first to greet Alfred, to introduce him to colleagues, and to acquaint him with the local attractions. Neyman had been at Berkeley since 1938 as director and builder of the statistical laboratory of the mathematics department. It would seem natural that these two men with a language and culture in common would become lasting friends, but it was not to be. Instead, over the years they became more like heads of warring fiefdoms.

In the first place, Neyman had not been in favor of Tarski's appointment; his choice for the position had been Antoni Zygmund, a mathematical analyst and yet another Pole.[1] Zygmund, a friend and colleague of Tarski's in Warsaw, was also a refugee, but since he already had a position at the University of Pennsylvania his situation was not desperate. More importantly, Griffith Evans, chairman of the mathematics department, had been impressed by the quantity and quality of the recommendations he had received on Tarski's behalf from many great mathematicians and philosophers. The Princeton mathematician Oswald Veblen had even urged Evans to apply for a special Rockefeller grant for Tarski, saying, "I hope you will pardon me for having interfered so much in this affair. My excuse is a double one: a) I have always had a particular fondness for the University of California, and b) I think Tarski to be an extraordinarily useful man."[2]

There were other antagonisms between Tarski and Neyman, most notably in their political outlook – Neyman being much farther to the left. But that in itself would not explain their antipathy, since Tarski had many friends with whom he strongly disagreed politically while still finding them congenial. Rather, the tension between them grew from their personal characteristics of ambition and competition; both were empire builders. Before too long, it was a standard joke in Berkeley to describe Tarski and Neyman as "Poles apart."

Tarski was now forty-one. Uncertainty and anxiety had been part of his daily diet from the moment he came to the United States, and the menu did not change much in Berkeley. He had come to this new, very different place – the Far West – as a lecturer with a one-year contract and only an implied promise that his reappointment would be looked upon favorably. The whole world was now at war; lives everywhere were disrupted, dislocated, and discontinued. Not surprisingly he had periods of depression and despair, especially about his family in Poland. In December 1942, Tarski unburdened himself in a long letter to Kurt and Adele Gödel:

Unfortunately, the conditions under which I work this year are even more difficult than in preceding years; and this concerns as much physical as psychological conditions. It seems to me that in the past I never worked so much.

This is for me the first year of regular lectures in English; as one might expect I am much absorbed by their preparation. This semester I teach almost exclusively "undergraduates" (with one exception – I have a seminar in logic where participants are exclusively professors and instructors of philosophy and mathematics; but the level is almost the same as in my other classes). Next semester I shall give a course in higher algebra for graduates; and it is desired that I – as a contribution to the war effort – give an extra course in cryptography, a subject about which I practically do not have any notion. The living conditions here are difficult; this is a typical "boom" (the well-known Kaiser shipyards about which you have surely heard). The population has greatly increased in the last months – many workers came here from the East. Prices are very high – it is much more expensive here than it was in Princeton in the spring. It was very difficult to find a pleasant place to live; but finally I succeeded with this. But anyway I encounter every day new "practical problems" which I have to settle and which take much time. And independently of this I still try to carry on scientific work. In the summer I started to write the article on the calculus of relations; I hope to finish this in a few weeks, but at the cost of great strain. It seldom happens that I go to bed before 3 o'clock, even though I have to lecture the next day.[3]

No matter where he was or what his situation, Tarski never failed to point out how overworked he was, but his problems and predicaments in Berkelely were all too real.

Adele Gödel responded to Tarski's litany of woes with a letter and a cake. By return mail he answered,

[your gift] affects my heart at least as strongly as my sense of taste.... What is special about the cake which I received today is that already its very form made me homesick – I have not even seen such for some 3 1/2 years, and would never expect to see this here – in Berkeley. It was really very kind that you have thought of me![4]

But a year later Tarski wrote, pleadingly, that he had not heard anything more from Adele and Kurt. Urging them to visit Berkeley, he said "it appears to me you could feel very well here .... It is a pity that it is so far from Berkeley to Princeton! I would like very much to see you again, to chat with Adele, and to discuss various questions with Kurt. I hope that at least I shall now have a few words from you." He enclosed stamped envelopes and paper so that Adele would have no excuse not to write and ended the letter on this sad note: "I receive now letters from Warsaw very seldom; and what I find there is not pleasant."[5]

University of California, late 1930s, aerial view from the west.

By the time Tarski arrived in Berkeley in September 1942, the war had begun to have a significant effect on the university.[6] The standard two-semester system had been replaced by three sixteen-week semesters in order to speed up instruction; non-essential courses were eliminated and new courses were added such as "spherical trigonometry and navigational astronomy" and "nutrition in peace time and war"; physical education courses were given more emphasis. Students were required to take at least one National Service Course to train for some form of emergency service. The most visible change was in the composition of the student body, since most of the men were called up for the armed forces or went to work in war-related jobs. Over the period 1942–1945, male enrollment at UC Berkeley dropped from over 11,000 to about 4300. Officer training programs accounted for a number of the men on campus, for whom the curriculum concentrated on basic engineering, foreign languages, and military studies; an Army Air Forces pre-meteorology program began in 1943. Fraternity and sorority houses were commandeered by the training units, and Army reservists marched to class. The annual Big Game football contest between Stanford and Berkeley was suspended, and instead they challenged each other to a war-bond contest with the goal of buying one B-25 bomber; in the end the students raised enough money to buy two bombers.

The list of courses for which Tarski was responsible at UC during the war years regularly included calculus, undergraduate algebra, and elementary mathematics for advanced students – this last designed for students preparing to teach mathematics in secondary schools. One semester he was put down for plane analytic geometry in the Navy program; there is no record that he ever had to teach cryptography as he had feared. Tarski was also listed for graduate courses in algebra and for the theory of functions of a real variable, both with the proviso that they were to be offered only if a sufficient number of students enrolled. In his second year Tarski was able to institute a seminar entitled Topics in Algebra and Metamathematics that was based on his own work.[7]

Bjarni Jónsson

Almost as soon as he got his bearings in Berkeley, it would seem that Tarski set out deliberately to re-create the atmosphere that had formed him as a student in Warsaw. By necessity, progress was slow because he was mostly teaching basic courses to undergraduates. Yet, being Tarski, he was already introducing his own subject whenever he could.

Bjarni Jónsson, who was to become Tarski's first Ph.D. student in Berkeley, was then an undergraduate. He had signed up for an algebra course with no idea that Tarski, the instructor, was someone of international reputation:

> The subject matter was quite traditional, mostly matrix theory. He adhered to the course description in the text, but occasionally made allusions to other topics. At the beginning of the spring semester I asked if I could sign up for a reading course with him, and when he asked what subject I had in mind, I told him that he had mentioned set theory and the foundations of mathematics in class and that I would like to find out more about these .... He suggested that we start with set theory, beginning by reading Hausdorff's classic *Mengenlehre*. This opened up a whole new world for me.

Bjarni Jónsson, Tarski's first student in the United States, at Sather Gate, UC Berkeley, 1946.

For a while I kept reading and reported to him once a week but then something happened that caused an abrupt change.[8]

Jónsson had discovered what he thought was an error in one of Hausdorff's proofs. Tarski was thrilled and impressed to find his student's objection valid and immediately promoted him to the more difficult problems contained in the list of theorems without proofs that he and his student Adolf Lindenbaum had published in 1926. "He asked me to prove as many of these as I could," Jónsson said, "and that is how I happily spent the remainder of the semester."

Thus was he seduced – as were future generations of students – by Tarski's enthusiastic recognition of his ability and his passion for the subject they now had in common. The following year, Jónsson enrolled in more reading courses and seminars with Tarski and had extensive contact with him outside the classroom. He said that he benefitted from the fact that Tarski had few faculty members to talk to and no other students who were interested in logic, set theory, or algebra:

At first we met in his office or we went for walks or sat over coffee near campus or I walked with him down to Shattuck and University, where he caught the street car home. Later we started meeting in his apartment. The late night sessions suited me quite well, for this was how I had always worked myself.

I don't remember when he started writing the book about cardinal algebras but I became quite involved in this project from the beginning. I read all the drafts of the manuscript, and he discussed with me what he was doing. I remember in particular one time spending an hour or more on the phone, standing in the hallway of the house where I had rented a room, while he described how he was going to start all over again with a completely new system of axioms. He was quite enthusiastic and obviously needed to talk, and at that moment I was probably the only person he could talk to about this topic.[9]

A poignant picture: the short, effusive Polish teacher confiding in the tall reserved Icelandic student as they walked all over Berkeley or as they worked late into the night in Tarski's apartment. One evening, Tarski shared his anxieties and told Jónsson "the things you must do in case anything happens to me." While Jónsson worriedly took these ominous forebodings to mean that his teacher was contemplating suicide, he was also marveling at the enormous enthusiasm and energy Tarski had for his work even in a time of crisis.

The material that Tarski was relentlessly working over with Jónsson dealt with algebraic properties of operations on finite and infinite cardinal numbers. With Jónsson's essential assistance, it would eventually be organized and prepared as a book, entitled *Cardinal Algebras*, that appeared in 1949. None of Tarski's other students or colleagues were drawn into this particular subject in the intervening years. But he did have people besides Jónsson to talk to in Berkeley about his many other mathematical interests. At the very outset, during the school year 1942/43, J. C. C. McKinsey visited UC under a Guggenheim Fellowship. In his proposal to the Guggenheim Foundation, McKinsey had planned to work on decision problems for intuitionistic and modal propositional calculi,[10] but in Berkeley Tarski drew his attention to a 1938 article he had written interpreting those calculi in topological terms. It seemed to Tarski that the core of that work would best be reformulated in algebraic terms, and he and McKinsey invested considerable effort in carrying that out. The result of their work during McKinsey's visit would be a long joint article, "The Algebra of Topology," which appeared in 1944 and would be the basis of further collaboration after McKinsey left Berkeley.

One would have expected Tarski to make intellectual contact with the resident logician Benjamin A. Bernstein – McKinsey's teacher from the

1930s – and the algebraist Alfred Foster, but he considered their interests and approaches old-fashioned. Personal relations with Bernstein and Foster reached a low point following an acrimonious debate that Tarski had with them over a point of difference that arose in an algebra seminar.[11] But Tarski needed people to connect with, no matter what the subject, so instead he took up with younger faculty, Raphael M. Robinson and Anthony P. Morse. Both worked primarily in mathematical analysis, but Robinson enjoyed attacking problems over a wide range of fields and Tarski eventually attracted him to some questions that proved to be useful to both of them.

Complaints

Tarski's complaints were a constant, a darker tone playing counterpoint to the bright enthusiam with which he worked on the long list of things that he wanted to do. As for the "practical problems" mentioned in his first letter to Gödel, he was very good at getting others to solve them. Secretaries, students, colleagues – all were pressed into service. His attitude, always, was "ask and ye shall receive." The problem of a pleasant place to live was handled by Sarah Hallam, the mathematics department secretary. She found him an apartment on Shasta Road in the Berkeley hills, a choice location.[12] Hallam – a pert, deceivingly prim-looking young woman in her early thirties – had been at UC since 1936, first as a half-time secretary while she finished a degree in mathematics. After completing her master's degree, she became the full-time administrator of the department in a career that was to last forty years. She liked her job because of the people and the variety of different things she was called upon to do, and she chose it over a teaching position at Reed College, where she had received her undergraduate degree. Tarski was, in her opinion, one of the most interesting people she'd ever met.

Sarah knew Alfred "from the beginning," as she put it, meaning as soon as he arrived in Berkeley. They met at the home of Tony Morse, who, with the enthusiastic help of his wife Mary, entertained frequently in an informal way. The Morses had invited Sarah and the newly arrived Tarski to their home in Walnut Creek, near Berkeley, and almost immediately thereafter Sarah was taking his dictation, typing his letters, and helping with those practical problems that Alfred hated to deal with and consequently handled ineffectively. She did not feel imposed upon because, besides being sympathetic to his plight, these were things she liked to do. She knew that Tarski was one of the leading logicians in the world, that he was a refugee, and that he was worrying about his wife and children in Poland.

Early in the game, Sarah noticed that Tarski could become so concentrated on a problem that he seemed to forget everything else. One day he asked Griffith Evans to help him get information that might aid in getting Maria Tarski and the children out of Warsaw; he had made a big point about how important it was. With considerable effort, Evans produced results the very next day, but to Sarah's astonishment, Tarski hardly seemed to remember what he had asked for. She was to see the same phenomenon many times. "When he was doing his work, which was a large part of the time, he was entirely into it; he didn't think about anything else; everything else took a back seat, even his family's escape."[13]

One more Tarski "trademark" was to shock Sarah Hallam in those early days before she became aware that he liked women *very* much. Not long after they met, Alfred "propositioned" her. Remembering the moment some fifty years later, she said: "He asked me at the beginning, long before Maria came. We were in my apartment after having been at the Morses. I was very naive at that point, and I was very shocked, I guess. He is an intense person and when he wants something he really goes after it, strong! But he accepted my 'no' and didn't pursue it."[14]

Gossip had it that Tarski did indeed "pursue it," chasing Sarah around the table in her mathematics department office. True or false, the story remains and, in all likelihood, he did behave inappropriately more than once. Sarah put up with it – as others did – because she liked him and found him interesting to talk to about art and politics and all manner of subjects that most people in the department didn't usually discuss with her. In the years to come – when Sarah became a close friend to everyone in the Tarski family, particularly Maria – she knew as much as anyone about Tarski's professional business and personal relationships.

In general, Berkeley people were good to Tarski, inviting him to their homes and doing what they could to cheer him up. At the Morses, he and Jónsson played bridge. (Card playing and especially bridge had been a regular activity in his bourgeois family milieu in Warsaw.) His new colleagues also took him on day trips to the scenic high points of California, something he appreciated greatly. These excursions became a habit with him – an essential part of his life – and once he was established, he went on an outing almost every weekend: for a hike in Muir Woods or to Pinnacles National Monument or to the North Coast. Later, when visitors came, Tarski repaid the early kindness shown him by taking guests on beautiful trips all over California and the Southwest.

The annual mathematics department picnic that brought together faculty, staff, and students was held in the spring of 1943 at Codornices Park. On that occasion McKinsey witnessed at first hand Tarski's cluelessness in practical matters, in an event that has passed into legend as "The Sanitary Napkin Story." Before the picnic, Alfred had insisted that, like everyone else, he be allowed to contribute something. To make it easy for him, he was asked to bring paper napkins. McKinsey was driving him to the party when Tarski, reminded of the promise he had made, suddenly said to him, "Please stop. I must buy something for the picnic." McKinsey waited in the car while Tarski went into a drugstore and asked the clerk for napkins.

"You mean sanitary napkins?"

"Of course, sanitary," said Alfred, thinking, what kind of a fool is this?

"A big box?" asked the clerk.

"Yes, yes, certainly, a big box."

And so it was that, to the stunned silence of the assembled members of the department and their families, Tarski proudly placed a giant-sized box of Kotex on the picnic table at the park.

ટન ટન ટન

In spite of the welcome extended to him, Tarski greatly missed the East Coast. In one of his letters to the Gödels he said that people in Berkeley are more disposed to *quatschen* [small talk or gossip] and that the atmosphere is friendly but not terribly serious. Except for McKinsey, he had not yet found people like Gödel, Carnap, Church, Curry, Erdös, and Eilenberg, with whom he could have a deep, fertile exchange of ideas. He was yearning for a trip east and brought up the possibility of a visit in the spring of 1944, adding realistically, "the trip is rather expensive and under present circumstances one can hardly count on any invitations to lectures which would cover the expenses of the trip."[15]

Two months later Alfred was still working on getting the crucial invitation, and in his next letter to Gödel he asked him outright to "try to arrange for me an invitation from the Institute ... even without any material obligation on the side of the Institute (if impossible otherwise)." He proposed to give lectures on the algebra of relations in the second half of March.

In case if the invitation is sent to me *in the next days by Airmail,* perhaps I shall obtain money from the university and that will make the whole undertaking somewhat easier. It would perhaps be still better if the invitation could be sent in the name of the Institute and Princeton University. [emphasis in original]

Hedging against disappointment, Tarski ended with a postscript: "Don't try too hard.... [I]t is not particularly important."[16]

But very obviously it *was* important to him and – to his own evident embarassment, for he was not used to being rejected – he could not let go. Explicitly blaming himself for making up his mind too late – and implicitly for asking Gödel, who was notoriously cautious and slow to make any kind of arrangement for anyone, including himself – Tarski wrote:

I see from your last letters that you have had much trouble in connection with my plans. I am awfully sorry!... It appears now as if everything was in vain.... I shall wait still a few days for the letter from Princeton University; however, in case I don't get one, I shall abandon my travel plans. I am very sorry since I really wanted to come to the East to see old friends once again and to talk over many things.[17]

Finally, facing the fact that neither the institute nor Princeton University was leaping at the suggestion that they pay for his travel to the East Coast in the spring, Tarski went to Bozeman, Montana. This was not as far-out a choice as it might seem, because McKinsey had gotten a job teaching at the college there after his Guggenheim Fellowship at Berkeley. In an almost triumphant voice, telling how well things turned out after all, Tarski wrote to Gödel:

Instead of coming to Princeton I finally went to Bozeman.... I shall give one or two lectures and stay two or three weeks. It is beautiful here – mountains all around – views which could remind one of the Niedere Tauren [in Austria]. We [Tarski and McKinsey] shall possibly write a paper together – a continuation of our article on the algebra of topology. What is new with you? Has your work on Russell already appeared? How is Adele? Soon I shall send you a few reprints.[18]

McKinsey had difficulties of his own in the 1940s. In spite of his excellent reputation as a teacher and researcher, he never advanced beyond the position of instructor at NYU. Anecdotal reports from students and colleagues indicate that his open homosexuality got in the way of a tenure appointment, and the spirit of the times on this matter makes it likely. Following his fellowship year in Berkeley, his position as assistant professor of mathematics at Montana State College[19] lasted two years. After that he went to Nevada, then to Oklahoma, and in 1947 to a research group at Douglas Aircraft Corporation in Los Angeles that was transformed a year later into the independent think tank for Cold-War research and development known as the RAND Corporation.

His situation was precarious everywhere and especially at RAND, where a high-level security clearance was required of all the research staff. Eventually McKinsey was asked to leave RAND because homosexuals were thought to be high-risk employees, the rationale being that they were more likely to divulge secrets under the threat of having their private lives revealed. The fact that Chen was completely open about his private life made no difference or, perversely, may have made matters worse.

The three weeks with McKinsey in Bozeman did wonders for Tarski. He loved talking to McKinsey, with whom he had perfect rapport; for him it was a tonic like no other. The additional blessing of high altitude (Bozeman is 5000 feet above sea level), clear air, and beautiful mountains gave him a further boost. In Bozeman, McKinsey and Tarski continued their substantial work on the algebra of topology, bringing out its connection with Brouwer's intuitionistic logic. Tarski wrote Gödel of their progress and Gödel responded with thanks for his two postcards and a letter, saying that he and Adele were really very sorry that he did not come to Princeton, adding hopes that he might still make the trip east.

Tarski returned to Berkeley feeling exhilarated and ready to begin teaching again. From then on his situation improved steadily. He was elected president of the Association for Symbolic Logic for the years 1944–1946. Perhaps in response to this added recognition of his distinction, in 1945 the university promoted him from lecturer to associate professor, leapfrogging over the usual intermediate rank of assistant professor; the following year he was promoted to a full professorship.

The End of the War in Europe

In mid-1944, the Soviet armies swept through Eastern Europe, routed the Germans from the Baltic States, and advanced toward Warsaw. In light of these encouraging events, the Polish resistance forces, mistakenly counting on the Soviet troops to support them, launched a full-scale battle and attempted to retake the city from the occupying Germans. The Warsaw Uprising, which lasted from August to October 1944, was a grievous disaster. The Soviets, camped on the other side of the Vistula River, held back while the Germans crushed the uprising. With the advances by the allies on all fronts, the German high command knew they were losing the war and, in revenge, Hitler ordered Warsaw to be razed. With systematic cruelty, Nazi demolition squads dynamited the public buildings, palaces, and museums and then went from street to street and house to house setting fire to the doomed city. The inhabitants of Warsaw who did not flee were deported to forced-labor or concentration camps or were murdered in the streets.[20]

As reports of the conflagration filtered out to the Western world in bits and pieces, Tarski was, for many months, beside himself with anxiety. Then, to his enormous relief, in early 1945 he received a letter from Maria with the marvelous news that she and the children were safe in Cracow. They had escaped Warsaw before the worst consequences of the uprising, and while they were there, the Russians occupied Cracow with almost no fighting. Now Tarski's main concern, once again, was how to keep in touch with Maria and how to get her and the children out of Poland and into the United States, hardly an easy matter. In the chaos following the end of the war that left six million inhabitants of Poland dead – three million of them Jews – those who survived were desperately seeking to find and rejoin their loved ones whose fate was unknown.

It was only after the war that Tarski learned the full details of how Maria and the children survived and of what had befallen the rest of his family.

In September 1939, when the Germans invaded and bombed Warsaw, Maria and the children were on holiday in the countryside and were caught completely unprepared. They returned to their flat in Zoliborz, which fortunately had not been damaged, packed their possessions, and moved in with Maria's sister, Józefa Zahorska, a widow who lived with her son in the same neighborhood. Jan was five and beginning school; little Ina was eighteen months old.

Living with Józefa in the early part of the war, Maria and the children were marginally safe because Józefa, like Maria, was not Jewish. It was not long before she and Maria became active in the Polish Underground, writing and distributing information and often acting as couriers, as they had done in the Piłsudski days. Maria also did a variety of relief work and for a while was in charge of a kindergarten.

Not surprisingly, Jan and Ina had different experiences and retain somewhat different memories of that period. Except for the continued absence of his father, there were few changes in Jan's life. For the first two years of the war he went to a private school; when it was shut down, classes continued to meet "underground" in the homes of the students. He was unaware of his grandparents' disappearance until several years after the fact. Before the German occupation they had moved to Otwock, a suburb of Warsaw, and their visits had become infrequent, so he did not notice the point at which they no longer visited at all. As for his father, Jan knew that they were all waiting to be reunited, and meanwhile they wrote letters. He recalled that at one point they were almost on their way, bags packed and ready to leave, when suddenly the plan changed. He knew nothing of his father's professional importance and was surprised when, just before the family actually left Poland in 1945, one of his teachers asked him: "Are you related to the famous mathematician?" Since it was well known that Tarski was a Jew, perhaps the teacher had not dared to risk the question earlier. Yet Jan himself had not worried about being identified as Jewish because – paternal grandparents notwithstanding – he didn't think he was.[21]

Ina Tarski, four years younger than Jan, was much more anxious and confused about the disruptions in her life. Maria, who was constantly worried that the children might be identified as Jews and seized by the Nazis, had placed her in an orphanage for girls in the countryside, just outside Warsaw. Ina did not understand why she was there since she knew she was not an orphan. It wasn't that she was afraid or unhappy but in retrospect, she said, it was a mystery to her:

I didn't have bad memories of that place; there were other girls to play with and I had a little friend I was very close to. We got packages from the Red Cross, so considering everything, we had good conditions. But I didn't know *why* I was there and for years afterwards I didn't know – in fact, not until many years later when my husband asked my mother why she put me there. She told him it was because we had been denounced as Jews, and she thought it was a safe place. In hard situations, my mother was a very capable person. She did what she had to do with no nonsense. She understood and sympathized with people who were weak but she herself was strong and effective.[22]

The Escape to Cracow

As soon as the Warsaw Uprising of 1944 began, Maria took Ina out of the orphanage, and she and the children fled to Cracow where friends found them a place to live in an abandoned castle. Ina remembered being very frightened because they were within sight and sound of a German army camp. "We had to hide and make it look as if the castle was still uninhabited. It was dark and spooky and no one else was around. We stayed there until the war ended and the Germans were gone."

Then they returned to Warsaw in the coal car of a freight train, Ina lying on top of the coal. Warsaw was a smoking wasteland, buildings burned to the ground, everything at a standstill. People were hitching themselves to carts to transport whatever could be salvaged. Maria and Jan walked, while Ina rode in one of those carts from the train terminal to Maria's sister's apartment house in Zoliborz. Miraculously, it was still there and so was Józefa. Those were days when the news of death and atrocity – who had been killed, and how – was relayed between families and friends; people already numbed and devastated by what had happened to them had to endure further heartbreak when they heard about the fate inflicted upon others. Although most members of Maria's closest family survived, almost all of Alfred's family was gone. His mother and father had been transported from Otwock to the ghetto in Warsaw, and from there they were sent to Auschwitz. With few exceptions, his aunts, uncles, and cousins were also murdered. His brother Wacław avoided being identified as a Jew and had managed to survive until almost the end of the war by hiding in a boat on the Vistula. Then, during the uprising, he went in search of his daughter, Anna. He was last seen in a train station, where he was captured and subsequently killed. His wife Tamara, a nurse who was active in the underground, was killed at about the same time when she went to the aid of a man who had been shot and was lying in an open square, crying for help. As she was attending to his wounds, a German sniper took her life.

Only those who have lived through such savagery know how they managed to do so, but if internal strength is a factor, Maria Tarski had it. In the face of calamity she kept her head, and although she did not know how or if she and her children could escape, she knew Alfred was working on it and she made sure that when the opportunity came they would be ready.

The Family Arrives

In September 1945, Tarski finally was able to make the trip to the East Coast that he had been longing for since 1943. He had invitations to lecture at several universities and at least some of his expenses would be paid. More importantly, this gave him the opportunity to go to Washington D.C. to speak personally with immigration officials about what could be done to bring his family to him quickly, now that the war was over. Coincidentally his friend Van Quine, who had served as an officer in the U.S. Navy during the war, was in Washington waiting for his discharge to be processed. They were able to spend a good deal of time together, and he was able to point Tarski toward officials who might help him. These initial efforts were disappointing. In his autobiography Quine wrote of Tarski "vainly pressing the bureaucrats to help bring his family out of [now] Soviet Poland." On the positive side, Quine and Tarski "had good discussions of logic and philosophy and some jolly evenings with Waves from my office. We got away to a logic conference in New Jersey... and talked of writing a book together, but nothing came of it. Late in October he came back for further vain efforts with the bureaucrats."[23]

Here is a perfect example of Tarski dividing himself: while he was tormented by his family crisis and what could be done to solve the problem, he was doing logic and squeezing in "some jolly evenings with Waves." Quine recalled Alfred's parting words as he left Washington with everything still up in the air about prospects for Maria and the children: "Grreet all gerrls," he said, meaning of course, the jolly Waves. (Incidentally, a few years later one of those "gerrls" became Quine's second wife.)[24]

The efforts with the bureaucrats in Washington had come to nothing, but the help Tarski sought from other sources was effective. Again Anders Wedberg played a major role. Wedberg had returned to Sweden in 1943, after the British defeat of the German U-boat fleet made it possible for him to return by a "safe conduct" convoy. His family was wealthy and influential; his father, Birger Wedberg, was known to everyone in the Swedish government. Combining his prestige as Supreme Court Justice with a promise to defray all costs for the stay of Maria Witkowska Tarski and Janusz and Krystyna Tarska, he applied for a Swedish entry visa for Maria and the children. In the letter of application, prepared by Anders, he outlined Tarski's situation from his August 1939 departure from Warsaw to his appointment in Berkeley and briefly summarized what had happened to Maria and the children, ending with "she is now in a desperate situation." Wedberg also explained that Tarski was now an American citizen (as of June 1945) and that the State Department had already authorized the American Embassy in Warsaw to grant the Tarski family a "nonquota" visa to America and that all the other necessary conditions were fulfilled. "Under the present conditions, however, it may be a long time before the visa will be issued." The Wedberg plan was to move the family from Warsaw to Sweden and then, as soon as matters could be arranged, from Sweden to the United States.[25]

This plan – put in motion in Sweden in September at the very moment Tarski was feeling frustrated in Washington – worked with the aid of several people "in high places" (as Maria later labeled them) pulling strings. Chief Justice Harlan Stone, the father of Tarski's colleague Marshall Stone, urged the U.S. Embassy in Poland to act, while Wiktor Grosz, a Defense Department minister in Poland's new Communist government, moved things along in Warsaw. Grosz was the husband of Alfred's ex-fiancée, Irena; both of them were loyal friends in spite of their political differences with Tarski. On 11 November 1945, two months after Wedberg's application, the Swedish government's Alien Commission granted an entry permission for Maria Witkowska Tarski and the children Janusz and Krystyna. Packing a few precious possessions and saying heartbreaking farewells to the family members who had survived, they flew from Warsaw to Stockholm – a momentous first flight for all of them. The distance in miles was not great but the contrast was dazzling. Jan Tarski recollected:

All Polish cities were dark at night but when we flew into Stockholm, it was all light and bright. We lived for a month at an elegant country estate [Bodal Mansion, the Wedbergs' home] and did all sorts of wonderful sightseeing and regained our equilibrium and composure. Mother had time to take care of the formalities of travel, our visas and tickets. My father had sent one hundred dollars but my mother changed sixty dollars worth to Swedish Crowns and sent it to her relatives in Poland. Then she had to explain how the money had "disappeared" to the Swedish authorities, which was a bit touchy.[26]

A month later, Maria, Jan, and Ina left Göteborg for America on the *Atomena*, a small freighter bound for Philadelphia. Jan remembered that there were six passengers, "three Tarskis, two Finns, and one Estonian." The weather was "not wonderful" but the children thought the trip was fun because the crew fussed over them and on Christmas Day they had dinner with the captain. Their first host in the United States was their Polish friend Antoni Zygmund, who lived in Philadelphia where the ship docked. The family stayed with the Zygmunds for a week to recuperate and acclimate before making the long trip across the country.

Alfred had remained on the East Coast until early November but, with all the variables and uncertainties of their arrival date, decided that he had to return to UC in time for the new semester and wait for them there.

When at long last they arrived in Berkeley on 6 January 1946, a big crowd, including reporters and photographers, was at the train station. Tarski had told his friends that his family was coming and many joined him to welcome them. The young philosopher Benson Mates, who had worked closely with Tarski, remembered the drama of the moment: "Alfred was wearing a rumpled greenish-black suit that looked as if he had slept in it; you could see the outline of his body. He paced back and forth until the train pulled into the station and then he ran along the platform as the train slowed, peering into the cars and searching for the faces of his loved ones."[27]

What made the event of particular interest to the press was that Maria and the children were among the first postwar refugees from Poland. The photograph in the *Oakland Tribune* shows a smiling Tarski holding Ina on his lap while she clutches her teddy bear; Maria is sitting close on one side and Jan is leaning toward him on the other. The caption reads, in part: "Separated in 1939, when Alfred Tarski, famous Polish mathematician, came to America to lecture, Mrs. Maria Tarski and their two children ... were reunited with him yesterday at Berkeley Station."[28] Ina, a baby in 1939, did not remember her father. She felt overwhelmed by the crowd, the confusion, and the strangers speaking an incomprehensible language in another unfamiliar city.

The family was taken in by Griffith Evans and his wife. They had a large house on two levels, and they gave the bottom half to the Tarskis until they found their own place to live. This was only the beginning of the generosity and sympathy offered in those early days of their life in Berkeley. Alfred and Maria would never forget it.

Nine months later, Tarski wrote a long letter to Heinrich Scholz, his German colleague in Münster who had helped logicians in Poland during the war. He gave Scholz a vivid account of how things were going with his family and brought him up to date on his own state of mind:

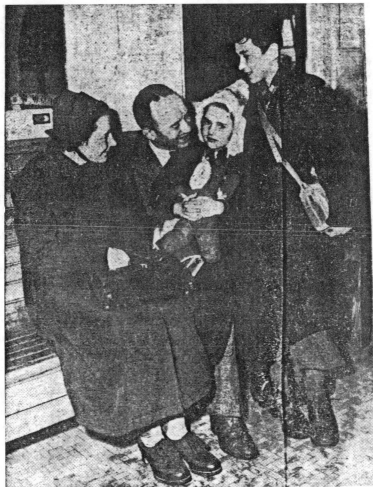

Separated in 1939, when Alfred Tarski, famous Polish mathematician, came to America to lecture, Mrs. Maria Tarski and their two children, Eva Christina, 7, and Janusz, 11, were reunited with him yesterday at Berkeley Station. With the help of "high people," Mrs. Tarski and the children escaped from Poland to Sweden in November.—Tribune photo.

Tarski family reunited at the train station in Berkeley; *Oakland Tribune*, 6 January 1946.

As you can imagine, I feel happy that after these gloomy years of the war my wife and children have succeeded in joining me. My children feel rather well. My wife has not yet recovered and I wonder whether she ever will, completely. Her stories about the occupation period and about her own experiences sound nightmarish. Except for my wife and children, practically my whole family

Tarski, Jan, and Ina in the garden of their house on Cragmont Avenue, 1947.

(about thirty people, counting only close relatives) was murdered by Nazis. My parents were killed in 1942. My only brother and his wife were atrociously murdered the day after the end of the Warsaw Uprising. The losses in my wife's family are also heavy. As you probably know, many of my colleagues and former students in Poland have been murdered besides those named in your letter – Pepis, Presburger, Mr. and Mrs. Lindenbaum to mention only a few; the death of the Lindenbaums has strongly affected me personally.[29]

On a more positive note, Tarski continued:

> But I am glad that Ajdukiewicz, Kotarbiński, and Mostowski have survived and at least for the time being, have some bearable conditions for their work. As you may have heard, Kotarbiński is now president of the University

of Lodz, and Mostowski is acting professor of philosophy of mathematics in the University of Warsaw.

My own situation is in general satisfactory. I have a full professorship here and hence a feeling of stability. California is undoubtedly a beautiful place to live, but I cannot deny that intellectually I feel somewhat isolated. I am not giving any courses in logic and there are very few people here who are interested in this domain. I am now working mostly in abstract algebra. My publication plans of which I wrote you in 1941 have not yet been realized.[30]

Less than two years later Tarski would be writing Scholz again, telling a completely different story – of students getting degrees, of seminars in logic, of his coming sabbatical and the plans he has for research, and, most importantly, of his plans to organize "a systematic study of logic and foundations in Berkeley." Fairly quickly, many more than a few people were "interested in this domain."

And how did that change come about? Even before the ink was dry on the first letter, the irrepressible Tarski embarked upon a campaign to create exactly what he needed to have, personally and intellectually: a place of *congregation* rather than isolation.

7

Building a School

When Solomon Feferman, an eager new graduate student in mathematics, arrived in Berkeley in 1948, he – like legions before and after him – didn't know exactly what he wanted to do. He was assigned to an office in a "temporary" wooden barracks building that remained there for a half-century more, a large space that he shared with other students and teaching assistants, most of whom were further along in their studies. Naturally, they gave counsel.

One of the older students, Frederick B. Thompson, had already chosen Alfred Tarski as his Ph.D. thesis advisor; his conversation was all about logic and Tarski, whom he idolized. Aware that young Feferman was listening attentively, Thompson became even more expansive and urged him to enroll in Tarski's course on metamathematics. The following year Feferman took his advice and fell under Tarski's spell.

I knew immediately that this was to be my subject and Tarski would be my professor. He explained everything with such passion and, at the same time, with such amazing precision and clarity, spelling out the details with obvious pleasure and excitement as if they were as new to him as they were to us. He wrote on the blackboard with so much force that the chalk literally exploded in his hand, but step by step a coherent picture emerged. Methodically yet magically, he conveyed a feeling of suspense, a drama that managed somehow to leave us with a question hanging in the air at the end of the hour.[1]

Feferman would have been surprised to learn that only two years earlier Tarski had complained of feeling isolated, that he wasn't giving any courses in logic, and that very few people were "interested in this domain." In Feferman's view, interest in this domain was vibrant and students were clamoring for Tarski's attention. Bjarni Jónsson and Louise Chin had already completed their degrees and had positions elsewhere;

Julia Robinson had just finished her thesis; and Wanda Szmielew would arrive in 1949. Following Fred Thompson, soon Anne Davis, Robert Vaught, Chen-Chung Chang, and Richard Montague were intensely engaged, and in 1950 Dana Scott would come upon the scene as an undergraduate prodigy breathing hotly down everyone's neck. Even before Leon Henkin was recruited to become a partner and organizer in the campaign to build a great center for logic in Berkeley that would bridge the departments of mathematics and philosophy, Tarski had his project well underway. Griffith Evans now understood fully what Oswald Veblen meant a few years earlier when he wrote, "I think Tarski to be an extraordinarily useful man." Aware of Tarski's charisma and his potential to add luster to the university as a whole, Evans gave him free rein to choose the courses he wanted to teach and backed him in the hiring of new faculty so that he could build the logic curriculum. Tarski brought Jan Kalicki, a young logician from Poland, as a visitor to the mathematics department in 1951 and, convinced of his value to logic, urged his appointment to the philosophy department in 1953.

Benson Mates, a newly appointed assistant professor whose specialty was in the logic of the ancient Stoics, was Tarski's ally in the philosophy department. He had written his thesis on a topic suggested by Tarski. Although he was not his official advisor, Mates had taken his courses and discovered that Tarski had a deep knowledge of Greek and the Greek philosophers. They became good friends with common interests, and when Tarski pushed to recruit logic faculty for the philosophy department, he used Mates as his emissary to chairman William Dennes. "People felt he was always on their backs, pressing his agenda," Mates said. "He had this European notion that everything was done by influence." Even when Dennes thought he didn't have the power to do what Tarski wanted, Tarski managed to get him to do it anyhow. Then Tarski would say to Mates, "You see, I told you!"[2]

"Two students of mine – both girls"

In June 1948, Tarski wrote to Heinrich Scholz giving a succinct summary of his *new* circumstances and his plans for the future; optimism glows on the page:

> In general my situation in Berkeley is improved. I shall now have more freedom in choosing topics for my courses and in organizing here *a systematic*

study of logic and foundations. I have even reasons to believe that for several years there will be some younger people around, employed by the University as research associates or assistants, whose main task will be to help me in research.

Two students of mine – both girls – have completed their work during the last year. One of them, a Miss L. H. *Chin* (probably the first Chinese woman-mathematician) wrote you a short time ago, on my request, to answer your question concerning the algebra of relations. Miss Chin has been helping me by preparing for publication *a book on relation algebras*; and I am sorry that our collaboration will be interrupted since she is going to leave Berkeley soon (she has been appointed assistant professor! at the University of Arizona). The other, Mrs. Julia *Robinson* (the wife of my colleague here, Prof. R. M. Robinson) has obtained some interesting results concerning definability and decision problems in number theory. Her paper will appear, probably in the Journal of Symbolic Logic; it will contain a short account of some unpublished results of Mostowski and myself which Mrs. R. has applied in her work.[3]

Tarski was proud but somewhat amazed that Chin, one of his "girls," had immediately found a university appointment; because of his competitive streak, he was in all likelihood a bit jealous. *He*, after all, had not become a professor until twenty years after his Ph.D. degree. Furthermore, he would be losing one of those valuable assistants whose main tasks, he thought, would be to help him with his research. This was the European model, where former students sometimes stayed on for years.

On the other hand, a small fringe benefit resulted from Chin's departure: whenever she drove from Arizona to California, which she did with some frequency in the years following her Ph.D. to assist him with various publication projects, Tarski asked her to make a detour via Mexico to purchase the bargain-priced tequila that he used for making the famous slivovitz served at every Tarski party. He was notoriously shameless about asking anyone near the Mexican border to cross over and buy tequila for him. A few years later, while visiting his student Jean Butler in Tucson, he asked her and her husband to drive him across the border; they would be allowed to bring back one bottle, duty-free, for each person in the car. Butler recalled, "He was very disappointed that I refused to take our children out of school for the day so he could have two more bottles."[4]

Tarski's other "girl" was Julia Robinson, destined to become famous for her contribution to the solution of Hilbert's tenth problem on Diophantine equations; she would also be the first female mathematician elected to

Julia Robinson, c. 1950. Tarski's student, later first woman president of the American Mathematical Society.

the National Academy of Sciences *and* the first woman to be president of the American Mathematical Society. Immensely gifted but modest, Julia was in Berkeley when Tarski arrived; she was married to Raphael Robinson, who had been her teacher and mentor as an undergraduate. After their marriage, Raphael continued to encourage and support her mathematical work in every way.

She began her studies with Tarski by auditing the seminar he gave on Gödel's incompleteness theorems. At one point he read a letter from Andrzej Mostowski, his former student in Poland, who raised the question of whether it is possible to define the addition of integers in terms of successor and multiplication. To her surprise, within a few days, Julia came up with a complicated definition. Immensely pleased, Tarski said the work was so original that it would do for a Ph.D. but, by the time Julia actually finished writing up the result, she had generalized and simplified it to such an extent that, in her words, "it became trivial"; she knew without asking that it wasn't enough for a thesis.[5] (It is now an exercise in Herbert Enderton's book, *A Mathematical Introduction to Logic*.)

Like all of Tarski's students, Julia Robinson found his way of setting results into an elegant framework exciting and inspiring, especially since "he just bubbled over with problems." Her appreciation is captured in Constance Reid's *Julia*:

There are teachers whose lectures are so well organized that they convey the impression that mathematics is absolutely finished. Tarski's lectures were equally well organized but, because of the problems [he raised], you knew that there were still things that even *you* could do which would make for progress.[6]

These problems sometimes came to Julia second-hand, after her husband Raphael met with Alfred at the Men's Faculty Club. Over lunch, they discussed various mathematical topics. Some of her best work was stimulated by Raphael's transmission of a problem posed by Tarski when she was not present. In particular, her thesis, "Definability and Decision Problems in Arithmetic," began as an attempt to solve a problem raised during an Alfred–Raphael lunch conversation; the question was whether one can give a first-order definition of the integers in the field of rationals. Julia showed by a quite difficult and novel argument that this is indeed the case; it follows that the elementary theory of the rational number field, like that of the arithmetic of the integers, is undecidable.[7] This result was far afield from the original problem Tarski had assigned her about relation algebras, but it was still connected to other things that Tarski had done.

House and Home

In a remarkably quick and economically astute move, Tarski purchased a house in the Berkeley hills shortly after Maria and the children arrived in 1946. Good houses were for sale at reasonable prices; with large loans and low-interest rates easily available, most faculty members became homeowners. Encouraged by his colleagues with offers of financial help and general advice, Alfred settled on a charming if slightly pretentious house at 1001 Cragmont Avenue; jokingly referred to as "the Tarski castle," it was high above the street and had large rooms suitable for entertaining, a big garden, an aviary, and a nice view. The many steps to the front door were the only drawback. Within a few months, the family was installed, the children were going to school, and Maria was doing her best to adjust to life in America. The writer Lensey Chao, who was then a mathematics student living nearby, recalled that as she and her family drove past the Tarski home they would often see him standing on his landing at the top of the stairs, appearing for all the world like the lord of the manor, gazing out with satisfaction toward San Francisco Bay and the horizon. They would wave, and Tarski would smile widely and wave back.

Life in this new house was organized with Tarski as reigning genius and Maria as minister to his needs, taking care of all household matters, leaving him free to do his work, to advance his field both personally and institutionally. He worked late and slept late; the children were taught to tiptoe around in the morning and not to disturb him. Although he did nothing of a practical nature around the house, he took a major interest in planning and planting the large garden, to which he and Maria regularly devoted their Monday afternoons. He was partial to rhododendrons, azaleas, and roses, and he had a special love for tropical plants like the feijoa. He also took charge of the family's social and recreational activities – deciding where they would go, whom they would invite for dinner, and on what occasions to have a large party.

The children went to the neighborhood schools and made their way as children do. Fitting into the California environment was easier for eight-year-old Ina than for twelve-year-old Jan. Ina was teased by her schoolmates for a short time but soon she had good friends; Jan felt like an outsider for years. In Warsaw, he had begun *gymnasium* and intellectually was ready for eighth-grade work, but in Berkeley he was placed in the sixth grade until he learned English. He was sent to school wearing the Swedish suit his mother had bought him in Stockholm. "I was something of a curiosity," he recalled. "When I finally appeared wearing American jeans there was a lot of comment about that."

Gradually Jan was moved up to the appropriate grade but the level of work was not very challenging, and this did not escape his father's attention:

> He was critical of just about everything. I heard over and over again about the higher standards in Poland, especially those in the school *he* went to. When I joined the Boy Scouts, he was negative about that too and about the way all my activities were shaping up. He had different ideas about how a boy should be brought up, but he never took the trouble to find out exactly what I did in Boy Scouts or in school. It was just a general reaction.[8]

Ina, by contrast, felt her father was focused on what he himself was doing and paid little attention to what was going on in her life. From her perspective,

> we were a traditional old-style European family with an authoritarian father and a mother who was supposed to be submissive – but would explode on occasion. As children, we were supposed to be polite, to be seen and not heard,

and do what the parents said. We had dinner together every night and talked about whatever my father felt like talking about, his day at the university or something on the news or something he heard someone say. It was all very proper but not comfortable. We led parallel lives but were not close. I didn't tell what happened at my school and he didn't ask; he was scarcely involved in my life. I did what was expected and there wasn't any friction until later when we began arguing about what I would or would not eat.[9]

Wanda Szmielew Arrives

Although contact between Julia Robinson and Tarski was friendly and properly professional, it was not personally close. She was not his research assistant and thus had no long, late-night sessions with him; as far as anyone knows, he made no seriously flirtatious overtures toward her. Over time she became aware of Alfred's reputation as a Casanova, but Julia, a tall, handsome woman, created just the right balance of engaged distance – maybe without even thinking about it – to discourage Alfred.

Tarski's next student, Wanda Szmielew, did *not* want to discourage him. Born in 1918, Szmielew had started working with Tarski in Warsaw before the war, and very soon they had established a personal relationship that included hikes in the Tatra mountains. She arrived in the United States in 1949 as the first Pole to come to Berkeley under his sponsorship; given the Cold-War political situation, this must have taken some powerful intervention in both countries. Before she came she had already solved an important problem – the decidability of the theory of Abelian groups – and this work was to be written up as her dissertation. She came with the goal of completing her thesis in one year, which she did.

Liberated, mysteriously beautiful, and independent, Szmielew set about doing what she wished with little concern for the reaction of others. Unconstrained by convention, she came to Berkeley, leaving her husband in Poland, and accepted Tarski's invitation to live in his house with his family as a guest for the year. On the face of it, this was quite reasonable. Wanda had a grant with limited funds, she had limited time, and she didn't have a car; so it made eminently good economic sense for her to be right there. Furthermore, she was going to be working closely with Tarski, whose habit it was to work into the early hours of the morning. Jan's bedroom, adjacent to Tarski's study, was made hers and Jan moved to a room in the basement. He welcomed the change at first – at fifteen

Sarah Hallam, Tarski's secretary, and Wanda Szmielew, his student, on the steps of the Cragmont Avenue house, c. 1950.

he was happy to have a secluded place to himself – but he was not happy when it became clear that the rapport between Alfred and Wanda went far beyond a shared interest in mathematical logic. Since Tarski was in the habit of doing as he wished at home without having to justify his actions, it is difficult to know whether he had any misgivings about creating this ménage. He had had mistresses before and would have others in the future; it was, after all, part of the culture in Europe – especially among intellectuals, artists, and freethinkers. No one made a fuss, particularly if there was some discretion. The problem in this instance was the blatant lack of it.

Why did Maria put up with this? The simple answer is: She didn't think she had much choice. In the United States she had no income of her own and, at first, no language competency; she was completely dependent upon her husband. Furthermore, she was accustomed to his excesses and willfulness and excused him because he was a "great man" whose status as genius gave him license. As she saw it, her role was to do everything to make him comfortable and happy so that he could do his brilliant work, and without making too much of it, she enjoyed the reflected glory of being his wife. "You can't judge him like you do other people," she later told her daughter, "because he is exceptional." Maria knew that Alfred was deeply attached to her; he needed her and, in his own way, held her in high esteem for her courage and strength. Her easy warmth with people made her a great asset to him socially; as more than one friend observed, she was a sympathetic person of great tact and intelligence. If she felt humiliated by his open love affairs, she was nevertheless aware of her own worth.

Still, she wanted more than that. Before coming to the United States, Maria had worked outside the home. She had been a teacher, had been involved in underground activities during the war, and had done relief work. She was a "giver" – she felt a satisfaction in making herself useful in the larger world. In addition, Maria decided that she had to have some means of supporting herself in case Alfred decided at some point to leave her – or she him. With that in mind she enrolled in a school of practical nursing, and when her training was completed she immediately took a part-time job. According to Ina, Alfred objected strenuously, saying he did not want her to work; he wanted her at home taking care of the household, the children, and (implicitly) him. She acquiesced but did not forget what she had learned.

A time would indeed come, ten years later, when even the saintly Maria would decide she had had enough: she would leave Alfred for several years and come very close to divorcing him. But in 1950, in the Cragmont Avenue house, she tried her best to live with a humiliating situation and keep her emotions under control. Periodically, though, there would be a crisis, a boiling over, usually precipitated by Szmielew's presumption of privilege and the expectation that Maria would take care of domestic tasks for her, as if Alfred's wife were the housekeeper rather than the lady of the house. At those moments, Ina (who was then twelve) recalled, Maria would blow up and there would be loud bitter arguments and recriminations, "shouting and crying; everybody in the family hearing everything." Eventually the heat would subside, and all would simmer down until the next crisis. But both Ina and Jan retained an everlasting resentment and dislike for Wanda Szmielew because of her insensitivity toward Maria as well as toward them.[10]

Meanwhile, among the logicians, gossip was rife about the nature of the relationship between Tarski and Szmielew, but few colleagues outside the intimate circle really knew what was going on. Independently, two of Tarski's male research assistants who were assigned to help Wanda revise her thesis and put it into a more mathematical form for publication had to spend long hours with her. They each felt she was making not-so-subtle sexual advances and were very uncomfortable about it. Perhaps they imagined it, perhaps not. But they were sure about one thing: Wanda Szmielew, the beautiful, heavy-lidded femme fatale, exuded mystery and created tension.[11]

The Parties

Although in his personal and professional life Tarski was self-involved and ego-driven, socially he was charming, personable, and genuinely concerned about the welfare of his friends and students. He loved stimulating company and intellectual excitement and sharing a good time with others. By nature, he and Maria were expansively hospitable and frequently invited guests to their home for dinner. They also gave many big parties for students, colleagues, secretaries, visitors, friends, and family. Maria did most of the hard work, cooking for days ahead of time, while Alfred was in charge of the drinks and the program of the party.

These gatherings at the house on Cragmont had a certain predictability, but that in no way diminished the fun. After a long climb up the stairs to the front door, a guest was greeted by a gleamingly effusive Alfred, cigarette in hand. A comment about the view was de rigueur and then a glass of slivovitz, flavored with fruits or berries from the garden, was offered. It was Alfred's concoction, a potent brew that he was very proud of. Some liked it; many did not, but either way it was almost impossible to refuse at least one drink. Well before all the guests arrived, the party was already at a high pitch.

Maria, friendly and informal, bustled and banged between the kitchen and the dining room, carrying quantities of food to the table: a large Polish ham and, almost always, *bigos*, a Polish stew made with prunes. Usually, Sarah Hallam, by then a close family friend, helped with last-minute details. Guests were served buffet style as Maria hovered, encouraging everyone to eat while Alfred plied the crowd with liquor. Whatever stiffness there might have been among the students who held him in awe was dissipated by the food and drink and the graciousness of host and hostess.

After dinner, the group reassembled in the living room and conversation flowed: about politics, art, flora and fauna (Tarski liked to display his knowledge of botany and zoology), and miscellaneous small talk about people in mathematics, logic, philosophy, and the world at large. The political discussions were heated, especially in the 1950s when McCarthyism was at its height. The loyalty oath at the university, the Rosenberg spy case, and the Alger Hiss–Whittaker Chambers affair were examined in detail: were Ethel and Julius Rosenberg guilty and, if so, how critical was the information they had passed, and did they deserve the death penalty? Was Alger Hiss guilty? Opinions differed and passions ran high. Tarski surprised his colleagues and students by being to the right of most of them on most issues; for example, he was *for* the death penalty for the Rosenbergs. As a fervent anti-communist, he believed the "red threat" to be more real than did most liberals. On the other hand he described himself as a social liberal when it came to domestic affairs; he was a member of the American Civil Liberties Union.

Almost every party included a musical interlude. When he thought the right moment had arrived, Tarski would step back from the center of a conversation, then move forward and back again in his signature lecturing posture, to announce loudly, "I think it is time for a little music." Directing an ingratiating smile toward his student Robert Vaught, he'd say: "Bohbb, please play something for us." Bob, lanky and handsome (he looked like the actor Gregory Peck), would roll his eyes, mutter something, amble to the piano, and play a romantic piece. Next it would be Richard Montague's turn. "Deeck, now you ... please" and Montague – slight, nervous, wearing glasses with heavy frames – would move quickly to the bench and launch into something con brio, accompanying himself in a high falsetto. His performance was always breathtaking; an accomplished organist, he had once considered making music his career.

For a few years before his move to Stanford, the logician John Myhill was in Berkeley and he, too, was called upon to perform. Known for his ability to improvise, he would say: "Give me a book, any book, and open it to any page and I'll do an opera." One memorable evening Maria Tarski handed him *The Kinsey Report*, considered sensational when first published, which Myhill proceeded to use as a libretto from a page opened at random; completely deadpan, he sang, in several voices, the statistics on male homosexuality.[12]

During the year that Evert Beth – Tarski's longtime friend and colleague from the University of Amsterdam – was visiting with his wife Corry, she too would be invited to sing. The rotund Madame Beth hushed the room, offering a Schubert song in her sweet soprano. On occasion, daughter Ina was asked to perform and obliged with her most recent short piano recital piece.

After the music, a new round of drinks would be served. All the flavors of slivovitz – plum, apple, berry, and juniper – had to be tasted and evaluated. The dining room was cleared, tables and chairs were pushed back, and the waltzing began. Maria's courtly younger brother Antoni Witkowski, recently arrived from Poland, cut a fine figure. As the evening wore on, a core of "bitter enders" remained, the ones who (like Tarski) enjoyed the late hours. Those were the moments of intimacy and occasional indiscretion as inhibitions melted, the time of *Brüderschaft*, of drinking with arms linked, when more songs were sung and Tarski would say to one and all, "Call me Alfred."

The contrast between the "call me Alfred" bonhomie at parties and the authoritarian "Professor Tarski" at the university, pacing around the mathematics department commanding attention and service, was striking. Every Tarski student had to find his or her way of interacting with these two personalities. In general, Tarski's expectations of what a student would do for him were broad. His model was that of the Herr Doktor Professor with a coterie of subservient, idolizing acolytes – an exaggeration perhaps, and eventually a fading phenomenon, but the essence was never forgotten. (Somehow, he himself had never served in the role of acolyte or devoted assistant.) Yet, for all his authoritarian ways, Tarski could be and often was an extremely sympathetic and loyal friend; he cared about his students, and he wanted to help when they had serious personal problems.

So the mixture of personal and professional was ever present, and negotiating the shifting currents was an art practiced with varying success by his assistants as they graded problem sets for his courses or helped him write up his research papers and project reports. He gave exact instructions for the way everything had to be done or formulated. Language was extremely important; since he wasn't a native speaker of English, he often asked for advice about the meaning of a phrase or word even though after much discussion he often favored his own first choice. The working sessions on research papers frequently involved spending late-night hours with Tarski in his smoky study at home, which some found agreeable but most did not; nonetheless, it was a rare student or "postdoc" who did not become accommodated to Tarski's routine.

More Students of the 1950s

Frederick Thompson, an earnest and upright young man with none of Tarski's bacchanalian tendencies, completed his Ph.D. thesis on "Some Contributions to Abstract Algebra and Metamathematics" in 1952. The primary subject matter was Tarski's newly created concept of cylindric algebras, and a basic question to be answered – not solved in the dissertation but thrown into greater relief – was whether abstract algebras of this sort could be represented concretely, as was the case with Boolean algebras. Thompson came as close as any American student to the ideal of the acolyte who wished to serve in every way. He happily undertook any task Tarski set him and invented some of his own. At the Cragmont house he painted the dining room and the breakfast room, tiled the bathroom, and changed fuses when the power went out. Fred claimed he loved doing each and every task and never felt Tarski was taking advantage of him, because "after all when I was changing a fuse we would be talking about Fermat's last theorem." Thompson also claimed it was he who finally succeeded, where others had failed, at teaching Tarski to drive a car with a manual gearshift. He may be taking more credit than was due him, for Louise Chin had also put in many brave hours at the task. His account reveals as much about Thompson as about Tarski:

> He [Tarski] was not catching on very fast, but I knew he had a superb intellectual bent, so one day when we were out practicing I said "Pull over to the curb." Then I pointed to the clutch and said "Do you know what that is?" He said, "Well, I know it is called the clutch but that is all I know about it." I said, "When you push that down it disconnects the motor from the wheels." Then I said, "Do it" and he did and the motor of course went faster. Next I had him push the brake and he asked, "What does that do?" I said, "That grabs the wheels and holds them, and if the motor is still attached that is going to make the motor stop too but if you push down the clutch which disconnects the wheels and then you push down the brake, then you can stop the wheels without stopping the motor." And I went on in that way, explaining each piece of equipment and what it did, and immediately he was able to drive. Nothing more was necessary. From there on he understood what was going on and he drove very well.[13]

That was Fred Thompson's assessment; most others cringed at the thought of Tarski at the wheel because he had the disconcerting habit of looking at the person he was talking to (and he was always talking) instead of the road. However, everyone did agree that Tarski loved to drive. Around town, he liked figuring out the best route from point A to point B and he particularly enjoyed finding odd ways to go. On weekends he organized excursions, taking the family and local friends to scenic spots a few hours away from Berkeley – sometimes in a convoy of cars – to places such as Pinnacles National Monument or St. Helena; when visitors came from afar, he planned longer journeys that involved several days of travel.

For Thompson, Tarski was "a towering figure, a major influence on every aspect of my whole life, not only my intellectual life, but all the rest." Indeed, years later Thompson separated from his first wife – a woman very much like him – and married a younger Polish woman from Warsaw who until 1962 had lived in Zoliborz, Tarski's old neighborhood. After his Ph.D. oral examination, he was told he had passed.

Everyone shook my hand, but as I held out my hand to Tarski he said, "Oh no, you have not received your degree yet. You must come to my house at seven this evening." I knew what was going to happen; there was going to be a lovely dinner party with the graduate students and the professors I had studied with, and every time I turned around, Tarski's hand would be in front of me with a glass of liquor in it. Before I went to Berkeley I had never had a drop of any alcoholic beverage, so I prepared by drinking a whole bottle of whipping cream and stuffing down a whole loaf of bread. At midnight when Tarski brought out the champagne to toast my degree, he said that since I was still absolutely sober, I had finally earned it."[14]

Yet, when the time arrived for drinking *Brüderschaft* with linked arms and the affectionate words, "now you can call me Alfred," Fred's surprising response was, "No, I'll never call you Alfred until I can stand beside you." His plan was to write a book and dedicate it "To Alfred", but somehow that never happened.

Working for a Ph.D. with Tarski was an unpredictable experience. Typically he suggested several problems, all difficult, and the student would choose one, as Thompson did. After two years of intense effort on a decision problem for ordinal arithmetic (a problem Tarski would assign again to future students), he was unable to solve it. Meanwhile he had obtained results on a number of smaller problems, primarily concerning cylindric algebras, which could be assembled together as a thesis. Thompson was unhappy that he had not solved a major problem, but he wrote up what he had and it was accepted. According to Thompson, Tarski urged him to do this, saying, "Let's get you out of here; you are ready to go out in the world." A job was waiting for him at RAND Corporation, where McKinsey had worked until 1951. He got the position because of Tarski and McKinsey's recommendation, although by then McKinsey had moved to Stanford, where the issue of his homosexuality (though known) did not prevent his appointment.

With his daunting standards for what was "thesis worthy," Tarski rarely pushed a student out of the nest. Typically he asked for more and yet more proof that the fledgling was ready to fly, demanding a solution to a major problem and insisting upon an exasperating exactness and completeness of exposition. Although it became common knowledge that doing graduate work and writing a Ph.D. thesis with Tarski was a harrowing experience, his personal magnetism and brilliance, and his passionate belief in the supreme importance of logic, were so strong that he attracted a steady stream of students eager to take up the challenge.

An Exceptional Case

Although usually Tarski's graduate students knew right away – sometimes after only a few lectures – that they wanted *him* as mentor, Robert Vaught chose him after losing his first advisor during a time of personal and general academic crisis. He had taken many logic courses from Tarski and, in Fred Thompson's view, was intimidatingly quick to grasp all the main ideas. But Vaught was also working even more intensively with Professor John Kelley in the field of functional analysis and was well on his way toward a Ph.D. under his supervision. The crisis arose when Kelley was fired because he refused to sign the Loyalty Oath of 1950.[15]

This extremely controversial oath was conjured up by the University of California administration to ward off the threat of investigation by the House Un-American Activities Committee of the U.S. Congress. During the Cold-War years of McCarthyite hysteria, the committee had been combatting the "red menace" by turning up a communist under every stone. In self-defense, the university administration decided to require all its employees to sign a special oath declaring that they were not members of the Communist Party or any other organization dedicated to the overthrow of the United States government, thus pre-empting the committee by saying, in effect, "we have cleaned our house ourselves." The decision created havoc among the faculty, as colleagues argued about the right thing to do. Professor Kelley and others on the faculty refused to sign on the principled grounds that every employee had already signed an oath of allegiance to uphold the Constitution of the United States *and* the Constitution of the State of California – an oath that specifically forbade any other oaths or tests such as this special one.[16]

Reports differ on Tarski's position. According to Kelley, "Alfred didn't think it was worth arguing about; as a practical matter, he thought one ought to just sign and forget it. He felt the same way about McCarthyism: he just didn't want to be bothered with taking a stand against McCarthy's tactics."[17] In the context of Tarski's hatred of the Soviet hegemony over Poland and his general anti-communist sentiments, this attitude would not be surprising. However, Leon Henkin, who (like others) refused a job offer at Berkeley in 1952 because of *his* principles regarding the oath, had the impression that Tarski was sympathetic to the faculty who refused to sign but did not join the protestors because he felt his personal circumstances precluded doing so. And he made no effort to talk Henkin out of his anti-oath position – which in itself was unusual, for Tarski seldom accepted a refusal without an argument. As it turned out, the oath was rescinded in 1953 and later declared unconstitutional by the California Supreme Court precisely because it had singled out one group of people and required them, uniquely, to sign a special "double" oath. And eventually Senator Joseph McCarthy was called to account for the outrageous excesses of his witch-hunting crusade.

In between, however, much damage was done. Kelley's dismissal was a major blow to the university and to the mathematics department, faculty and students alike, for he was a first-class mathematician and engagingly warm. Fortunately, he was offered a position at Tulane University that he accepted quickly; he also secured a fellowship for Bob Vaught so they could continue their work together (they had already written a joint paper) and Vaught could finish his Ph.D. with him as advisor. But Vaught was in a troubled mental state; he had had several periods of deep depression followed by manic episodes. Thesis or no thesis, he did not see how he could follow Kelley to New Orleans, a place where he had no family or friends. Sick with anxiety, he remained in Berkeley and turned to Tarski, with whom he had already established an intellectual and personal bond. It was to be a tortured journey. In despair, he attempted suicide several times, was committed to the county hospital, and later given electric shock treatments against his will. Furious, he demanded a legal sanity hearing and

Robert Vaught at the Tarski Symposium, 1971.

summoned witnesses, his friends and professors, to testify on his behalf, putting everyone in a terrible bind because Vaught expected his friends to support him – to say he was fine and that there was no reason for him to be locked up. But most of his friends felt he was indeed a danger to himself and perhaps to others, particularly since one of his suicide attempts had been to drive his car into a tree.

Vaught never forgot his professor's testimony at the hearing, recalling that Tarski spoke for about ten minutes and in the end said, "It isn't that he is so exceptionally good in mathematics [but of course he was!] but I admire his judgment," and then he started to cry.[18]

Throughout Vaught's many years of mental illness, depression, and mania, in and out of various hospitals, Tarski remained a staunch ally even though Vaught, in his disturbed state of mind, sometimes insulted and attacked him verbally – as he did others. When Vaught was sent to the state institution in Napa, a two-hour drive from Berkeley, Maria Tarski, profoundly sympathetic, visited him frequently – as often as two or three times a week – sometimes asking her friend Celina Whitfield to drive her. Alfred said he couldn't bear to see him in the hospital (and no doubt felt he couldn't afford to take the time), but he cared deeply. He attempted to understand the mysteriousness of the illness, helped Vaught seek out doctors and psychiatrists, was interested in his treatment, and, above all, encouraged him as a student and saw him through a successful Ph.D. degree. With a research fellowship rather than a teaching assistantship, Vaught was not obliged to teach and Tarski did not pressure him to help with his own research as he did others. Nevertheless, Vaught did have his share of late-night sessions with his mentor and incidentally noticed how large Tarski's pupils were without realizing why. Only in retrospect did he become aware that Alfred was taking stimulants to stay awake.

After finishing his doctoral work in 1954, Vaught was hired at the University of Washington but eventually, with Tarski's strong backing, returned to Berkeley and remained there as a professor of mathematics until his retirement. With the advent of new medications to treat his illness, his manic-depressive episodes were controlled. In an article on Tarski's work written for the *Journal of Symbolic Logic* a few years after Tarski's death, Vaught expressed his personal and professional feelings for his professor and friend:

Tarski stands with a few others like Aristotle, Frege, and Gödel as one of the greatest of all logicians. As a man, Alfred Tarski was warm and sensitive. In person he was somehow incredibly alive and alert in a quite unique way.[19]

After the loyalty oath was rescinded in 1953, John Kelley returned to Berkeley and Leon Henkin accepted the position that he had refused while it was in effect. Henkin could not have come at a better moment to help Tarski build the foundation of his empire in logic.

INTERLUDE IV

The Publication Campaigns

D IFFICULT AS THE YEARS 1939–1946 had been, Tarski continued his research and publication in a variety of areas, though now – with one notable exception – entirely in the English language. All of the resulting articles planted seeds that were to flourish abundantly in later years. Nineteen forty-one saw the publication of a groundbreaking paper on the calculus of relations,[1] a subject to which he was to return later and promote to the end of his life (see Interlude VI, Algebras of Logic, which follows Chapter 13). In 1943 he published an article on fields of sets and large cardinal numbers with the itinerant Hungarian genius, Paul Erdös; they were to revisit this subject two decades later when remarkable progress on the concepts involved had been made by Tarski's students William Hanf and H. Jerome Keisler.[2] Tarski's visit to McKinsey in Montana resulted in the publication in 1944 of an important article on the algebra of topology, which they would connect with intuitionistic and modal logics a few years later.[3] Also in 1944, Tarski published an expository article on his theory of truth in which he took the opportunity to clarify his goals and respond to critics; in subsequent years that article was widely anthologized and translated into a number of languages.[4] Then, in 1945, in homage to his Polish colleagues who had survived the war and had renewed publication of *Fundamenta Mathematicae*, he sent part two of an article on ideals in set fields; the first part had appeared in 1939 just prior to the Nazi invasion of Poland. Both parts were written in the "international language" of German, rather than Polish.[5]

Launching the Campaign

Beginning in 1946, with his position established at Berkeley and his family life reasonably settled, Tarski undertook – with a sense of great urgency –

the publication of a series of books and monographs that would present in a systematic and detailed fashion his main achievements of the past brought up to date. At the outset, the easiest thing to do was to prepare a second edition of the *Introduction to Logic and to the Methodology of Deductive Sciences*; the dedication read, "To my wife".[6] That book was very successful and, in the following years, went into two further editions and was translated into many different languages: first in Russian in 1948; followed quickly by Spanish, Dutch, and Hebrew; and later on by translations into French, Bulgarian, Swedish, German, Italian, Czech, Georgian, and Serbo-Croatian. The first two chapters would even be reprinted in the popular collection *The World of Mathematics*.[7]

After the *Introduction to Logic*, all the books and monographs that Tarski wrote, either individually or with collaborators, were at the advanced level and intended for graduate students and researchers. Incidentally, in all his dealings with publishers, Tarski drove a hard bargain; in his archives at the University of California there are reams of letters discussing terms of publication and insisting on the best possible royalties. He would refuse to grant rights if his conditions were not met.

The Decision Procedure for Algebra and Geometry

As noted in Interlude II, Tarski (along with many others) considered the completeness and decision procedure for elementary algebra and geometry to be one of the two most important research contributions in his entire career – the other being his theory of truth. Yet, though the work had been done by 1930, the details were not published until 1948. This is partly accidental because in 1939 Tarski did finally prepare a monograph, describing his method, under the title *The Completeness of Elementary Algebra and Geometry*. This was to appear as the first in a new series, *Métalogique et Métamathématique*, for a Parisian publisher, but the invasion of France by Germany in 1940 disrupted the publication process. As Tarski later wrote: "Two sets of page proofs which are in my possession seem to be the only material remainders of that venture." In 1948, the RAND Corporation became interested in Tarski's decision procedure for elementary algebra. J. C. C. McKinsey, who was then working for RAND, may have suggested that the procedure could be programmed for computer calculations of the optimization of strategies in the theory of games, but implementation of that would require first writing up its theoretical details in full. He was handed the job of preparing Tarski's aborted 1939 work for publication, though what he actually ended up doing was revising it entirely under Tarski's supervision. In final form the monograph was put out as a RAND report with a new title: *A Decision Method for Elementary Algebra and Geometry*.[8] Three years later, a second edition was published by the University of California Press and thus became publicly available.[9] Finally, a lightly edited version of the original 1939 page proofs was brought out in 1967 in France.[10] Tarski's perspective had clearly changed in the meantime, with the different titles reflecting a change of primary aims from completeness to decidability.[11]

Cardinal Algebras and a Tribute to Lost Colleagues

Fresh on the heels of the RAND report, Tarski's next major publication of the early postwar period was *Cardinal Algebras*; completed with the assistance of Bjarni Jónsson in 1948 and including a jointly written appendix, it appeared in 1949.[12] The dedication to the volume reads, "To the memory of my friends and students murdered in Poland during the Second World War". In his preface, Tarski explained how he was recouping and building on a group of results in general set theory obtained twenty years before. After acknowledgments to Chin, Jónsson, and McKinsey, Tarski wrote:

It would be impossible for me to conclude this introduction without mentioning one more name – that of Adolf Lindenbaum, a former student and colleague of mine at the University of Warsaw. My close friend and collaborator for many years, he took a very active part in the earlier stages of the research which resulted in the present work, and the few references to his contributions that will be found in the book can hardly convey an adequate idea of the extent of my indebtedness. The wave of organized totalitarian barbarism engulfed this man of unusual intelligence and great talent – as it did millions of others.

The main aim of *Cardinal Algebras* was to isolate in algebraic form a number of results about finite and infinite sums of cardinal numbers that could be proved with*out* using the Axiom of Choice. Put in this abstract form, they could then be applied to other mathematical systems besides cardinal numbers – as was shown by Tarski's joint work with Jónsson in the appendix. That connected to their monograph, *Direct Decompositions* *of Finite Algebraic Systems*, on which Jónsson and Tarski had collaborated two years earlier and which incorporated the main part of Jónsson's doctoral dissertation.[13] Its main purpose was to generalize to finite algebraic systems of an arbitrary kind the well-known theorem of group theory according to which every finite group has, up to isomorphism, a unique representation as a direct product of indecomposable subgroups.

Algebraic Logic

Hardly had these books been brought to publication when Tarski set about advancing his ideas for the algebraic treatment of logic through the use of relation algebras that he had introduced in 1941. Louise Chin made particular progress on that in her 1948 Ph.D. dissertation, and Tarski used the opportunity to combine her work with an updated exposition of his approach to the subject in a jointly authored monograph, *Distributive and Modular Laws in the Arithmetic of Relation Algebras*, that appeared in 1951.

The axioms for relation algebras can account for the logic of only dyadic (binary, or two-placed) relations, and that only incompletely, whereas modern logic in general deals with polyadic (many-placed) relations. It was to put the latter in mathematical terms that in the late 1940s Tarski introduced the idea of cylindric algebras, so called because the operation of "cylindrification" (projection on a component) on polyadic relations corresponds to the logical operation of existential quantification; this approach was developed initially in collaboration with Louise Chin and Frederick B. Thompson and was to become a major subject of research in the Berkeley school. Those efforts would lead two decades later to the first part of an opus, *Cylindric Algebras*, authored by Tarski together with Leon Henkin and Donald Monk;[14] the second part would not appear until after Tarski's death.[15] These developments are elaborated further in Interlude VI.

Undecidable Theories: An Unexpected and Unexpectedly Successful Book

Complementary to the notion of a *decidable theory* is the notion of an *undecidable theory*, by which is meant a theory for which no effective step-by-step procedure can be found to tell whether or not a statement is provable from its axioms. The first main undecidability result was found by Alonzo Church in 1936; he showed that the system of *Principia Mathematica* is undecidable if it is ω-consistent. A year later, J. Barkley Rosser strengthened this result considerably by showing that the theory called Peano Arithmetic (for the natural numbers under addition and multiplication) is undecidable and that the same holds for any consistent extension of it. Julia Robinson's 1948 thesis established – by interpreting the property of being a natural number in the system of rational numbers – that the elementary theory of rational numbers under addition and multiplication is undecidable. A year later in several abstracts (one with Mostowski), Tarski announced the undecidability of the elementary theory of groups and of some other algebraic structures, together with an abstract stating a rather general method for proving a wide class of theories undecidable. But these results did not attract much attention, and it was generally thought that the Tarski school was almost entirely dedicated to working on decidability results. Most prominent in that direction after Tarski's pioneering work on algebra and geometry was Wanda Szmielew's proof in her 1949 dissertation that the elementary theory of Abelian (commutative) groups is decidable.

It was thus quite unexpected when, in 1953, Tarski published the short book *Undecidable Theories* with Andrzej Mostowski and Raphael Robinson as co-authors.[16] It was divided into three distinct parts: the first and the third written by Tarski alone and the second written jointly with Mostowski and Robinson. The main new concept of the first part of the book is that of a theory being *essentially undecidable*, which means that the theory together with all its consistent extensions is undecidable. Tarski showed that this property is preserved under the interpretation of one theory into another. Another principal method of establishing undecidability of a theory would be to obtain it from a known undecidable theory by deleting finitely many axioms from the latter (without restricting the language). In order to combine and apply these methods as widely as possible, one would need to start with an essentially undecidable theory having a finite number of axioms and open to a wide variety of interpretations. In the 1940s, Tarski and Mostowski had found one such example, which was later considerably simplified and whittled down through the work of Raphael Robinson. The resulting system is a very weak fragment of Peano Arithmetic, and Part II of *Undecidable Theories* is devoted to showing its essential undecidability. Finally, in Part III, Tarski applies his general methods in combination with Robinson's theory to prove the undecidability of the elementary theory of groups (as he had announced four years earlier). This constituted a foil to Szmielew's decidability result for the elementary theory of Abelian groups. Since the theory of Abelian groups simply adds the commutativity axiom to the theory of groups, the latter is an example of a theory that is undecidable but not *essentially* undecidable.

Unlike *Cardinal Algebras,* which appealed to a rather limited audience, *Undecidable Theories* attracted widespread interest among logicians; besides being short it was easy to read and its elegant, widely applicable, and powerful methods led to a great deal of subsequent research. To some extent the difference in reception also lay in the fact that the former book more or less finished off a subject, whereas the latter opened up a whole new series of lines to follow.

Logic, Semantics, Metamathematics

When Tarski visited England in 1950 to deliver the Shearman lectures, his old friend J. H. Woodger proposed to prepare, in English translation, a collection of his major contributions to logic, semantics, and metamathematics published before the Second World War – a proposal that Tarski happily accepted. Out of Woodger's substantial effort came, six years later, the volume *Logic, Semantics, Metamathematics.*[17] Included in it, of course, is the great paper, "The Concept of Truth in Formalized Languages" – a translation of a translation, from Polish through German. Other articles of note are the two papers resulting from Tarski's Ph.D. dissertation, several papers on fundamental concepts of metamathematics and the methodology of deductive sciences, one on definable sets of real numbers, and a famous and influential one on the notion of logical consequence. There are also three articles written jointly: one on sentential calculi with Jan Łukasiewicz; one on logical operations and projective sets with Kazimierz Kuratowski (related to that on definable sets of reals); and one with Adolf Lindenbaum on invariance of logical notions under permutations of the universe of individuals. The volume ends with an article on the sentential calculus and topology that was to lead directly to Tarski's later work with McKinsey on algebras of topology. The article with Lindenbaum was to form the basis of a striking but controversial proposal that Tarski made in a 1966 lecture he gave in London under the title "What Are Logical Notions?"[18] Years later the philosopher John

Corcoran urged Tarski to publish the text of that lecture and took on the task of editing it.

For the first edition of *Logic, Semantics, Metamathematics*, Tarski assisted Woodger with points of translation and also took the opportunity to make some additions and corrections. A more substantial revision of the translations with further additions and corrections was made in the second edition, again edited by John Corcoran during the last few years of Tarski's life.[19] According to Corcoran's preface, for their work together Tarski prepared more than fifty single-spaced pages of changes.

Ordinal Algebras

Tarski's final book in the remarkable decade of publication that had begun in 1946 was the slim volume entitled *Ordinal Algebras*.[20] Its aim was to derive on an abstract algebraic basis a substantial number of properties of isomorphism types of binary relations under the operations of ordered addition and converse. Tarski's algebraic approach to these operations was similar to that taken in the book *Cardinal Algebras*. As in that previous work, the addition operation is extended to finite and infinite sequences of objects, and the algebra is thus infinitary in a suitable sense. Tarski used this framework to recapture many of the joint and individual results that had been announced without proof in a paper with Adolf Lindenbaum thirty years earlier. In addition, the book included a number of new results and benefitted from the inclusion of two appendices by Chen-Chung Chang and Bjarni Jónsson (respectively) giving further applications.

What Next?

Having thus brought much of his prewar work into definitive monograph and book form, Tarski was now ready and eager to undertake several new major publication projects, including books on relation algebras, cylindric algebras, set theory, and geometry. These projects would keep him busy to the end of his life.

8

"Papa Tarski" and His Students

IN *The Best Man*, Gore Vidal's play about presidential politics, a character says: "A lot of men need a lot of women and there are worse things, let me tell you." To the extent that is true, it is not unusual that Tarski was a ladies' man who needed a lot of women; what is unusual is the number of female *students* he had – probably more than any other mathematics or philosophy professor of his time – graduate students and postdoctoral researchers whom he mentored and sponsored. Of his first seven Ph.D. students, four were women who wrote their theses in mathematics.

While he was busy seducing them, or trying to, Tarski took women seriously. The seduction was mental as well as physical. As with all his students, he encouraged women to think, write, edit, and prepare papers for publications. He was flirtatious and aggressive, but it was rarely a one-way affair. He had predatory instincts, but his "victims" were at times willing and may even have taken the initative. There were those who saw Tarski as an overbearing little man, but others found him very appealing. One woman who attended his course on the foundations of mathematics but never spoke to him said, "people laugh at me but I thought he was one of the sexiest philosophers I had ever seen. A big ego, yes, but a mind that made him terribly attractive and interesting. It was like watching a great stage performance. He was magnetic."[1]

Indeed, there was a frisson about him that made it attractive to risk an encounter. A Mills College student who attended his seminar was warned that he would try to seduce her on the spot. When he behaved like a "gentleman," she told a friend, "I kept asking myself,... but what is wrong with me?"[2]

Anne C. Davis (Morel), c. 1950. Tarski's student.

Anne C. Davis (later Morel)

Between Fred Thompson and Robert Vaught, Anne Davis added her name to the growing list of Tarski's Ph.D. students as well as to the list of women with whom he became personally involved. Slight in build, Anne had an appealing, quizzical manner – a way of wondering out loud about the right thing to do, both mathematically and socially. She had come to Berkeley as a mathematics graduate student in 1942, when Tarski himself had just arrived and was teaching mainly undergraduate courses. Not finding a special subject that engaged her, she left and joined the Waves to help in the war effort. Two years later she returned, married to Alan Davis, also a mathematician who had a job teaching at the University of Nevada in Reno. The marriage – according to Delos Morel, who would later become Anne's second husband – began on shaky ground, and almost immediately she was looking for a way to extricate herself.

When Anne re-entered graduate school in 1946, she discovered Tarski and logic. He welcomed her to his classes and seminars, and she chose him as her advisor. Dividing her time between Reno and Berkeley, Davis joined the circle of Tarski students of the late 1940s to the mid-1950s. The Tarskis invited her and her husband, when he was in Berkeley, to come along on their Sunday excursions, which usually included Sarah Hallam and Louise Chin. Anne and Wanda Szmielew became friends during the latter's first stay in Berkeley, and she and C. C. Chang, a younger student with whom she shared an office, were "buddies," as he put it. To most others she seemed straitlaced, but that view changed as rumors circulated about her relationship with Tarski. The gossip was not lost on C. C., who worried "whether people weren't talking about me and Anne too, since as office mates we were often together in close quarters. Of course I knew nothing was going on between *us*."[3] Years later, Sarah Hallam, who was the soul of discretion, revealed that she knew exactly when Alfred and Anne became lovers. She and Anne had taken a trip to Mexico in 1950. "Suddenly Anne wanted to cut it short and return to Berkeley. It dawned on me then that she wanted to do it so that she could be with Alfred. Their affair was just beginning – right after Wanda Szmielew returned to Poland."[4]

Perhaps more than most students, Davis looked to Tarski for guidance and approval, asking for advice on problems and approaches to their solution. Her dissertation in set theory, on the arithmetic of order types, followed a line from Cantor, Hausdorff, and Sierpiński to Tarski and Lindenbaum; it was very much in the Tarski idiom, and that pleased him. Although she proceeded slowly, he liked the seriously careful way she worked. Problems in her personal life and the distance of the commute between Reno and Berkeley impeded her progress, so they corresponded or spoke by telephone. But the many letters she wrote him were often touchingly personal rather than mathematical.[5] Although at first she addressed him as "Dr. Tarski," by 1949 it was "Dear Alfred"; she asked about his teaching load, his travels, and his health. After his return from

London (where he gave the Shearman Lectures in 1950), Davis wrote, "I'm sorry you did not have more time abroad but I am selfishly glad that you are back.... I look back to last semester when I saw you more frequently. It seems to me there is never enough time to talk but perhaps if I were seeing you now, you would be bored." Her letters are intimate but self-deprecating. She reveals her feelings, then says: "This does not sound the way I want it to."

When her thesis was nearly finished, the tone of her correspondence shifted. Since she was writing about the state of her work, the salutation changed to "Dear Professor," an ironic opening but appropriate to the major content of the letter. The conclusion, too, suggests a more confident intimacy than before: poignantly she describes her problems with her husband and follows with, "I think of you and miss you. I am very unhappy ... please write to me. It would give me so much pleasure to get something in your handwriting." The letter closes, "With my love to you," and a postscript: "I am reading *Emma* [Jane Austen] and I love it but I'm not sure you would like it"; then she repeats, "I miss you."

"Dear Boss," Anne writes in the next letter, "If you come by tonight after 8:30, I'll be home and awake."

Davis completed her Ph.D. in 1953 and was hired as an instructor at UC for two years. In 1955 she and Alan divorced, and, according to Delos Morel, Alfred asked Anne to marry him. Morel, a lawyer, and Anne had been friends for many years, and it seems she had been confiding in him, too. He was recently separated from his wife, who had been Anne's friend as well; Morel and Tarski also considered themselves friends. Morel's account of the evening of Tarski's proposal smacks of a scene from Feydeau:

One evening I dropped by Anne's apartment on Russell Street and Tarski was there. A few minutes after I arrived, Anne excused herself and went upstairs. So there I was and there *he* was ... it was very awkward but we made some kind of conversation. After a time, Anne returned and rather quickly, Tarski left. Then Anne told me he had come to propose marriage, adding that he wanted to have a child with her. I don't know whether she answered directly at the moment or later but she said she did not want to marry him. In the weeks that followed, she frequently saw him driving by her apartment and on several occasions when I came to see her, I saw him sitting in his car parked across the street from her house. From this I assumed that he took her refusal very hard.[6]

So, after years of involvement, instead of falling into Alfred's arms Anne turned to Delos. In 1957 she married him and changed her name to Morel; the following year, they had a daughter. She held positions at UC Davis, spent a research year at the Institute for Advanced Study in Princeton, and ended up as the first female full professor at the University of Washington. Whether or not Tarski really would have divorced Maria to marry Anne if she had accepted him is unknowable, but he certainly was very fond of her and was very hurt when he lost her. He needed an intimate, affectionate, and preferably sexual relationship with a woman, preferably a younger woman – someone he could counsel and inspire, someone who admired and flattered him. He needed constant reaffirmation not only about the greatness of his work but also of his whole person, including his sensual appeal. The "Papa Tarski" sobriquet was only half a joke. Being a father-professor, a powerful man, was an essential part of his attraction, and the position of father-professor enhanced his own feeling of power. He depended upon and respected Maria (and dominated her, too), but he was a romantic and always in search of a new passion.

The Morels as a couple and the Tarskis remained friends, and when they came to California they would visit with Alfred and Maria. In 1960, the year she was at the Institute, long after their affair had ended, Anne again wrote to Tarski about her unsettled feelings, confiding that she found it difficult to make friends in Princeton: "This sounds rather like a letter to father from which attitude I have at times tried to break away. At the moment, however, meaningful human contacts are so few that such analysis is not in order." She added, "this sounds self pitying which is not the case. My day to day existence is pleasant ... and my daughter is a happy, lovely, young creature."[7]

But over time, as Morel learned of Tarski's subsequent affairs with other women and the effect it had upon them, her attitude toward him changed. The feeling of closeness changed to harsh disapproval, for she felt he was taking advantage of his position of power in a way she now viewed as unacceptable, and in retrospect it inevitably soured her view of their past relationship.[8]

Jean Butler was one more woman of the 1950s who was more briefly involved with Tarski as a Ph.D. student. "I backed into math because my husband was a mathematician," she said; "I took courses from Tarski and I was bowled over by his brilliance and power." When her husband, a young mathematics professor, moved to Arizona and then to UCLA, she continued to work on her own, sending her results to Tarski and consulting with Richard Montague at UCLA. Although she presented her work at the Cornell meeting that would be held in 1957 and was part of Tarski's entourage there, in the end he disappointed her by showing her work to another logician, who took her material and ran with it. Butler had thyroid disease and also (like Morel) marital problems; as a result, she quit mathematics for a long period. When she was ready to return in 1967, Tarski did not give her the financial and personal support she needed. He had apparently lost interest in seeing her through her Ph.D. She went instead to the University of Washington, where, after starting all over again, she was successful in completing a Ph.D. in graph theory with Victor Klee.[9]

By and large, whatever the level of personal intimacy – and there had been something of that nature between Butler and Tarski – he was usually helpful and encouraging to his female students, perhaps in some instances giving them more leeway than the men. But that was not the case with Butler. The knotty circumstances are too difficult to untangle, but her complaints are among those that contributed to Anne Morel's jaundiced view of her teacher and former lover.

Chen-Chung Chang

"What is it like to be Tarski's student?" C. C. Chang asked Fred Thompson when he was worrying over who to chose as an advisor.

"He treats you like a son," was the unhesitating response.[10]

But what kind of a father was he? Initially every student found him friendly, approachable, and encouraging. However, once a student entered his seminar and began to participate, the smile vanished and an authoritarian, demanding, unyielding father-professor took center stage, one who could and would mortify a floundering student by pouncing on every error, large or small. In his insistence that presentations be done exactly as he thought right, he showed no mercy.

Chang had no difficulty doing things right; his experience as a Tarski student was almost all positive. A smooth, elegant young man, he came to the mathematics department at UC in 1949 after completing undergraduate work at Harvard; like Feferman, he was seeking a subject and a mentor. At Harvard he had taken a logic course from Quine but, eschewing the

Chen-Chung Chang at the Tarski Symposium, 1971.

usual reverence, Chang said, "Quine was such a dull lecturer. He came to class with a big stack of three by five cards and essentially read his book to us, card by card – the same book we were using as the text for his course."

In Berkeley, Chang enrolled in Tarski's class on universal algebra and was amazed at how different it was from Quine's course. In the first place, Tarski *never* read his lectures. He was completely concentrated on what he was talking about; he was enthusiastic and had an intense desire for students to understand, but most importantly, Chang recalled, "Tarski offered several directions in which a person could go. You didn't have to just stay in algebra; you could do model theory. You could do set theory or metamathematics. When I took his seminar I found I could get into it right away... I could steer my own boat."

Early in his career Tarski had established the habit of inviting students or colleagues to his home to work with him into the wee hours of the morning. Some were suited to the task, others not, but no matter. A Tarski student had to get used to those night sessions and get used to his urgent phone calls to come to his house as soon as possible. Richard Montague's famous remark after one such call was "Oh dear, I've got to go now. Papa Tarski wants me to come over and lick stamps." No one ever addressed Tarski as "Papa Tarski" to his face but the claim is that he knew of his nickname and, being ignorant of the mockery, rather liked the affectionate sound of it.

Obviously, Tarski wanted much more than stamp licking. He wanted assistance with the details of a paper or a book in progress; or help with the notes or the bibliography; or a close scrutiny of a manuscript with the idea of honing it to perfection; or help with a project report. Any of these tasks could require endless revision.

Although Chang took six years to finish his doctorate, it was, for the most part, easy sailing. What was difficult was the work Tarski asked him to do during his graduate student years as research assistant. Chang's description of an all-nighter with "Papa Tarski" illuminates the scene perfectly.

They would start about 9 P.M. when Tarski was just getting going. He always smoked, and he kept the door of his study closed so the smoke would stay in the room because he thought that made him concentrate better. "It was awful for me," Chang said, "because I had asthma, but what could I do? I was his student. I wasn't really a night person either and after a while it was a struggle to keep my eyes open. Around 2 A.M. he'd ask me if I wanted some coffee and I'd say yes. Sitting at his desk, with the door closed, he'd scream, 'Mariaahh, Mariaahh,' as loud as he could. If there was no answer, he'd repeat it, sometimes three or four times until Maria finally opened the door, half asleep, saying 'Yes, Alfred?' He'd ask her to bring us two cups of coffee and she trudged into the kitchen to make the coffee and bring it to us. I've never seen anything like it before or after."

Alfred never faded; at 4:30 A.M. he was still going strong. Chang would leave the house as the sun was coming up. "I'd see the sunrise from the top of those high stairs. I knew he would be going to bed, to sleep until noon but I would have to be at the department first thing in the morning to give the manuscript to the typist because he was always in such a hurry."[11]

Chang was a meticulous young man about whom a friend once remarked that his white buck shoes never got dirty. So, the physical condition of the manuscript made him very uneasy. As he described it, Tarski had a habit – from the old days in Poland – of being frugal with paper. He would take a sheet, fold it in half so it became like a little four-page book, and he'd write in pencil on all four sides. If he made a mistake, he wouldn't scratch it out. Instead, he'd erase and erase. Sometimes he erased whole paragraphs and wrote over the erasure so that the paper looked grey and smudged with charcoal. Then he'd say, "It doesn't have to be clean; it just has to be clear." And while he was doing all that erasing, Chang would stand there horrified, thinking, "I am going to have to tell her [the typist] that this awful looking *thing* has to be typed," while Tarski kept shaking his head, erasing and repeating, "doesn't have to be clean, just has to be clear."

Yet, embarrassed as he was about the "awful looking thing" the paper had turned into, Chang, along with almost every student of the Tarski school, was everlastingly grateful for having been imbued with that passion for clarity. "You look at our writings and you see his influence," he said, "and we pass that on to our students."

On only one harrowing occasion did Chang refuse to follow his teacher's lead. In November of 1953, the promising thirty-one-year-old Polish logician, Jan Kalicki, was killed in an automobile accident while driving from Berkeley to a logic meeting in Los Angeles. With him in the car were Tarski, in the right-hand passenger seat, and Mrs. Tarski and C. C. Chang in the rear. Chang had perfect recall of the awful moment, and of Tarski's panic in contrast to Maria Tarski's composure and compassion.

Kalicki was driving seventy miles an hour, talking to Tarski and turning his head to look at him as he spoke; he didn't see a curve in the road until the last minute and overran it; then as he tried to turn back, the car rolled over. It rolled three and a half times and landed on its roof. We were upside down but Jan was thrown out of the car and it had rolled over him.

Tarski was screaming "Get us out. Get us out!" and I was thinking: Hey, if the gas tank leaks.... It was a two door car but somehow Maria and I crawled out of the back seat before Tarski got out. Maria really impressed me; she was the calmest one. Jan was lying on the ground, blood coming out everywhere, from his nose, his eyes, ears, and mouth but he was breathing. Tarski and I didn't know what to do – we were just walking around in circles. Maria, though, went up to him and lifted his head, wiped away some blood and tried to talk to him.[12]

Eventually help came and the group was taken to a nearby hospital. Maria and Alfred were bruised and shaken; his coat was torn but neither was seriously injured. C. C. had a deep leg wound that required many stitches. Jan Kalicki was dead.

The accident occurred soon after their departure, and the news reached Berkeley almost immediately. Benson Mates recalled that, only moments before leaving the university, Kalicki had said he would be driving his car and mentioned, in a humorous way, that he was supposed to wear glasses but did not – because he thought he didn't look good in them. "Kalicki left and I went home," Mates said, "and almost as soon as I got there, the phone was ringing and Tony Morse was telling me that Kalicki was dead. I couldn't believe it! I have no idea whether his near-sightedness had anything to do with the accident but I do wish I had said something to him about the glasses."[13]

Kalicki had been a student in the underground university in Warsaw during the German occupation, held master's degrees in mathematics and philosophy from Warsaw University, and had managed to get to England in 1946 to do a Ph.D. at the University of London. Tarski, recognizing him as a promising young researcher and an excellent teacher, invited him to California and helped him launch his career. As a newly appointed assistant professor of philosophy specializing in logic, he was an early link in the chain of logicians that Tarski would see appointed henceforth. A modest young man, extremely well liked, the irony of his sudden death on a tranquil California road after surviving the war in Poland was inescapable.

It came as a surprise to Chang when, a few days after the crash, a lawyer representing Tarski came to interview him about the accident and suggested that he join in a suit for damages. Initially Tarski's suit was directed against Kalicki's insurance company but, in the course of the investigation, falsifications in Kalicki's application were discovered and so the policy was null and void. Tarski then decided to sue UC because he and the others had been traveling on university business, but Chang, usually compliant, particularly where Tarski was concerned, refused to participate in the suit.

"It was something I did not want to do," he said, "I wasn't going to be in it. Even though I realized there was some kind of pressure from my Ph.D. professor and I wondered if I shouldn't go along, I just couldn't. That is the only time I did not do what I thought he wanted me to do."

In the end, Tarski dropped the suit; perhaps Chang's stance had an effect or perhaps Maria may have had some influence. Kalicki's wife is said to have been furious; she refused to speak to Tarski for many years because she thought it was his original action against the insurance company that had caused them to cancel, thereby leaving her uncompensated for the loss of the car. For his part, Tarski had no idea that there were irregularities (Kalicki had lied about a previous accident and the fact that an earlier insurance policy had been cancelled) or that there would be a financial loss to Mrs. Kalicki.

A Sabbatical

For Tarski it had been a long haul, from 1942 to 1955, getting himself and his program in logic established in Berkeley, and when his colleague Evert Beth of Amsterdam urged him to spend a year abroad, dividing his time between the Netherlands and France, Tarski agreed. Maria remained in Berkeley in the new house they had bought at 462 Michigan Avenue. It had a fine view of the bay, a large garden, and there were no stairs, making entry to the house much easier. Ina was still in high school; Jan, by then, was a student at UC and living on campus.

Beth and Tarski had known each other since the late 1930s, meeting first at a congress in Paris and the following year at a smaller conference in Amersfoort. Directly influenced by Tarski, Beth turned away from his earlier philosophical interests in Kant and intuitionism and began to work in logic and the foundations of mathematics. Like Tarski, Beth was a small man with big ideas and vision, a successful and energetic promoter of logic. At one point, on his own initiative, he posted a sign on his door declaring it to be the office of The Institute for the Study of Foundations of Mathematics and Logic, and in this way brought it into existence.[14] A few years later, he and a colleague founded the Netherlands Society for Logic and the Philosophy of Science, this time with official backing.

Amsterdam was a good base for Tarski to spread his influence in Europe and a charming city to visit for its architecture, museums, canals, and flowers, all of which appealed to him. In that pleasant atmosphere, he and Beth made their plans to ensure that logic would be at the forefront of the newly formed international organization: Logic, Methodology and Philosophy of Science. As congenial colleagues, they agreed on almost everything, down to the pleasure of smoking cigars. Evert's wife Corry, a large enveloping woman, did her part handling practical matters and making Tarski comfortable. Alfred responded by being effusively charming and attentive. At a later date, he and Leon Henkin sent her a note on a postcard that read: "Dear Corry, It is midnight. We are sipping champagne

Meeting of the Association for Symbolic Logic, Amsterdam, 1 September 1954. From left to right: *first row*, Tarski (2nd), Abraham Robinson (4th), Evert Beth (5th), Robert Feys (6th), Willard Quine (7th), Leon Henkin (8th), Józef M. Bocheński (9th), Arnold Schmidt (10th); *second row*, Corry Beth (in white, behind Tarski), Haskell Curry (behind Beth and Feys), Frits Staal (behind Henkin), Richard Jeffrey (at right end); *on the steps*, Paul Bernays (next to window), Georg Kreisel (top left), Marjorie Quine (next to Kreisel), Hao Wang (2nd from right).

and thinking of you." She treasured that card and told her friends about it so many times it became a standing joke among Dutch logicians' wives, but she never stopped being flattered to be remembered in this way.

While logic was thriving in the Netherlands, in France almost nothing was happening in that field. In fact, there was general disdain for the subject. On a mission to alter this view, Tarski gave a series of five lectures in English on the theory of models at the Institut Henri Poincaré in Paris. Prior to those lectures he had attended an international colloquium sponsored by the Centre National de la Recherche Scientifique (CNRS), where he was an active participant and attempted to promote his cause.

Paulette Février, a philosopher with a research appointment at the CNRS, and her husband Jean-Louis Destouches, a physics professor at the University of Paris, were Tarski's hosts. Both were engaged in research on the foundations of quantum mechanics. Destouches and his teacher, the Nobel physicist Louis de Broglie, had visited Warsaw in the early 1930s and were impressed by Tarski even then. It was Destouches who was instrumental in inviting him to Amersfoort in 1938. Février later recalled her husband predicting at that early date that Tarski would rise to the pinnacle of his field. "That was why," she said, "he tried to invite Tarski to come to Paris as often as possible because he was aware of the value of his teaching for his students and collaborators and because he liked him."

In addition to being a young scientific admirer, Paulette Février – small, blonde, and vivacious – was his ministering angel in Paris, finding him an apartment and helping him to buy whatever he needed or wanted. Once, it was precious stones that he planned to bring back to Amsterdam to be set; another time, he was looking all over the city for crystal glasses to match a pattern that he knew; and, always, he needed a good supply of Kola Astier, a stimulant that could be bought over the counter in France. She continued to be his "supplier" ever after, sending packages from France whenever requested. Tarski loved her company and, like almost every woman who became a close friend, Paulette Février found it necessary to point out that there was only badinage between them, and that this was because Alfred knew how much she and Jean-Louis loved one another. Nevertheless, like Corry Beth, she was thrilled to be the object of his attention and enjoyed his way of being charming and romantic in a "Mittel-Europa style," which she defined as an exaggeration of French style. (A younger Polish colleague corroborated this description, calling it "a kind of chivalry" typical in the interwar years in Poland.)[15] His visits were always exciting; he brought gifts for her and her daughters. He came with a suitcase full of exotic presents, including Indian cotton material, which Paulette made into dresses that the girls called "robes Alfred." For "Paolette," as he called her, Tarski always brought perfume or else, when they went on their shopping trips, he would say, "First, Paolette, we must buy you some *Ma Griffe*."[16]

During his year abroad Tarski went, for a short time, to Warsaw – his first return since his fateful departure in 1939. Encouraged by the slow softening in the hard-line Moscow-dominated Polish government that had begun following Stalin's death, Tarski decided he could visit Warsaw with reasonable safety. Until then, because of his open and well-known criticism of the regime, he had not risked a return. Irena Grosz, his friend and former fiancée whose husband was a highly placed government official in Poland, arranged a visa for him. There seems to be almost no account of his thoughts and feelings during this visit to his birthplace, but one can easily imagine a gamut of emotion welling up after seventeen years of absence: sorrow, anger, pain upon hearing again the unbearable stories of how his loved ones had died. Perhaps he also felt guilt for having been absent during the worst of times – even though, rationally, he knew it had not been his fault.

His brother's daughter Anna Tarska had survived the war, and Alfred at least had the consolation of seeing her as an adult and meeting her husband. Also in 1956 he visited some of his closest friends, especially Wanda Szmielew, Andrzej Mostowski, Karol Borsuk, and Tadeusz Kotarbiński, the teacher whom he revered. By then, Szmielew had a daughter, Aleksandra, who was two years old. Tarski would form an attachment to her as she grew older, and he always asked about her in his letters to Wanda. Tarski had professional business to attend to as well; he was determined to do all he could to bring Polish logicians to the United States and especially to Berkeley for conferences and for longer stays. It was at this time that he set the wheels in motion for Szmielew and Borsuk to attend the Axiomatic Method Symposium planned for 1957 and for Mostowski and his student Andrzej Ehrenfeucht to come to Berkeley the following year.

The Program in Logic and the Methodology of Science

On his return from the sabbatical in 1956, Tarski, always looking for ways to solidify the importance of logic at Berkeley, may have been prompted to action by witnessing the success of Jerzy Neyman in transforming the statistical laboratory into a separate department. Neyman had been promoting this move since the early 1940s, but it was not until Charles B. Morrey, Jr., succeeded Griffith Evans as chair of the mathematics department that it garnered the requisite internal support. The redesignation of Neyman's group as the statistics department was formally approved by UC president Robert Gordon Sproul in July 1955. Morrey had trumpeted the importance of both Neyman and Tarski to Sproul in the summer of 1954 when the two were in Amsterdam as invited speakers for the International Congress of Mathematicians. Only five Americans were honored in the same way and, said Morrey, "Our university is the only university in the entire world from which two speakers have been invited."[17]

While Tarski was as ambitious as Neyman to have Berkeley be the leading center in the world for *his* subject, he could not think of forming a department devoted entirely to logic. What he could do, though, was make use of logic's genuine connections with philosophy in order to strengthen the case for a special doctoral program. In a letter dated 13 August 1956, a formal proposal was submitted by Tarski and a group of his colleagues to the dean of the graduate division to offer a Ph.D. degree in a new field of study: logic and the methodology of science. Joining Tarski in support of this proposal were Leon Henkin and Raphael Robinson in mathematics as well as Ernest Adams, Benson Mates, and John Myhill in philosophy. In addition, Tarski's friends Yuen Ren Chao in oriental languages and Victor F. Lenzen in physics signed the proposal, which stated:

> In recent years the University of California at Berkeley has become one of the most important centers of logical and methodological studies in the world. The prospect of doing advanced work in these fields under expert guidance has attracted students from all parts of the United States and even from Europe. If these students choose to work towards an advanced degree, however, they are forced to decide between two alternatives, neither of which is adequate for their purposes.[18]

The argument was that the requirements for a Ph.D. in mathematics were so onerous that students specializing in logic had to forego training in relevant subjects such as semantics and philosophy of language, philosophy of science, and philosophical logic. Likewise, students in philosophy who were interested in logic had to spend so much time on such subjects as ethics, metaphysics, and the history of philosophy that they did not have time for the necessary mathematical training. The proposal would replace these with a combined system of requirements with the anticipation that "students obtaining their degrees under this plan will be especially well qualified to teach logic and related topics in both philosophy and mathematics departments, and that they will have been prepared, more adequately than has hitherto been the case at any university, to make original contributions to the advancement of the subject."[19]

The proposal was approved in May 1957; the first student entered the program in 1959 and graduated in 1964. Since then, there has been a steady succession of Ph.D.s in logic and the methodology of science. Most have gone on to academic positions, and many have attained international recognition for their contributions to the field.

Solomon Feferman and Richard Montague

Aside from the trauma of the automobile accident and his smoke-induced asthma attacks, C. C. Chang's six years as a graduate student had been uncomplicated compared to others of the fifties – so much so that he asked Tarski if could stay on for an extra year as a postdoc because he felt he had more to learn from him. Flattered, Tarski responded warmly to the request and secured him a position as a lecturer. Feferman and Montague, his last students of the decade, were of a different mind; they left Berkeley as soon as they could, well before the last details of their theses were completely worked out.

Feferman, shy and boyishly handsome, was a graduate of the California Institute of Technology. He began his thesis work by obediently undertaking two problems proposed by Tarski: one on cylindric algebras and the other on the decision problem for the ordinal numbers with addition – the same problem that had stumped Fred Thompson. Feferman solved the first but obtained only a partial result on the second. Tarski said that this wasn't enough to meet his criteria for a dissertation, so Feferman tried harder. Thus far his experience was typical, but then fate in the form of the military draft intervened. He was inducted into the United States Army and sent to Fort Monmouth, New Jersey, to serve two years as a lowly corporal in a research division of the Signal Corps. There, between stints of kitchen police and long days spent in secret calculation on how many rocket missiles the United States would need to launch in order to intercept a hypothetical Soviet missile attack on New York or

Solomon Feferman in the Colorado mountains, c. 1950.

Washington, he began to work on problems connected with Gödel's incompleteness theorems.

Returning to Berkeley in the fall of 1955, Feferman discovered that Tarski had already left for his European sabbatical and had delegated Leon Henkin, newly hired at UC, to be his substitute advisor. Thus it was that Feferman regularly discussed with Henkin the details of his progress on the new direction he had chosen, while Tarski was kept up to date in a more general way by correspondence. When it appeared to both Henkin and Feferman that at last he had done something that was clearly "thesis worthy," the whole thing was given to Tarski, who had by then returned to Berkeley, for his approval. To Feferman's dismay, instead of saying "excellent!" Tarski hemmed and hawed.

Perhaps he was irked that the subject was not the original one *he* had suggested and not in any of his own main directions of research; instead, it sharpened and generalized the method of arithmetization used to prove Gödel's incompleteness theorems. Perhaps the old rivalry Tarski felt with Gödel over those theorems was awakened. In any event, he decided not to decide on his own whether the work was sufficiently important and instead asked Feferman to send a summary of the results to Andrzej Mostowski in Poland. This took more time and created more tension. To Feferman's relief, Mostowski found the results novel and interesting and strongly encouraged Tarski's approval. However, because of the delays, the final details and all the necessary requirements were not completed until the following year (1957) when Feferman was already an instructor at Stanford University, appointed on the premise that he would soon have his doctorate.

Richard Montague, on the other hand, did exactly what "Papa" wished. His Ph.D. thesis on the axiomatic foundations of set theory dealt with problems close to Tarski's heart (it would be years before he launched his seminal work on the formal semantics of natural language), yet he too had plenty of aggravation. A prodigy, small and brilliantly quick, Montague was one of the students who idolized Tarski and took up his problems and systems with manic enthusiasm, to the latter's obvious satisfaction. Among friends he labeled Tarski "despotic," but he never stopped wanting to please him – to be the apple of his eye, his favorite. Oddly, Montague's worship and desire to do the "right thing" had a negative effect on the final preparation of his thesis; his method and style were so heavily based on Tarski's that proofs had to be carried out to perfection, exactly as or better than Tarski himself would have done. Furthermore, the priority issue that was so important to Tarski also reared its head; documenting the footnotes and establishing attribution for who solved which theorem first in 1927 or 1928 in Warsaw became a major stumbling block.[20] All of this final "nitpicking" effort took place while Montague was at UCLA, for he too had accepted a position offered him in the philosophy department on the premise that his Ph.D. would soon be awarded. It took two years for all the formalities to be completed, but through it all, he continued to do everything he could to please Tarski.

Although most of Tarski's disciples took a page from his book, insisting on precision in their own students' work, none was more like him in this respect than Montague. Nino Cocchiarella, a Montague student, labeled him a "little tyrant ... who refused to be on a Ph.D. committee unless he could be chair so that he could dictate how things had to be done." Cochiarella recalled having to write a fifteen-page, painfully detailed logical definition in order to satisfy him, even though he could have explained it more informally, yet adequately, in a few paragraphs.[21]

Not all of Montague's problems as a student were Tarski's doing; in his private life he was in constant trouble of his own making. Like J. C. C. McKinsey, he was openly and flamboyantly homosexual. Seeking excitement, he would cruise the gay bars of Oakland and San Francisco; when he allegedly seduced a minor, the boy's parents brought legal charges against him and the case went to trial. Fortunately for Montague, whose external manner was exceedingly proper, there was a hung jury. An impressive array of character witnesses – including Tarski, Benson Mates, and William Dennes, chairman of the philosophy department – testified on his behalf and persuaded the district attorney not to retry the case, promising that he would seek psychiatric help and vouching that he would act "properly" on campus. Unfortunately, when he went to UCLA Montague continued to take what his friends thought were crazy risks in terms of the company he sought; no one could persuade him otherwise. Perhaps Tarski also urged caution but perhaps not, for he had had his own sexual adventures. In any case, he was never judgmental or pompous about sexual morality – he saved that attitude for logic.

Leonard Gillman (a "Ph.D. by Mail") and Dana Scott

Montague and Feferman, Tarski's tenth and eleventh Ph.D. students, were awarded their degrees in 1957, after which there was a hiatus until Jerome Keisler's thesis was completed in 1961. But there were two good students in the 1950s that Tarski dearly wished he could claim as his. One was Leonard Gillman, whom Tarski later called "my Ph.D. by mail"; the other, Dana Scott, left Berkeley for Princeton – to everyone's astonishment – after his first year in graduate school.

Gillman had been a graduate student at Columbia University, but during World War II he left his Ph.D. research and signed up as a mathematician for the Navy. He continued in this capacity until 1950, when he was offered a sabbatical year on the condition that he go to MIT. Regretting that he couldn't return to Columbia, his home university, he nevertheless accepted the opportunity. His plan was to write a thesis in game theory and to do it quickly, but before he got underway a neighbor happened to give him *Leçons sur les nombres transfinis* by Wacław Sierpiński. Gillman said, "I plunked down in a chair and read it from cover to cover. It read like a novel and I was hooked." Gillman put aside game theory and decided to risk working in set theory, a field completely new to him, and soon obtained what he thought were good results. He had no set theorists to advise him or evaluate his work, but boldly he sent his paper to the *Annals of Mathematics* journal and it was accepted. For further confirmation, he simultaneously sent the work to Tarski, with a letter explaining his situation. Tarski responded promptly, saying, "Yes, your theorem appears to be original but it follows easily from work Paul Erdös and I published in 1943." Furthermore, Tarski showed Gillman a way that his results could be strengthened by using that earlier work. Gillman was not at all thrilled with the second part of the response because it gave him the upsetting feeling that, instead of advising him as a student, Tarski was "horning in"; indeed, in the exchange of letters that followed, Tarski indicated he was thinking in terms of a joint paper. This troubled Gillman even more, for he had assumed this work would be a major part of his Ph.D. thesis.

In the end, after further correspondence and what appears to have been some pressure and diplomacy on the part of Gillman's former professors at Columbia (including Eilenberg, who knew Tarski well) about the impropriety of Tarski's proposing himself as a co-author, Tarski backed off. Appropriately, Gillman accorded him detailed acknowledgment in his thesis and, in his revised paper, thanked him for his "encouraging interest and generous advice." The two finally met in person in 1952, at a conference in Michigan and again at one in St. Louis. There, in a café, Leonard recalled, "We had a pleasant chat about set theory over coffee and cigarettes, and he told me that he thought of me as 'my Ph.D. by mail'." And so, in every list – official and unofficial – of Tarski's Ph.D. students, Gillman's name appears.[22]

During the period of turmoil over credit for who did what, the Gillman family acquired a puppy and irreverently named him "Tarski." When asked why, Mrs. Gillman explained, "a dog is something that brings both joy and pain." The fact that the Gillmans had a dog named Tarski did not go unremarked in the wider logic community, and in time it gave rise to many tales of "the dog named Tarski" that were embroidered into the Tarski legend.

Dana Scott's association with Tarski began when, as a very young undergraduate, he found his way into Tarski's graduate seminar in set theory. In his sophomore year, while working at his part-time job in the UC Berkeley library, Scott chanced upon the *Journal of Symbolic Logic*. Thumbing through it, he spotted an article by Jan Kalicki that, to his delight, he could understand; then he realized this "Kalicki" was the teacher in his theory of equations class. Scott was also taking a philosophy course from Benson Mates. From his two professors, Scott began hearing about Tarski and the excitement he generated and one day, on his own, he decided

Dana Scott and David Kaplan at the Tarski Symposium, 1971.

to attend Tarski's seminar – perhaps not realizing or perhaps not caring that it was for graduate students. In no time, Scott became a regular participant, for he was immediately recognized by everyone present as outstandingly brilliant. Like others before him, Scott was captured by Tarski's charisma and his ability to communicate a passion for his subject. He too had found what he was looking for. A few years later a story went the rounds that if Tarski proposed a problem and Scott came up with a strategy for arriving at a solution, Tarski figured it was not hard enough to assign to anyone as a thesis problem.

In this seminar and other classes and lectures, Scott met the whole circle of students and faculty inspired by Tarski in the 1950s, which – since J. C. C. McKinsey's appointment at Stanford – had expanded to include faculty and students from that university as well: McKinsey, his student Jean Rubin and her husband Herman, Patrick Suppes, and Donald Davidson among them. A joint Berkeley–Stanford Logic Colloquium was organized and met at least once a month and was always followed by a party. In this atmosphere of professional and social camaraderie, a strong sense of community developed – exactly what Tarski had been seeking to create as a replacement for the stimulation, interaction, and ferment of the Polish scene that had nurtured him.

Scott, a large and seemingly even-tempered young man, and Richard Montague, small and volatile, became best friends; they lived in the same house and had common interests other than mathematics. Montague, a graduate student, had taken more courses and at first knew more, but he was only two years older than Dana, having himself been something of a prodigy. Their friendship did not preclude an intense if mostly unspoken rivalry; vying for Tarski's approval, they constantly tried to outdo one another. In the set theory course, Tarski would walk into class, write a theorem on the blackboard, then step back and ask, "Now, how shall we prove this?" The students were supposed to deliberate about it; but often before he could get "how shall we prove" out of his mouth, Montague would be waving his arms and hands and shouting out the solution, while from the back of the room Scott would calmly ask, "Well, do you want it with or without the axiom of choice?" – meaning he had *two* ways of proving it.[23]

Thrilled as he was at having such stellar students, Tarski's delight was not unmitigated. He was visibly annoyed at always having them right on his back and sometimes, as soon as Montague and Scott started their act, he would say, "Wait, wait a minute. I want to do this one myself." Also, for their benefit, one day after he presented a theorem he said, "Look – you know – I first proved that when I was thirteen years old."[24] He was obviously ambivalent about these young upstarts, who were so much on top of everything. There was, it seems, a *three*-way rivalry afoot.

Why, when he was doing so brilliantly, did Scott leave Berkeley after just one year as a graduate student? The precipitating event would appear to be a small matter, except for the personalities involved.

Tarski had asked Montague and Scott to proofread the manuscript of *Logic, Semantics, Metamathematics*, a selection of his papers from 1923 to 1938 that had been translated by the biologist J. H. Woodger, a longtime admirer of the Polish school of logicians in general and of Tarski in particular. Woodger felt that the importance and scope of the Polish contributions to logic was insufficiently appreciated in the English-speaking world and thought he would be "performing a public service as well as acknowledging a debt to his Polish friends" by preparing an English edition.[25] This magnanimous offer was gratefully accepted by Tarski, but the book took years of work and the effort of several long-suffering research assistants before it met his standards. As Scott explained:

Dear old Woodger... one of those amazing English eccentrics, had worked incredibly hard to translate all these things, but not being a mathematician and only an amateur logician, he bungled a whole lot of it. So Richard Montague and I had to do the proof-reading of those articles [there were twenty-seven]. We had done about half of them when Richard went down to UCLA to begin teaching and I was left to finish the proof-reading by myself, which I could not bring myself to do. I procrastinated and procrastinated. Tarski would phone the place where I lived and I would tell my friends to say, "No, he's not here now." He got very angry with me for falling down on the job – and he stayed angry for a long time.[26]

In fact, Tarski was so infuriated that he fired Scott without telling him. Later, when the tempest was over, Scott – in an amazingly frank letter to Tarski – said he felt he had been placed in an impossible position: that essentially he had been forced into the proofreading job because he could not say no to Tarski. Scott wrote:

You of course are conscious of the great power you have over your students, but I do not think you always admit it. The fact that your genius has put you in a position of power is a good thing but when you use this power in order to get things done, it must be admitted all around on everyone's part. This is especially the case for those who are as immature as I was at the time this business started.... You have no idea how I was destroyed on the spot when C. C. [Chang] told me you had gone to Kelley and asked for a new assistant. If only you had come to me first, I would have understood! I knew why you were doing it, but the bitterness of your not telling me to my face was more than I could stand. It seemed as if you had rejected me and entirely withdrawn your support. My only reaction was to run away and try to show that I didn't really need you.[27]

Scott ran to Princeton, where he was welcomed with open arms. By the following summer, however, there was a rapprochement between him and his now former teacher at a mathematics meeting in Ann Arbor; and over time, a great warmth developed between them – so much so that a few years later, when Dana married and he and his wife Irene had a child, Alfred was asked to be her godfather. Not surprisingly, Tarski strongly regretted losing his star student to Alonzo Church. As Scott's reputation grew, he would say to him in front of a group, "I hope I may call you my student." He had, after all, suggested his thesis problem.

With their friendship intact, Scott was left with no lasting regrets. Though at first he had been devastated to lose his privileged position as assistant and favored pupil, the wound healed quickly; it was a relief to have escaped the tyranny of the all-night sessions and what he called "the slave labor." As a Californian who had never been east of Reno, Princeton for Scott was a whole new world. Although he thought the lectures of his new advisor, Alonzo Church, did not compare to Tarski's, Scott found his other contacts among students and visitors exciting. He met and was strongly influenced by Michael Rabin, a young Israeli; Georg Kreisel, visiting from England; and Stephen Kleene, a former Church student who had returned to Princeton for a sabbatical year. Nevertheless, Scott acknowledged Tarski's influence as seminal: "He was the first major researcher I ever got close to. That's a big thing – somebody intellectual right away. I had some good teachers, but he was *the teacher!*"

9

Three Meetings and Two Departures

What a conference! There has been nothing else in logic remotely comparable.   Anil Nerode[1]

Cornell: Summer 1957

IN JULY 1957, at a month-long Institute in Symbolic Logic at Cornell University in Ithaca, New York, eighty-five scholars – ranging from graduate students to the most distinguished professors in the field – gathered together for a groundbreaking series of talks and discussions that dramatically changed the landscape of modern logic. "Its proceedings are full of exciting new beginnings... [The] list of participants[2] is an incredible *who-will-be-who* of young logicians."[3]

The conference was unusual for its length and breadth; the speakers represented all branches of mathematical logic and, for the first time, a large number of computer scientists took part. Most of the participants lived in the college dormitories and ate in the communal Cornell Hotel School dining room. It was like being at a summer camp, with informal parties, picnics on Cayuga Heights, and hikes to waterfalls. For weeks the green, hilly campus buzzed with talk about model theory, recursion theory, set theory, proof theory, many-valued logics, and, significantly, the logical aspects of computation; there was also the usual discussion about whose work was most important. There was novelty, rivalry, conviviality, and scandal.

Although Alfred Tarski was one of the principal organizers of this Summer Institute, the inspiration came from Paul Halmos, a debonair, "brash" (he called himself) mathematician from the University of Chicago. Halmos had a growing interest in algebraic logic, and during a 1953 visit to Berkeley he cultivated a close personal connection with Tarski, who helped expand his knowledge of the field. In his memoir written thirty years later, Halmos told how he got the ball rolling:

There weren't many conferences, jamborees, colloquia in those days and the few that existed were treasured.... I decided it would be nice to have one in logic, particularly if it were at least partly algebraic. I had no stature as a logician, I had no clout, I wasn't a member of the in-group; all I had was the brass (willingness to stick my neck out) and the drive (willingness to do the work).[4]

With the assistance of others – including Tarski, Stephen Kleene, Willard Quine, and Barkley Rosser, all of whom had the necessary stature – Halmos was successful in obtaining the sponsorship of the American Mathematical Society and funding from the National Science Foundation.

That done, the matter of where to hold the meeting was discussed. Tarski wanted it in Berkeley – his turf; Rosser was equally adamant in wanting it at Cornell, *his* university. After contentious correspondence, Rosser, who had already built a relationship with the scientific establishment outside the academic world and with government funding agencies, prevailed by arguing that it would cost less since the majority of participants would be coming from the East Coast. Reluctantly, for by now it was rare that he did not get his way, Tarski acceded.

The issue of who would be invited to speak also created heat. There was quick agreement about the most prominent senior scholars, but discussion – mostly by correspondence – about who to choose among the up-and-coming younger crowd went on for many months, with each of the main organizers giving preference to his own disciples. On this score Tarski did very well; about one fourth of the speakers were under his influence in one way or another, and many of them gave two or three talks. In this way, he succeeded in positioning himself as the leading man of the occasion. Because the reclusive Gödel, whose name was first on the invitation list, had declined to attend, there was no direct challenge to Tarski's assumption of that role.

Competition and Appreciation

The competition that existed from the outset was played out in the subtext of lectures and discussions. All the prominent logicians had their protégés. Alonzo Church, whose list of brilliant students rivaled Tarski's, could proudly point to Martin Davis, Simon Kochen, Michael Rabin, Hartley Rogers, and Dana Scott (although Tarski also counted Scott as his). Stephen Kleene of Wisconsin and Barkley Rosser of Cornell were Church's students in the 1930s. (Church's most famous student, Alan Turing, had died in 1954.) Kleene was accompanied by his own pupils, John Addison – who was later to ally forces with Tarski – and Clifford Spector. Adding to the interconnections, Addison was married to Church's daughter, Mary Ann. Rosser's links outside of academia brought computer scientists to the meeting, one of whom was former student George Collins, now a researcher at IBM. Collins was working on implementing part of Tarski's decision procedure for algebra for machine computation.[5]

From Harvard, William Craig, Burton Dreben, Henry Hiż, Hao Wang, and a very young Charles Parsons were all associated with Quine, the first American to have known Tarski. Anil Nerode, then a student in Chicago who would later join Rosser at Cornell, was urged by Halmos to attend; he wrote that "it was heady stuff for me ... to meet all the great names (except Gödel) from Tarski and Kleene to Quine and Church."[6] Tarski's crowd included Evert Beth, Jean Butler, C. C. Chang, Solomon Feferman, Leon Henkin, Richard Montague, Julia Robinson, Raphael Robinson, Dana Scott, and Robert Vaught.

As closely interrelated as the world of logic was, these scholars had been working on a variety of subjects with methods that were not universally familiar. It was as if they spoke the same language but in different dialects, and part of the excitement of the meeting was learning to understand these dialects and how they might be adapted to one's own work. The beauty of the conference at Cornell was that, for the first time, logicians of every stripe came to grips with what their colleagues had been up to; the opportunity to have face-to-face contact and lengthy discussion with individuals who had previously been only disembodied names was exciting, and the general feeling of exhilaration was enhanced by a combination of appreciation and competition.

The Wild Cards

Two intriguing men, Abraham Robinson (not related to Raphael Robinson) and Georg Kreisel, came to the Cornell Institute unfettered by a link to a mentor. Both would soon have enormous influence on their younger colleagues; Kreisel in particular was something of a pied piper and seemed deliberately to seduce the students of others. Both men were European, like Tarski, but much younger than he and more recently arrived in the United States. Robinson, born in Germany, had lived in Israel, France,

England, and Canada, and he would return to Israel once more for a few years before settling in the United States.[7] Kreisel – Austrian born, educated in England, and a frequent visitor to France – had a position in Reading, England. At the invitation of Kurt Gödel he had spent the previous two years at the Institute for Advanced Study in Princeton. Largely self-taught in logic and less bound by tradition, neither Robinson nor Kreisel owed allegiance to a single methodology. Both had worked in applied mathematics in England during World War II, and as a result their style was much more experimental and free-wheeling than Tarski's. Also, each of these men in his own way exuded intellectual self-confidence; neither was afraid to lock horns with Tarski. Polite to his face, Kreisel was unrestrained in his behind-the-back mockery of Tarski's methodical "no 't's left uncrossed, no 'i's undotted" style. *His* approach was impressionistic and intuitive; he made dramatic statements about major ideas, lightly sketched in the details of a proof, and let the listener or reader figure out the rest – or more often not. Naturally, this "hand waving" and lack of precision irritated Tarski greatly. Moreover, Kreisel's work was in proof theory and constructive mathematics, subjects that did not interest Tarski and about which he knew almost nothing. A black mark in Tarski's mind was the influence Kreisel was having on some of his students, particularly Feferman and Scott.

Robinson, on the other hand, worked on problems concerning the applications of model theory to algebra, which interested Tarski very much; in fact, they were in direct competition with each other in that area. He too had a certain looseness of presentation that annoyed Tarski, but his contributions could not be ignored. In 1960, Robinson would make his mark internationally with the creation of non-standard analysis as a means to restore in a coherent way the use of infinitesimal quantities in mathematical analysis. Though of great intuitive value, the use of these supposed numbers had long since been discarded from the foundations of analysis because of their inconsistent properties. It was through the application of model theory that Robinson was able to show how their use could be regulated logically, thus combining their heuristic value with a firm foundation.

A surprising connection between Tarski, Robinson, Kreisel, and Henkin was revealed at the Cornell conference. This was related to the seventeenth problem that David Hilbert had raised in his famous list of twenty-three open mathematical problems. Hilbert conjectured that every polynomial of real numbers that never takes on a negative value is representable as a sum of squares of ratios of polynomials. This conjecture was proved correct in 1927 by the algebraist Emil Artin. Two years before the Cornell conference, Abraham Robinson showed how to use Tarski's famous work on the completeness and decidability of axioms for the real numbers to sharpen Artin's result by putting uniform bounds on the number and degrees of the polynomials whose squares occur in the Artin representation. At Cornell, both Henkin and Kreisel gave talks entitled "Sums of Squares," which further showed how such bounds could be calculated by a computing machine – at least in principle. Henkin did this by a more careful reworking of Robinson's model-theoretic method combined with Tarski's decision procedure. Kreisel, instead, demonstrated how it could be done by applying proof-theoretic methods to Tarski's axiom system for the real numbers. Few – Tarski included – were able to follow Kreisel's argument, both because of the relative unfamiliarity of the methods he used and the sketchiness of his presentation. In fact, Kreisel never wrote down the details, and it took some thirty years for them to be worked out in full by others.[8] In contrast, Henkin's presentation satisfied Tarski; it was "clean and clear" and did not require elaboration. Still, the unexpected connection between proof theory and model theory aroused considerable interest.

The Coach

During the question periods that followed the talks, Tarski was inevitably the first to spring up to make a remark about the subject at hand. Rare was the time that he did not have something to say about attribution or priority – about who had anticipated the result and who had worked on such problems earlier. His stance was judgmental; he felt it his duty to comment on the style of the presentation and the content, almost as if he were conducting his own seminar. In one instance he went so far as to assert that he did not find the direction of the speaker's work at all worthwhile and then proceeded to discuss his own early work, which was only distantly relevant. More than a few members of the audience were disturbed by Tarski's autocratic posture, but only Hilary Putnam, then a young philosopher at Princeton, had the gumption to respond, saying he thought these critical remarks were inappropriate and should be reserved for Tarski's autobiography. Putnam expected Tarski's students to rise to the defense of their teacher but they did not. His conclusion was that they were not displeased to hear him criticized.[9]

On the other hand, the philosopher William Tait, then an insecure graduate student at Yale, had a different view: he felt Tarski was a positive force at the meeting, and Tait was swept up by his enthusiasm and wide range of interest. "Most of the more senior people [at the conference] were rather inaccessible to students, or so it seemed to me. The two exceptions, to whom I have always felt grateful, were Paul Halmos and Tarski. Both seemed to welcome interaction with students and I spent a number of evenings in their company.... Speaking with them helped me lose my sense of being an alien and gave me confidence about my own work."[10] Between these opposing views was Henry Hiż's characterization of Tarski as "The Coach."

These disparate reactions are typical. Tarski at his best was encouraging, inspiring, and even lovable; at his worst, harshly critical, egotistical, and overbearing; it was all in a day's work for him. Although he himself seemed to have an everlasting need for appreciation and adulation – for being the best or the first – he was remarkably insensitive to the feelings of others.

Gödel's Absence, Kreisel's Presence

Tarski may have dominated the discussion sessions, but no one at the Cornell conference created more of a stir than Georg Kreisel. As a student in Cambridge (England) in the 1940s, he attended Wittgenstein's lectures on the philosophy of mathematics and had regular discussions with him. Kreisel was only twenty-one and still an undergraduate when Wittgenstein declared him to be "the most able philosopher he had ever met who was also a mathematician."[11] But unlike others in Cambridge at the time, he "was not the stuff of which disciples are made."[12] Never one to mince words, Kreisel later wrote: "Wittgenstein's views on mathematical logic are not worth much ... because what he knew was confined to the Frege–Russell line of goods."[13]

At the Institute for Advanced Study in Princeton during the two years before the Cornell meeting, Kreisel cultivated a close personal and professional relationship with Kurt Gödel, a privilege the reclusive genius accorded only a few of his colleagues – and Kreisel made sure everyone was aware of it. A lightning-quick thinker with a devastating wit, Kreisel had a reputation for outrageous behavior, but he also knew how to be courtly and charming. In Gödel's company he was sympathetic and appropriately deferential while revealing his own intellectual power. Beyond

Georg Kreisel, Los Altos Hills, California, c. 1959.

mathematics and philosophy, which was, of course, their major topic of conversation, the two Austrians had much in common. Both hypochondriacs and insomniacs, they commiserated and shared the details of their ailments and their strategies for combatting sleeplessness.

In the course of the conference, Kreisel gave a talk on a previously unpublished result of Gödel's: a constructive functional interpretation of the system of intuitionistic arithmetic. The audience was impressed because it was generally assumed that Gödel had ceased work in mathematical logic after his stunning results of the late 1930s concerning the axiom of choice and the continuum hypothesis. Here was a not-so-subtle show of what Kreisel knew that others did not – a clear bit of one-upmanship that Tarski, already more than a little put out at the attention Kreisel was attracting, did not appreciate.

In the personal realm, Kreisel caused an even greater sensation when he brought Verena Huber-Dyson – the attractive, cosmopolitan wife of the renowned physicist Freeman Dyson – to the conference as his companion.

Verena Huber-Dyson, self-portrait in a mirror,
Princeton, New Jersey, c. 1956.

A mathematician with a Ph.D. from the University of Zurich, she and Dyson had met while doing postdoctoral work at the Institute for Advanced Study and subsequently married. By the time Kreisel arrived on the scene, she had three children and was committed to family, home, and hearth; mathematics was supposedly forgotten.

She and Kreisel became close friends during his stay in Princeton, where he was a frequent visitor at the Dyson home. By 1957 their friendship had developed into a full-scale love affair. Not surprisingly, her presence in Cornell as Kreisel's "friend" was gossip topic "A" of the conference and, indeed, elsewhere in academic circles for years afterward. Ironically, it had been Dyson who had urged Gödel to invite Kreisel to the Institute for Advanced Study, for they had been friends and fellow students at Trinity College in Cambridge. "As mathematicians, Kreisel was a deep thinker and I was a craftsman," Dyson once wrote.[14]

Whatever mischief and personal pain Kreisel may have caused by running off with his friend's wife – and there was much – Mrs. Dyson, as he referred to her in public, acknowledged the "liberating effect" of the two years they were to live together. (They did not marry, but after she and Dyson divorced, Kreisel, exercising his perverse sense of humor, introduced her as "My wife, Mrs. Dyson.") "For me," she wrote, "it was a time for reflection and realignment that ultimately returned me to a self of deepened awareness."[15] Although her relationship with Kreisel brought her emotional upheaval, it served to return her to mathematics and to a professional rather than a domestic life in the academic world.

Tarski would have noticed Mrs. Dyson under any circumstances because of his eye for "girls," especially one as alluring as Verena, but the fact that she was Kreisel's girl, Dyson's wife, and a mathematician made it impossible for him to resist flirting, and indeed he made a flamboyant impression upon her. Almost as soon as they were introduced at the first reception, he recited, in German, the whole of Goethe's "Heidenröslein," a poem about a wild young man plucking a beautiful fresh rose even though the rose warns she will prick him and hence he will always think of her in sorrow and pain. Eyes gleaming, blushing, and obviously pleased with himself, Alfred told Verena that this, of all Schubert's *Lieder*, was his favorite. Verena said, "I felt as if he were offering me a huge bouquet of flowers."[16] Four years later, when her affair with Kreisel had run its course and she was on her own in Berkeley, Tarski would recite the "Heidenröslein" and other poems to her again and again.

Enter the Computer Scientists

The Cornell logic conference was the first to include many speakers from the emerging field of computer science, the theoretical foundations of which had been laid in the 1930s by Gödel, Church, Turing, and Emil Post. Especially through Stephen Kleene's work, recursion theory – the subject of what can and cannot be solved by computing machines – became one of the main fields in logic.

The connections between the theory and application of computation had begun toward the end of World War II, when the first large-scale electronic digital computers were built. At that point, the hardware had to be programmed by hand for each kind of application, a long and arduous task. John von Neumann, who had done important work during the war, demonstrated how to circumvent the lengthy process by using Turing's concept of a "universal computing machine," which allows programs to be treated as software and stored in computer memory.

By 1957, companies such as IBM and Remington Rand were producing the first generation of commercial electronic computers, and the high-level programming language FORTRAN had been established as an industry standard; that made possible the relatively routine translation of mathematical formulas for scientific computations into machine-ready programs. Some – but by no means all – logicians were quick to grasp the implications of these developments. At the Cornell meeting, Rosser gave a talk on the relation between Turing machines and actual computers; Church gave a series of talks on the logical synthesis of switching circuits for computer hardware; and Abraham Robinson spoke on theorem proving as done by man and machine. Among the younger contributors, Michael Rabin and Dana Scott spoke about finite automata, and Martin Davis talked about his implementation on the "Johnniac" computer (at the Institute for Advanced Study) of a decision procedure for the arithmetic of the integers under addition – a procedure that had been discovered in 1930 by Tarski's student Mojżesz Presburger in his Warsaw seminar.[17]

On the industry side, IBM and some of the other companies employed a number of researchers with backgrounds in mathematics and logic, and these people turned out in large numbers at Cornell, both to listen and to speak. There were fifteen talks given by researchers from IBM, many of them demonstrating the utility of FORTRAN-like programs for solving problems of potential interest to logicians. In particular, George Collins's talk on the implementation of parts of the decision procedure for the algebra of real numbers on an IBM 704 should have caught Tarski's attention, because this work linked Tarski's theory with possible practical applications. But Collins reported, "He didn't show any appreciation for my work, either then or later. I was somewhat surprised and disappointed."[18] Nor did Tarski show any special interest in Michael Rabin's work then or later in computer science, though he valued Rabin highly for his contributions to logic and even tried to hire him.[19] Tarski was always on the lookout for talent, and the Cornell conference was a prime hunting ground. Rabin and others would later visit Berkeley for a year, and Addison and Craig were to become permanent faculty members.

It certainly is surprising that – despite his own recognition of the importance and systematic pursuit of the decision problem for various algebraic theories – Tarski did not evince the least bit of interest in the practical computational applications of those problems for which a decision procedure had been found.

A veritable cascade of meetings in logic followed the Cornell conference. There would be meetings with national and international mathematical associations as well as similar meetings with philosophical societies and computer science organizations. After Sputnik, during the years of competition with the former Soviet Union, money flowed toward any endeavor that might possibly give the United States a boost in the sciences, and scholars were more than ready to take advantage of this bonanza. Academics would spend their summers plotting a course from one conference to another. "Will I see you in Amsterdam? In Paris? Bucharest? Where will you stay? Are you going to Jerusalem? Warsaw? Mexico? Stanford? Berkeley?" Presenters and pilgrims followed one another around the world reporting on new work, meeting new colleagues, exchanging ideas and methods, drinking and eating together, jousting with old friends and enemies, and making new ones. It was a heady time for all.

A Symposium of His Own

Although Tarski was the dominant figure in Cornell, he had been one of several organizers and had been forced to make concessions. The first meeting he could truly call his own, the "International Symposium on the Axiomatic Method, with Special Reference to Geometry and Physics," was held six months later in Berkeley from 26 December 1957 to 4 January 1958.[20] Although the planning for this conference had been simultaneous with the one for Cornell and there was some overlap between the two, its focus, as the title indicates, was quite different. The participants now included a select international group of mathematicians, physicists, and philosophers of science in addition to the logicians. Moreover, Tarski was in complete control.

The choice of geometry as one of the pillars of this conference was natural; when Tarski was teaching mathematics in high school, geometry was one of his main subjects. This stimulated him to rethink its logical foundations, and in doing so he arrived at an elegant new axiom system for Euclidean geometry that – in its precision and all-around economy – was an improvement on Hilbert's earlier axiomatization. What further distinguished Tarski's system from Hilbert's was that it was formulated without the use of set-theoretical notions. Tarski called his version of geometry *elementary* (which does not mean that it is simple). He was so pleased with this new approach that in 1926/27 he gave a course of lectures about it at Warsaw University.[21]

After a lapse of thirty years, Tarski returned to this subject and in 1956/57 gave a course on the foundations of geometry at UC and, through his renewed attention to the subject, quickly interested a number of students and colleagues in expanding the work in various directions. Among these, Dana Scott dealt with problems concerning the concept of dimension in Euclidean geometry, while Wanda Szmielew took up the axiomatics of systems of non-Euclidean geometry and questions of completeness and decidability for them. Raphael Robinson and Halsey Royden examined various possible choices of primitive notions for elementary Euclidean and non-Euclidean geometry. Royden, on the mathematics faculty of Stanford University, drove fifty miles each way to attend Tarski's lectures that year. Meanwhile, there was a resurgence of interest in the foundations of geometry elsewhere, so the time was propitious to bring together people working actively on the subject.

Physics as the choice for the other main pillar of the Axiomatic Method conference was less obvious from Tarski's background. Though he had attended courses on physics after completing his Ph.D., unlike Gödel he never showed an active interest in the field. But the high degree of mathematization of physics, coupled with Tarski's conviction that logic and the axiomatic method was to be a central part of the methodology of all science, pointed to physics as an obvious candidate for advancing his general program. The axiomatization of physics can be traced back to Newton's laws for classical mechanics and his development of its consequences on the model of Euclid's *Elements*. The axiomatic approach was replaced by a more free-wheeling use of mathematics in the subsequent developments of physics, but there was a return of interest in that approach toward the end of the nineteenth century through the work of Ernst Mach and Ludwig Boltzmann (who had a centrally important influence on the Vienna Circle). At the beginning of the twentieth century, Hilbert included the axiomatization of the physical sciences among his list of the most challenging mathematical problems. Hilbert and his students Hermann Weyl, John von Neumann, and Richard Courant took up that challenge in the following decades, especially for the kinetic theory of gases, Einstein's relativity theory, and the newly emerging and in many respects puzzling quantum mechanics.

Tarski had no involvement in these developments but had close personal connections with colleagues who did, and that is why the axiomatic method in physics came to be the second main component of the 1957 Berkeley conference. First and foremost, Patrick Suppes, a philosopher of science at Stanford with broad interests in logic, attended Tarski's seminars at UC in 1951. The following year he was invited to give a course in Berkeley on the philosophy of science, and Tarski's students Richard Montague and Dana Scott attended. Meanwhile, McKinsey had been appointed to the Stanford philosophy department and almost immediately he and Suppes began collaborating on the axiomatic foundations of classical mechanics in a form that came closer to meeting Tarskian standards of logical rigor than anything done previously. After McKinsey's death in 1953, Suppes continued this work by moving on to relativistic mechanics, in part with his colleague Herman Rubin, while Suppes's Ph.D. student Ernest Adams worked on the axiomatics of rigid-body mechanics. Adams was appointed to a position in the UC philosophy department in 1956, thus cementing the connections between Berkeley and Stanford.

Tarski was chair of the organizing committee for the Axiomatic Method conference and responsible for its overall conception, but the detailed work fell to Leon Henkin and Patrick Suppes. Henkin remembers it as the first of several meetings that filled him with so much anxiety that his doctor feared he was developing a serious case of stomach ulcers.[22] As Tarski's partner, right-hand man, and secretary of the organizing committee, he felt responsible for every detail of the arrangements even though he had the help of others. Along with the usual matters of logistics surrounding any conference – finding sponsors, inviting speakers, arranging lecture rooms, choosing hotels, and deciding upon what sorts of excursions and parties would be best to offer during nonworking hours – there were maddening Cold-War difficulties to deal with because several scholars from Eastern Europe were among the invitees. Hampered by State Department restrictions on all sides, in the end only two Poles (Karol Borsuk and Wanda Szmielew) and one Romanian (Alexandre Froda) were granted entry to the United States from the Eastern bloc. The problems surrounding this meeting were only a taste of what was to come in preparation for the larger, much more ambitious international conference for

Dorothy Wolfe, secretary of the Axiomatic Method conference, c. 1957.

Logic, Methodology and Philosophy of Science to be held at Stanford University in 1960.

Accompanying Henkin's organizational problems was his nervousness about Alfred's personal relationship with Dorothy Wolfe, the young and attractive secretary of the symposium. In the course of preparation for the meeting, Ms. Wolfe, formerly a philosophy student, and Tarski met frequently enough for her to be captivated by – in her words – "his mind and his power." He in turn warmed to her intelligence, good looks, and, not least, her enthusiastic appreciation of his genius. The result was a liaison that was the talk of *this* conference.

Once underway, the Axiomatic Method meeting proceeded as planned. There were nine full days of lectures given by distinguished and younger scientists in three sections. The first, devoted to the foundations of geometry, included such speakers as Paul Bernays, Karol Borsuk, Arend Heyting, Dana Scott, Wanda Szmielew, and Tarski himself. In a second section, topics in the foundations of physics, from classical mechanics to relativity theory and quantum mechanics, were discussed by Percy Bridgman, Jean-Louis Destouches, Paulette Février, Pascual Jordan, and Patrick Suppes, among others. And in a smaller but very diverse general

The Axiomatic Method conference, UC Berkeley, December 1957. From left to right: *first row*, Leon Henkin (1st), Herman Rubin (2nd), Paulette Février (5th), Alfred Tarski (6th), Wanda Szmielew (7th), Marshall Stone (8th), J. H. Woodger (end); *second row*, C. C. Chang (1st), Griffith Evans (behind Tarski), Jean-Louis Destouches (behind Szmielew), Julia Robinson (3rd from end); *third row*, Dorothy Wolfe (behind Chang), Patrick Suppes (3rd), Dana Scott (4th), Paul Bernays (5th), Raphael Robinson (7th); *fourth row*, Thomas Frayne (1st), Ernest Adams (2nd), Bjarni Jónsson (end); *fifth row*, Benson Mates (5th), Karol Borsuk (7th). (On the steps of Wheeler Hall, where Tarski had his office and taught his classes.)

section, papers ranged from the axiomatic treatment of functions in the calculus to the foundations of genetics to the use of the axiomatic method in the development of creative talent. Here the speakers included Tarski's old friends from the 1930s, Karl Menger and J. H. Woodger.[23]

When Tarski was involved in a conference, the social program was on a grander scale than usual because he thought it important to bring people together in a relaxed atmosphere. He liked good food and drink as well as beautiful scenery, and because he assumed others did too, he went out of his way to assure that visitors, especially foreign ones, had a "California experience." On the official level, there was a reception at the home of the vice-chancellor of the University of California. More personally, the Tarskis gave a gala New Year's Eve party at their home for which (duly noted in the program of the conference) Almaden Vineyards donated California "champagne." Before and after midnight, there were plenty of other Tarski-type libations on hand. More California wines were donated for the formal banquet a few days later at the landmark Claremont Hotel, a huge, white, Victorian architectural marvel. There were smaller, intimate gatherings too in the homes of the Berkeley faculty as well as a conference-wide excursion to Muir Woods and Mount Tamalpais.

Among those who enjoyed the California experience in a very personal way were Février, Destouches, Szmielew, and Borsuk. In letters written forty years later (when she was eighty-three), Février described, in exquisite detail, a three- or four-day automobile trip they took to Death Valley – with Alfred as driver – after the meeting was over.

We gathered at night in the Claremont Hotel Parking lot. Alfred was very tired and had some difficulty remembering where his car was. At last we started. He asked me to sit beside him in the front. After a while he fell asleep. The car was going very fast. He awoke suddenly and put on the brakes; the car went on two wheels but he succeeded in calming it down. He stopped and after a deep silence said, "Paolette, light me a cigarette."[24]

It was a terrible moment, and no doubt Alfred thought of Kalicki and that fatal accident a few years earlier. But, as Paulette related, they continued on their way. The rest of the tour was perfect: Bakersfield, Tehachapi, Mojave Desert, Mustard Canyon, Borax Fields, Zabriski Point, Dante's View, Scotty's Castle and – on the way back – the Sierra Nevada Mountains. The atmosphere in the car was merry and witty. They spoke in

Karol Borsuk, Paulette Février, Tarski, and Wanda Szmielew, starting out on a trip to Death Valley, January 1958.

three languages (Polish, French, and English) and teased each other; they exchanged places in the car and occasionally Jean-Louis drove. Sometimes, but not often, they spoke of their work. In Death Valley, Tarski drank a Coca-Cola and proclaimed, "Look: I am an American!"

In the hotels they took two rooms: one for Jean-Louis and Paulette; and one for Wanda, Alfred, and Karol. Fully aware of Tarski's reputation as a Lothario, Février suspected that there was more than flirting that passed between Alfred and Wanda as well as between Karol and Wanda, but she refused to take up that topic. "It was part of their private life," she said, "and we [Jean-Louis and Paulette] considered that we did not have to inquire about it. So I will tell you facts and only facts about our trip together in Death Valley. My attitude in this respect may seem a little old-fashioned (it is French in spite of our reputation about such matters)."[25]

Tarski was less discreet; he was happy to be known as a lusty man – a lover of many women – and was not loath to call attention to this trait. To his credit he believed, at least intellectually, that the principle of free love applied to his woman friends, too. Karol Borsuk, four years younger than Tarski, was Tarski's long-time friend and mathematical colleague from Warsaw; they had known each other since student days. Wanda Szmielew was Borsuk's assistant in Warsaw and his close friend as well. Whatever their shifting relationships may have been, they were like family, with ties

"Look: I am an American!" – Tarski drinking a Coke.

as strong as blood bonds, and their friendship and loyalty to one another endured until the end of their days.

Maria Leaves

In the summer of 1959 while visiting Berkeley, Dana Scott brought Irene Schreier, his bride-to-be, to the Tarski home to introduce her. Dorothy Wolfe – a small, dark-haired woman – was there, too. Irene recalled, "I immediately assumed she was a new maid in the house because after a few minutes of polite chit-chat Mrs. Tarski turned to Dorothy and started showing her where the linens were kept, where to put the silverware and the dishes, and, in general, how to run the household." It took Irene a while to figure out that Maria was on the verge of leaving Alfred, that Dorothy was not being hired as a housekeeper, and that she would soon be moving into the house with Alfred.[26]

Most people thought that Maria left Alfred because of "that secretary who helped organize the Axiomatic Method symposium," but in fact it was not such a precipitous move. She had considered divorcing or at least separating from Alfred for a long time. Both Jan and Ina had moved out of the house in their first year at UC; they were living on their own and supporting themselves by working in the university library and at odd jobs. The affair with Dorothy Wolfe simply acted as a catalyst for a change that had already been in motion. But what was surprising to others was that she left *him* rather than the other way around. It appeared, finally, that Maria had had her fill of catering to Alfred. Her attitude was: let someone else do it – I'll even show her how – but it's time for me to be on my own.

Married for thirty trying years, separated by war and traumatic circumstances and then reunited with great difficulty, they were to be riven not so much by his infidelities as by his indifference to her desire for a life beyond serving his needs. According to Ina, her father was "frantic" when Maria told him she was leaving, and he pleaded with her not to go. "She was supposed to move out one weekend and she didn't; then the next weekend and she didn't," Ina said, "and finally on the third weekend, when I was out of town, she did it."[27]

Always competent, Maria planned her new life well. She bought a large pleasant house at 2811 Regent Street, within walking distance of the UC campus, and set it up as a rooming house with several bedrooms upstairs and a large living room and a fireplace downstairs. Meals were not served but food was provided for tenants to make their own breakfasts. Jan Mycielski, later a professor of mathematics at the University of Colorado, lived there in 1961 when he first came to the United States. He felt the place had a very congenial atmosphere, but "Maria was not around much because she was working."[28] That, too, was part of her plan. She found a nursing job in a nearby convalescent home and walked to work every day. At fifty-seven and without American credentials, she did not have a wide choice of positions to select from, but this was the job she had trained for and even worked at for a short time a decade earlier, until Alfred insisted that she quit. The work suited her; she was happy caring for people who needed help. Always a giver more than a taker, it pleased her now to be doing something worthwhile in a world that was larger than her own home.

But if Maria's idea was to completely separate herself from Alfred and live an independent life, her actions belied that notion, for to whom did she rent her rooms? – to visiting Polish scholars who had come to Berkeley at Tarski's invitation, usually with financial support that he had helped secure. They would come for a summer or a semester or a year, to do research with him on one or another of his many projects, to participate in lectures and seminars, and to partake of the ferment of ideas in logic. And what could be better for them than to stay at Maria's? – the ideal landlady, who spoke their language and could smooth the way and help when problems arose. Among the many fortunate ones were Jerzy Łoś, Leszek Pacholski, Helena Rasiowa, Czesław Ryll-Nardzewski, Lesław Szczerba, and the aforementioned Mycielski.[29] Thus inevitably she stayed in contact with Alfred.

Apparently unwilling or unable to shed her natural hospitality, Maria began inviting friends to lunch on her days off from work, and of course many of these friends were also Alfred's friends. Then, to their astonishment and in fairly short order, Alfred himself became a frequent lunch guest. Robert and Marilyn Vaught were present one day when he came walking into the house with flowers. Vaught recalled, "I just about dropped dead! We all thought they were in the middle of divorce proceedings." Instead it seemed Tarski was learning not to take Maria for granted and was turning on the charm. According to Lesław Szczerba, Alfred would always bring flowers, as if he were courting, and his manner was similarly seductive. Even Dorothy Wolfe was sometimes invited to lunch while she was living with Alfred. Maria bore her no particular grudge and, as had happened before, Alfred's current "amour" was added to the circle of friends.

Ina Tarski and Andrzej Ehrenfeucht

Just as Maria was in the process of separating from Alfred, Ina, who had just graduated from UC, announced her intention to go to Poland. There she planned to work toward a master's degree in sociology at Warsaw University. But this was not her only reason. The year before, Andrzej Mostowski and Andrzej Ehrenfeucht had passed through the necessary bureaucratic hoops and made their way to Berkeley. Mostowski, highly respected as a scholar, had already done important work in set theory and definability theory; as a person he was witty, warm, and extremely well liked. His student Andrzej, a young man with great potential as a researcher, was also charming and ruggedly good-looking. Both of them fit into the growing logic community in the San Francisco Bay area and were invited to parties and dinners at the Tarskis. Ina, too, came to the parties from time to time, although she was no longer living at home. So there

Andrzej Mostowski at San Gregorio Beach, California, c. 1958.

she met Ehrenfeucht and, immediately, they fell in love. The relationship quickly became serious and all was wonderful, except that he had to return to Poland at the end of the academic year. Also, Andrzej had a wife and two children – but that, apparently, was less of a problem because he and his wife were already estranged.

Nevertheless, Maria was shocked that her daughter would be following this "married man" to Poland. Ina's response was that she wasn't doing it for love alone. "I might have chosen to go even if Andrzej were not there. People do travel." Alfred, too, was vehemently against the idea, but for different reasons. It wasn't that he disapproved of her liaison with Ehrenfeucht – quite the contrary, he recognized all his virtues; instead he argued that for her the choice was poor academically. "You will not learn anything. You will be wasting your time and you will be thought of as a rich American." (She was hardly that, especially since Tarski was not supporting her.) He reiterated his negative views about the communist

Andrzej Ehrenfeucht at the Tarski Symposium, 1971.

government and the lack of academic freedom. Since he was so much against her going, he said he would not finance her trip. Ina said she would pay her own way – since she had been doing that all along. Tarski even enlisted Dorothy Wolfe to help convince his daughter not to go. Ina recalled going out to dinner with Dorothy and Alfred and having heated words with them both about the issue, but she stuck to her plans. After she left, her embarrassed mother told some of her friends that Ina and Andrzej were secretly married, but Alfred, the "truth" man, aghast at her face-saving falsehood, insisted she call them and say it wasn't so, and Maria complied.

Ina had many reasons for wanting to return to Poland; she wanted to see and understand the place where she had lived under dire circumstances for the first six years of her life. Significantly, two of her boyfriends prior to Ehrenfeucht had been Jewish, and both had grandparents who had come from Minsk – her mother's birthplace. Ina said, "I was definitely attracted to Jewish men. What's funny is that Andrzej was not Jewish but he had a grandmother from Minsk." Here was another matter that Ina and her father argued about a few years later. She said she felt Jewish and identified herself as a Jew. Heatedly, Tarski disputed this, saying, "No. It is impossible. You cannot be Jewish without your mother being Jewish. That is how Jewishness is determined." Ina continued to insist that she felt partly Jewish and certainly more Jewish than her father. It was, of course, an argument that went nowhere. Tarski was stating a "law" and his daughter was discussing her state of mind and heart.

On her way to Warsaw, Ina stopped in Paris; there she stayed with Paulette Février and Jean-Louis Destouches, whom Tarski had written to ask for help. According to Paulette:

> A.T. wrote me that Ina wanted to go to Poland against his will. He had to let her go without helping her, but asked me to welcome her in Paris and take care of her clothes and shoes because she left Berkeley without any satisfactory equipment for a winter in Poland. She stayed some days at rue Thénard [the Destouches-Février apartment] before flying to Poland. Later I met her and her husband in Jerusalem; they were not yet married, meeting everywhere difficulties to obtain the right papers.[30]

Whether it was Alfred himself or Maria who prompted him to contact Paulette, both were concerned about Ina's comfort. And most likely they were in touch with their Warsaw friends, too, asking them to keep an eye out for her. She stayed two years and ended by agreeing somewhat with her father's prediction about the academic level of her sociology courses, acknowledging that the program was very political. But she got what she wanted, her master's degree and a life with Ehrenfeucht; when eventually they married, both parents approved. "My father liked Ehrenfeucht very much; he said it was a good choice. I think it is one of the few things I did that he approved of," and Maria, too, became very fond of her son-in-law.

A Meeting in Warsaw

In the midst of all these major domestic changes, Tarski was preparing to go to Warsaw for a symposium in early September of 1959, the third meeting in two years in which he was centrally involved. Significantly, this brought together a sustantial number of Poles and Americans as well as representatives from a dozen countries, both East and West. The theme of the meeting, infinitistic methods, was very much in tune with Tarski's approach to the development of metamathematics by full use of the tools of set theory. Naturally, a majority of the American participants were from the Tarski school.

For most of the Americans, this was a first voyage to Poland, and the experience produced mixed emotions. So much of Warsaw had been destroyed and so much was still waiting to be rebuilt, though now Soviet style. Even to the casual visitor it was evident that the economy was in terrible straits, goods were scarce, and life was governed by restrictions. Nevertheless, as Dana Scott reported, it was thrilling to be at the source, the place where Tarski had been formed and where he had laid the ground for almost all of his future work. Although an appalling number of Polish logicians were murdered during the war, more than a few of Tarski's teachers and close colleagues from the old days had survived and were present: Tadeusz Kotarbiński, Kazimierz Kuratowski, Wacław Sierpiński, Jerzy Słupecki, Maria Kokoszyńska-Lutmanowa, Andrzej Mostowski, and of course Wanda Szmielew. Perhaps even more gratifying was the sense that, in spite of hardship, logic was alive and thriving again.

During the war when Warsaw University was shut down under the German occupation, an underground university had sprung into existence. Clandestine classes were held in private homes; bravely and stubbornly, teachers continued to teach and students continued to attend. Mostowski had been one such active teacher, and his wife, also named Maria, had been a student. Later she was a major force in rebuilding the mathematics library and in this effort Tarski was one of her greatest allies abroad, urging colleagues to donate books and journals to replace what had been destroyed and to provide up-to-date literature. Tarski's dedication to that project was emblematic of his abiding concern for Poland and its future.

The Borsuks in Berkeley

Following the Warsaw Infinitistic Methods meeting, the Borsuks came to Berkeley. Karol had been in Berkeley at the Axiomatic Method conference in 1957; now he came for a longer stay and brought his wife Zofia (Zosia) with him. This was very unusual and implied influence at a high level, because at that time Poland rarely allowed a man and wife to leave the country together; rather, the policy was to keep a "hostage" at home to ensure a return. But there they both were in September of 1959 as a welcome addition to the mixture of Tarski's colleagues and friends. Zofia Borsuk recalled: "Tarski spent a lot of time with us. He helped us a lot and was a very loyal friend." She meant that he helped them both at home and abroad – in Poland by stimulating the American community to help the Poles, and in America by being a charming and diligent host.

He took us to Mrs. Tarski very often, when he went to lunch on Regent Street, and he encouraged us to visit her without him too. Once we went to dinner at his house on Michigan Avenue and Dorothy Wolfe was there. She cooked. At Christmas time he took us on a big trip all over California, to Death Valley, to the desert, to San Diego, and to the Mexican border, but because we did not have visas, we could not cross over so we turned around and came back along the coast. The trip lasted ten days and we stopped many times. Both Maria and Dorothy went with us. Tarski and Dorothy did the driving. We all felt very close.[31]

As always, Tarski showed his visitors the sights he loved. The itinerary was similar to the one he had arranged in 1957 with Borsuk, Szmielew, and the Destouches-Févriers. This time Alfred was able to carry off the feat of having his wife (only partially estranged, it was now apparent) and his girlfriend with him on the journey, like one big happy family. In any case Tarski was happy – at least part of the time.

In the journal Zosia Borsuk kept during her visit to America, she noted that "Tarski is very volatile: sometimes gay, smiling, full of interest, full of enthusiasm, full of projects, and then there are times he seems pessimistic, morose, in a bad humor. When he is discontent, you can see it. His face changes a lot. I don't know why or what causes it." But New Year's Eve was a "gay, smiling" moment:

Alfred insisted that we had to stay up until midnight to drink to the New Year. And he and Maria toasted the day that they had become American citizens. They were so full of enthusiasm about that. "It was a very important day for us," he said. It surprised me because they were *so* Polish and such patriots.[32]

It was not a contradiction for Tarski to revel in the freedom of being an American, to drain his Coke and say "Look: I am an American!" yet remain Polish to the core. Until the end of his life he was an outspoken advocate for Poles everywhere and went out of his way to protest restrictive policies in Poland. He followed the political scene closely; he was a devoted reader of *Kultura*, the dissident magazine published in Paris that gave the news of what was *really* happening in Poland, and is said to have financed the cost of publishing one whole issue.[33] But he was infuriated by the communist politics of the country and the anti-Semitism that, like an ever-present fungus lurking underground, periodically burst forth with renewed growth. A few years later he would vow never to return to Poland, not even for a visit. It would be another twenty years before his hopes for Poland's future were renewed. In 1981 Tarski saw *Man of Iron*, the Andrzej Wajda film about the Solidarity strikes in the Gdansk shipyards, and left the theater moved to tears at the thought that the upheaval in Poland would inevitably lead to change for the better.[34]

In early 1960 Ina left for Warsaw, having completed all the necessary financial and bureaucratic arrangements, and the Borsuks also returned to their home. Wanda Szmielew came to Berkeley again but did not live with Alfred this time, perhaps because Dorothy Wolfe was with him – although by then Dorothy was already easing her way out of her relationship with Alfred. Perhaps Wanda's presence was a factor or perhaps, as it was rumored, Dorothy was attracted to another man. There seemed not to be any rancor about her departure. She had brought a rug with her when she moved in but didn't take it with her when she left; ever after, it was known to the family as "Dorothy's rug."

10

Logic and Methodology, Center Stage

THE FIRST INTERNATIONAL CONGRESS for Logic, Methodology and Philosophy of Science was held at Stanford University in August of 1960. Occupying the vacuum created by the demise of the Unity of Science movement, it was the culminating event, on an international scale, of a long process of reorganization of communities of the philosophy of science and of logic that took place in the fifteen years following World War II – a process that involved many competing interests and personalities. Alfred Tarski was the pre-eminent winner in that competition, for the organization of the Stanford congress and its many successors to come was stamped with his view of logic and methodology as being at the center of systematic scientific thought.

After the Unity of Science Movement

Emigration, death, politics, and philosophical conflict account for the failure of the Unity of Science movement to maintain its position at the core of the international organization in logic and philosophy of science. The leading figures who were still alive had long since dispersed, mostly to America, as part of the general intellectual and cultural flight from Nazism in Germany and Austria in the 1930s. In the United States, an effort was made to continue the movement through the fifth International Unity of Science Congress at Harvard in 1939 and the sixth, in Chicago, two years later. That meeting, held three months before the United States entered the war against Germany, Italy, and Japan, proved to be the last. Quine's joking characterization of the Harvard congress as "the Vienna Circle in international exile, with some accretions" applied to the Chicago conference, too.

The Vienna Circle had already disintegrated in the mid-1930s after Hans Hahn died and Moritz Schlick was murdered by a deranged student on the steps of the University of Vienna. Otto Neurath, the main energizer and promoter of the Unity of Science movement and a noted socialist, fled to the Netherlands when the Austrian fascists came into power. After the Nazi invasion of Holland, Neurath managed to make his way to England but died of a heart attack in 1945.[1] The mantle of leadership could have passed to Carnap, who had gone to Chicago well before the war. He was the foremost practitioner of the Circle's doctrine of logical positivism, but he was not a dynamic leader. What was needed was someone with Neurath's personality: endlessly enthusiastic and enterprising, with the determination to see his projects into existence. There wasn't any one like him around, at least not in the Unity of Science group.

In addition to the personal losses, the movement in exile in the United States was absorbed in a less programmatic and more diffuse development of the philosophy of science, partly due to critiques of its basic tenets – most trenchantly by Quine – and partly due to new influences. In particular, it had to relate to an American philosophical tradition of empirical philosophy stemming from the pragmatism of Charles Sanders Peirce, William James, and John Dewey that originated in the latter days of the nineteenth century. Though this school of thought also granted empirical science a privileged position in the unfolding of human knowledge, it did not make as sharp a distinction as the logical positivists had done between "meaningful", directly verifiable statements and "meaningless", metaphysical statements. Another difference was that the pragmatists did not give the new developments in mathematical logic a central position. The leading postwar representative of the American tradition in the philosophy of science was Dewey's student Ernest Nagel, the John Dewey Professor of Philosophy at Columbia University.

There was also a significant political dimension to the movement's decline, connected with a major change in perspective. In addition to its program of cleansing philosophy of unreason, logical empiricism had also been in the business of social enlightenment, and its leaders, especially Neurath and Carnap, had allied themselves with socialist causes. When the first representatives of the movement came to the United States before the war, they were welcomed to the milieu of leftist intellectuals centered in New York. But after the war (in the view of the scholar George Reisch), "the movement died because its methods, values and goals were broadly sympathetic to socialism at a time when America and its colleges and universities were being scrubbed clean of red or pink elements. The apolitical logical empiricism of the 1950s ... was a new-born child of the cold war."[2] Like many others, Carnap was subject to political pressures during the McCarthyite red-scare period of the 1950s. When Carnap moved to UCLA in 1954, the FBI – which had been investigating Philipp Frank as an alleged promoter of communism in the United States on the basis of an unfounded rumor – began compiling a file on Carnap, too. They found frequent occurrences of his name in issues of the American Communist Party newspaper, the *Daily Worker*, in support of humanitarian and internationalist causes, especially for peace. Carnap may not have been investigated directly, but many of his friends and colleagues were questioned. Earlier, Sidney Hook, another prominent student of Dewey's and a fervent anti-communist, had vehemently warned Carnap against being tagged as a communist sympathizer. On the occasion of Carnap's publicly signing plans for a meeting of the Cultural and Scientific Congress for World Peace, an alleged communist-front organization, Hook wrote Carnap that "anybody who is still a sponsor by the time the party-line begins to sound off at the Congress, will be marked for life as a captive or fellow traveler of the Communist Party."[3]

Whatever the exact balance of forces were that led to the decline of the Unity of Science movement, its exhaustion as such was widely recognized. Seizing the opportunity, Tarski and his colleagues eagerly stepped into the void. The organizational task that they faced at the outset was to bring American and European logicians and philosophers of science together in an umbrella organization that would do for those fields what the Unity of Science movement had done in the interwar period, but now in a way devoted solely to "objective" issues, without subscribing to the comprehensive scientific world view promoted by the movement in the 1930s that had included the rational transformation of society.

Organizational Jockeying[4]

In Western Europe soon after the end of World War II, a number of new societies for logic and philosophy of science came into existence, both as national and as international organizations. Most prominent among the latter was the International Union for the Philosophy of Science (IUPS), founded in 1950 by Ferdinand Gonseth, a mathematician in Zurich. Since the late 1920s Gonseth had been developing an approach to the foundations of mathematics and the philosophy of science in terms of an "open," dialectical philosophy that rejected the possibility of absolute foundations. Despite the handicap of severely impaired eyesight and the burden of teaching elementary courses at his university, he wrote prolifically, expounding his views, and in the 1930s he organized discussion meetings called the "Entretiens de Zürich"; later, with colleagues, he launched a new journal, *Dialectica*. Gonseth became president of the IUPS, which subsumed various societies of logic and philosophy of science that had arisen in Western Europe in the previous years. Though his union was broadly representative in composition, there were complaints that Gonseth ran it in a high-handed way.[5]

Meanwhile, Tarski was generating interest within the Association for Symbolic Logic (ASL) for the formation of a broader international organization for logic and the philosophy of science. Though the association itself was officially international and had many active foreign members, its center of gravity was in the United States and it was largely regarded as being American. In pushing to ally the ASL more broadly, Tarski and his colleagues hoped to put it in a better position to garner financial support for conferences, journals, publications, and other activities. Not surprisingly, Gonseth's group was also in pursuit of the same benefits. At that time, an expected source of funding for both was the United Nations Educational, Scientific, and Cultural Organization. However, UNESCO could not be approached directly; there was a hierarchy under it, branching along two main lines: the International Council of Scientific Unions and the International Council for Philosophy and Humanistic Studies. The Association for Symbolic Logic was already placed under the former through the International Mathematical Union, but that limited the types of conferences and activities it could pursue and did not represent the interests of its membership among the philosophical logicians and philosophers of science.

Gonseth's society *was* included under the philosophical branch of UNESCO, but Gonseth wanted desperately to be allied with the scientific branch; toward this end, in the early 1950s he courted the ASL to join forces in order to take advantage of the logic group's recognized strength and prestige. Tarski's Dutch colleague Evert Beth belonged both to the logic association and to Gonseth's group and was thus a natural go-between. The problem was that he, Tarski, and others in the association were put off by Gonseth's authoritarian way of conducting matters and by his lack of logical rigor. For that reason, Tarski and Beth discouraged the proposed alliance. Gonseth then tried to get his organization into the International Council of Scientific Unions (ICSU) on its own, but – as a result of behind-the-scenes maneuvering by Father Bocheński on behalf of the ASL – that too was quashed. However, the ICSU made a counter-proposal: entry would be approved if Gonseth's International Union for the Philosophy of Science would join with the International Union for the History of Science, which was already under its aegis. The historians of science were not at all happy with this proposal; but the decision had already been made higher up, and they were told that they would not continue to receive support from ICSU unless their union joined with the philosophers of science in this way. At the same time, power was wrested from Gonseth as president of his own organization. In 1953 a putsch by Beth and his friends took place: the Dutch logician Arend Heyting was made president; the position of vice-president was taken over by Tarski's friend, the philosopher of physics Jean-Louis Destouches; and Evert Beth assumed the role of secretary.

Tarski vs. Gonseth re Methodology

In 1953, Gonseth gave a lecture at the Colloque International de Logique in Brussels. Tarski attacked him head-on during the discussion period, brutally dismissing his ideas:

I must admit that I do not see, in the exposition of Professor Gonseth, *one single problem* which could be treated and settled by rational methods. When I hear, for instance, that there is an essential difference between mathematics and the natural sciences, my first tendency is to resist this opinion ... whether it concerns the origin of the disciplines involved, or the methods of inquiry applied in them, or perhaps the methods of organizing and establishing the results obtained; also under what conditions the differences can be called *essential*. Finally, I arrive at the conclusion that as long as these points are not clarified, no serious discussion of the problem is possible.[6] [emphasis in original]

In response to another participant in the same discussion, Tarski went on to remark:

It would be more than desirable to have concrete examples of scientific theories (from the realm of the natural sciences) organized into deductive systems. Without such examples there is always the danger that the methodological investigation of these theories will, so to speak, hang in the air. Unfortunately, very few examples are known which would meet the standards of the present-day conception of deductive method and would be ripe for methodological investigations; I can refer, however, to some recent attempts in this direction – to the work of J. H. Woodger in the foundations of biology and of J. C. C. McKinsey and his group in the foundations of physics. The development of metamathematics, that is, the methodology of mathematics, would hardly have been possible if various branches of mathematics had not previously been organized into deductive systems.[7]

In other words, a precondition for the methodological study of the sciences would, in Tarski's view, be their presentation as axiomatic deductive systems, and the paradigm for that was the axiomatization of various parts of mathematics and its study by the methods of metamathematics. All of this was directly opposed to Gonseth's anti-foundational, open view of science.

For many years Tarski and Carnap had held the ideal view that the sciences ought to be systematized in axiomatic deductive form. As the philosopher of science Michael Friedman puts it, Carnap – in his famous 1934 work *The Logical Syntax of Language* – had articulated the program of logical analysis as the principal enterprise of philosophy, "simply as a branch of logical syntax: specifically [that] of the language of science."[8] But Tarski had not always been as sanguine about its applicability. However, by the time of his criticism of Gonseth in 1953 he could, at least, point to the work on axiomatization of physics by McKinsey and Suppes and the group around them.

Even so, Tarski remained equivocal about the role of logic in the methodology of the physical and other sciences. In the preface to the proceedings of the 1957 Berkeley conference on the axiomatic method in geometry and physics, he wrote (with his co-editors Henkin and Suppes) that "much foundational work in physics is still of the programmatic sort, and it is possible to maintain that the status of axiomatic investigations in physics is not yet past the preliminary stage of philosophical doubt as to its purpose and usefulness. In spite of such doubts, an increasing effort is being made to apply axiomatic methods in physics."[9]

Tarski was not the only one to speak of mathematical and scientific methodology, and the use of 'methodology' as a key word for the kinds of research programs he had in mind does not originate with him; it goes back to the Polish philosopher Kazimierz Ajdukiewicz, though Tarski perhaps construed it in more specific terms along the lines of Carnap's project for the logical investigation of scientific language and theories as formal objects of study. The word 'methodology' has a ponderous sound to some ears, and its intended scope is not clear; but, since it was Tarski who emblazoned it as the emblem on his shield, the recurrence of the word in the organizational activities that he promoted is unavoidable. It had already become part of the name of the interdepartmental program in logic and the methodology of science that Tarski and his colleagues inaugurated at Berkeley in 1957. (In recent years, 'methodology' has become a vogue word, often misused as a pretentious substitute for 'method' in scientific and technical contexts.)[10]

A Marriage of Convenience

Following the directive of the International Council of Scientific Unions, in 1955 the international societies of the historians of science and of the philosophers of science joined to create a new entity, the International Union for the History and Philosophy of Science (IUHPS) under the ICSU umbrella. The parties to this marriage of convenience took care to demarcate themselves, respectively, as the Division of History of Science and the Division of Logic, Methodology and Philosophy of Science (DLMPS) within the new union. With the removal of Gonseth, the Association for Symbolic Logic had joined the IUPS the year before and was now represented with significant voting power in the new division for logic and methodology. At the same time, it managed to retain its place under the International Mathematical Union, thus enjoying the best of both worlds.

Between the years 1955 and 1960, the presidential position of the new DLMPS rotated between Alfred Tarski; Jean Piveteau, a paleontologist from the Sorbonne in Paris; Robert Feys, a Belgian philosopher of logic; and Arnold Schmidt, a German logician. Tarski's friends Jean-Louis Destouches and Evert Beth were appointed secretary and treasurer/adjoint secretary, respectively. Still, Gonseth was not totally out of the

Patrick Suppes at the Tarski Symposium, 1971.

picture; over Tarski's objection, Gonseth's backers saw to it that he was given the title of "Honorary President." In 1960 the terms of officers were lengthened to four years each, with Stephen Kleene becoming president for 1960–1963, Kazimierz Ajdukiewicz vice-president, Patrick Suppes secretary-general, and the Dutch mathematician Hans Freudenthal treasurer and adjoint secretary, putting Tarski's people firmly in the saddle. To start things off with a big bang, a major congress was proposed for Stanford in 1960.

The 1960 Congress

A grand idea in breadth and depth, the congress planned for Stanford in 1960 was almost derailed. Patrick Suppes, secretary-general designate of the division and the leading philosopher of science at Stanford, was the point man. In 1959, late in the planning stages, he told Tarski that he was having difficulty raising funds and was under pressure from the National Research Council and the National Science Foundation to abandon the idea of a separate international congress of the division and instead join the historians of science in *their* efforts to organize an international congress at a later date in the United States but not necessarily at Stanford.

Outraged at the proposal, Tarski wrote Suppes:

My reaction to the idea of holding an international congress jointly with the historians of science is decidedly negative .... A congress for the whole IUHPS [International Union for the History and Philosophy of Science] would be a gathering of people with very few common scientific interests uniting for some administrative, and not scientific, reasons. In particular, logicians would be engulfed in a sea of men who have entirely different approaches in their research and who apply entirely different methods, and I do not see what logicians could gain by participating in such a congress.[11]

Tarski wanted to stick with the Stanford plans for territorial reasons, too. "The work of our group in the San Francisco Bay Area [will give] a guarantee of a high scientific level for the proposed congress ... [and] if we join the historians of science in their efforts, the common congress will be held somewhere in the East." (Long gone was Tarski's feeling that the West Coast was intellectually inferior.)

Tarski prevailed in his opposition to a joint meeting, and the National Science Foundation, along with the American Council of Learned Societies, was persuaded at the last minute to help fund the 1960 congress at Stanford. Ironically, in the end, no financial support for this or succeeding meetings of the division was obtained from UNESCO, but its titular support gave the organization the desired international status.

The breadth of the congress was assured by the composition of the organizing committee, which was headed by Ernest Nagel as chairman and Alfred Tarski as vice-chairman and included leading scholars from biology, economics, logic, mathematics, philosophy, physics, and statistics. The plans for the meeting were ambitious in their scope and set a pattern that has largely been followed in succeeding congresses of the DLMPS. Invited lectures and contributed papers were distributed through eleven sections, the first three of which were designated for Mathematical Logic, Foundations of Mathematical Theories, and Philosophy of Logic and Mathematics. The next two were entited: General Problems of Methodology and Philosophy of Science; and Foundations of Probability and Induction. Following that were four sections on the Methodology and Philosophy of Biological and Psychological Sciences, Social Sciences, Linguistics, and Historical Sciences. The final section was labeled History of Logic, Methodology and Philosophy of Science.[12]

Despite the wide scope, more than a third of the invited lectures were in the first three sections. Thus the "Logic" of "Logic, Methodology and Philosophy of Science" was placed front and center – in contrast to the prewar Unity of Science meetings, where it had a secondary position. The first three sections of the Stanford congress had among its invited speakers such notables as Stephen Kleene, Abraham Robinson, Paul Bernays, Alonzo Church, Arend Heyting, Georg Kreisel, and Tarski himself, as well as a number of Tarski's students and co-workers. But while logic now took pride of place, methodology and philosophy of science commanded the major portion of the program. As in the Unity of Science meetings from 1935 to 1941, this was construed broadly to include the physical, biological, and social sciences.[13] Leaders in the latter field among the invited speakers were the economists John Harsanyi and Leonid Hurwicz, the psychologist Ernest Hilgard, the social scientist Paul Lazarsfeld, and the linguist Noam Chomsky. Other distinguished participants were the philosophers Karl Popper and Rudolf Carnap and the physicists Henry Margenau and John Wheeler. Berkeley and Stanford were well represented, as Tarski foresaw, yet the meeting was by no means insular; participants came from all over the world, including several from countries behind the Iron Curtain.

The list of speakers and disciplines generated palpable excitement; new personal contacts were made and new interdisciplinary sparks generated. These were enhanced by the many dinners and parties that were arranged informally as well as by several group excursions that were organized for the weekend.

"This country is so beautiful"

It was remarked that, at every opportunity during the Stanford congress, the Soviets and others from communist countries jumped into any group posed for a photograph and threw their arms around the most prominent Americans present. The "opportunists" explained quite frankly that they feared they were on a "hit list" back home and hoped that such evidence could be used as a form of protection to show that they were known in the West; Józef Bocheński was sure he was "number ten" in Poland.[14] Another concern for participants from Iron Curtain countries was that the U.S. State Department had set a limit of a fifty-mile radius from San

Francisco beyond which they were not allowed to go. (This was a tit-for-tat response to similar restrictions on the movement of Westerners in the Soviet Union.) A special tour to the Monterey Peninsula had been planned for one of the free days of the congress, including stops at the old Carmel Mission and Point Lobos State Park on the Pacific Ocean. Since the farthest point to be reached was nearly a hundred miles south of Stanford, special permission was obtained to extend the allowed radius temporarily. Two tour buses were hired and filled to capacity. At Point Lobos, with its spectacular scenery of craggy cliffs covered by windswept monterey pines and twisted cypress, paths are well-marked and there are strict warnings not to wander beyond their limits. But at the most famous Cypress Point, the excited scholar-tourists broke the rules and scrambled all over to see the view and take photos. When it was time to regroup and get on the buses for the return to Stanford, all the Russians were missing. After a long wait, ad hoc search parties and a ranger were sent out to scour the various paths and call out for them, to no avail. Now the tour leaders began to worry: Could one or more of the Russians have hidden so as to escape those of their company suspected to be KGB agents, or were they themselves agents? Worse yet, could one of them have fallen off the edge of a cliff to injury or death? Finally, after an interminable wait, one by one the Russians appeared as if out of nowhere, with hardly a word of explanation or apology except to say: "It's your fault! This country is so beautiful, we could not resist taking one photograph after another and it was impossible to stop."

After 1960, the LMPS congresses continued to meet every four years, with an occasional exception. They were held in Jerusalem, Amsterdam, Bucharest, London (Ontario), Hanover, Salzburg, Moscow, Uppsala, Florence, and Cracow. In August 2003, the first congress of the twenty-first century took place in Oviedo, Spain. Tarski's stamp on the organization and its meetings is indelible, and even during his lifetime he was gratified with the success of his vision.

11

Heydays

I consider the year I was in Berkeley to be the absolute heyday of mathematical logic.   Michael Rabin

IN THE TEN YEARS between 1950 and 1960 Tarski's influence became manifest, perhaps beyond his own wildest dreams, but it was the following decade that people labeled "the heyday." During the Cold War between the United States and the Soviet Union – and the concomitant arms and space race – money became available at a level never seen before for research in the sciences and mathematics and even for basic research in pure logic, which had no apparent connection with the problems of the day. The humiliation of the Soviet success in putting the *Sputnik I* satellite into space in 1957 – and four years later launching Yuri Gagarin, the first man to orbit the earth – exacerbated the competition. The United States was determined to catch up at any cost, and money flowed for conferences, symposia, summer meetings, and travel. There was money for new faculty, staff support, and visitors, as well as funds to establish new interdisciplinary organizations and to support new research projects. It all seemed to be there for the asking, and Tarski was one who knew how to ask or to get others to ask for him.

With the hiring of Leon Henkin (in 1953) and his former students Robert Vaught and Dana Scott (in the late 1950s), Tarski began to assemble a fabulous group of logicians at Berkeley in the mathematics and philosophy departments. He explicitly stated his goal to have logicians constitute ten percent of the mathematics faculty, and by the mid-1960s, with considerable help from his cohorts, he made it happen. His desire to build a home – a palace, really – for logic was realized, and there was a community of scholars living in it, as vital as the one he had known during his early years in Warsaw. His earlier complaints about being isolated in Berkeley were most certainly a thing of the past.

Kurt Gödel and Alfred Tarski, Princeton, New Jersey, 1962.

Driving himself and inspiring others, Tarski had arrived at a new level of fame and power. The program for students in logic and methodology at Berkeley was off to a strong start and he had presided over the first International Congress of Logic, Methodology and Philosophy of Science, but for Tarski "arrival" was never the end of the journey. There was always a further destination; there was always more that had to be done. His curriculum vitae of the sixties – when *he* was in his sixties – reads like a travelogue that would have exhausted a lesser man. Going from conference to conference and responding to individual invitations to lecture, he circled the globe and crisscrossed the American continent, making himself accessible wherever he went. He was never too busy to talk to people about his ideas and listen to theirs and, as a result, waves of logicians swept into Berkeley. Students, postdocs, faculty visitors, new faculty – all had much to gain from Tarski. Of course, whenever he could, Tarski enlisted these scholars to work on one or another of his many projects in model theory, set theory, foundations of geometry, and algebraic logic – the fields that were his major areas of interest at the time.

John Addison at the Tarski Symposium, 1971.

Newcomers: Faculty

William Craig, a gentle, modest man, seemingly unaware of his ethereally handsome appearance, came to Berkeley in 1960 as a visitor to replace Leon Henkin, who was on leave. Craig, one of Quine's students, had attracted Tarski's attention at a meeting in Cambridge even before receiving his Ph.D. He made another strong impression upon Tarski at the Cornell meeting with the presentation of his interpolation theorem, which was eventually recognized to be one of the most fundamental results in the metatheory of first-order logic. That same year John Myhill left the UC philosophy department to take a professorship at Stanford; Craig accepted an offer to replace him and became a permanent member of Tarski's team. He was a regular participant in the logic seminars and in later years took up the subject of algebraic logic, one of Tarski's main interests, but put his own stamp on it.

Next, John Addison, a man of incisive intellect and brilliant organizational skills, joined the mathematics faculty as an associate professor. Addison had been an undergraduate at Princeton where Alonzo Church held sway and a graduate student in Wisconsin under Stephen Kleene's tutelage. Noted for his substantial results about definability in higher-order languages, he too spent a year as a visiting professor and was subsequently offered a tenured position. Addison did not immediately jump at the opportunity because he had had a vision of building a logic center at the University of Michigan, where he, Roger Lyndon, and Clifford Spector already formed a nucleus. But those plans were sadly deflated when Spector died of leukemia in 1960. Two years later he accepted the UC position because it was clear to him that the stimulating atmosphere there and the presence of many logicians, both the permanent faculty and the visitors, would challenge him to do his best work.[1]

Addison's appointment created a bridge between the Kleene school of logic, mainly concerned with recursive function theory and hierarchies of definability, and the Berkeley school, which was concerned with set theory, model theory, and algebra. He was installed in an office between Dana Scott and Bob Vaught and had excellent interaction with both. His first assignment was to teach introduction to metamathematics, the course Tarski had created fifteen years earlier. As a measure of the exceptional interest in logic in those days, sixty-five students enrolled in the course and close to eighty students were in the logic and methodology program. Addison, too, was a regular attendee at Tarski's seminars; he thought Tarski was an "utterly superb" seminar leader and took him as a model for his own teaching, but it didn't quite work for Addison. He found he was incapable of applying pressure to students in the same way Tarski did, because "basically I'm not that kind of person."[2]

Not all the moves in this game of musical chairs for logicians were in the one direction. John Myhill, who had been at Berkeley since 1954, went to Stanford. There logic was having its own heyday in the sixties, especially through the cooperation of the mathematics and philosophy departments via joint appointments and its support in the former by Halsey Royden and in the latter by Patrick Suppes and Donald Davidson. Georg Kreisel visited frequently and would become a regular faculty member in 1964. Among the up-and-coming young logicians were Solomon Feferman and William Tait, both of whom had arrived in the latter part of the fifties. In mathematics, the young analyst Paul Cohen turned to questions in logic and obtained spectacular results about set theory in 1963.

When, after four years at Stanford, Myhill left for Buffalo, Dana Scott, wanting to build a program on his own rather than in Tarski's shadow, took his place. After Scott left Berkeley, Tarski still had to work at meeting his ten-percent goal for logicians in the mathematics department. Toward this end he swept Addison up into his plan. With his steely affability, he

Haim Gaifman at the Tarski Symposium, 1971.

was the ideal accomplice for this task and, like Tarski, once he got hold of an idea he never let it go. Another matter that would occupy Addison for several years was helping with the organization of an international symposium on the theory of models that was to take place in Berkeley in the summer of 1963. In later years, Addison exerted his influence on the mathematics department at Berkeley by assuming – twice – the demanding and commanding position of chair.

Newcomers: Students and Visitors

"Those were the heydays of Tarski. The group in logic and methodology was going strong and mathematics was too. The whole thing was flourishing." In a nutshell, Haim Gaifman thus described the scene at Berkeley and his reasons for wanting to do graduate work there.[3] Gaifman, an Israeli, had studied with Abraham Robinson in Jerusalem and after that was working toward a Ph.D. in the foundations of probability with Tarski's old friend and colleague Rudolf Carnap at UCLA. Serendipitously, Gaifman came upon and was intrigued by a problem in Boolean algebra that Tarski had posed. He dropped probability, focused on algebra, and was happy to find an interesting solution that he promptly sent to Tarski. Impressed, Tarski invited him to speak in Berkeley and, as Gaifman reported, "After my talk, Tarski didn't waste a minute. On the spot he invited me to come work with him and then he got me a research assistantship."

Carnap seemed to have no anger or irritation about Gaifman's leaving UCLA to become Tarski's Ph.D. student. "An angelic man, he let me go and we remained on the best of terms with no hard feelings whatsoever," said Gaifman. He was unaware or unconcerned about rivalries between teachers and of how much Tarski wanted outstanding students. Had Carnap lured away one of his brilliant students, Tarski is not likely to have been "angelic" about it.

Gaifman completed his Ph.D. in 1962, in only two years – an impressive record, even for one who already had a master's degree. Unintimidated by Tarski's demanding nature, he did not work on any of his projects and had remarkably few late-night sessions with him. He just did his own thing. Once he was asked to pick someone up at the airport. Gaifman said, "I resented that and let him know I didn't think that was my job. Tarski gave me the kind of grin that indicated that I didn't realize what he could demand if he wanted to; but he accepted what I said in good humor. I think he thought I had a lot of chutzpah to say what I did, but he took it as Israeli gall and bad manners."[4]

Michael Rabin, who would become the rector of Hebrew University as well as a professor of computer science at Harvard, was another Israeli who was attracted to Berkeley in the early sixties. As enthusiastic as Gaifman, he, too, used the word "heyday" to describe the general atmosphere: "There were so many people from all over the world. In fact I consider the year I was in Berkeley [1961/62] to be the absolute heyday of mathematical logic in Berkeley. I considered it to be an enormous privilege to be invited there. It really was the Mecca."[5]

Not everyone agreed with Gaifman and Rabin about the ideal time. The "greatest years" were in the eye of the beholder. The best time to be in Berkeley was when *you* were working with Tarski – under his spell, as it were. Steven Givant, Tarski's close colleague and friend of the seventies, commented: "What great power Tarski had, to make everyone think that *their* decade with him was the high point."

At the Cornell meeting in 1957, with one of his flash "I can get anyone I want" decisions, Tarski had offered Rabin an assistant professorship in Berkeley. Rabin declined then, but three years later he asked if the offer was still open. He came as a visiting associate professor, a position that took some effort for Tarski to secure because of Rabin's relative youth. Discussing what had impressed him about Tarski personally, Rabin said:

Michael Rabin at the Tarski Symposium, 1971.

I was twenty-nine years old and he was sixty, so to me he looked very ancient – that's the natural order of things, sixty doesn't look old at all to me now – but I was so impressed by his intensity, by how interested he was in *everything*, in logic, in mathematics, in research, in new ideas. I explicitly said to myself, I wish when I am his age I will still be as much alive intellectually and scientifically and as much connected.[6]

In the seminar on decidability and undecidability in which Rabin took part, he observed Tarski in action, conducting his course in the European tradition of the old master. While students presented their work, he interrupted with questions, corrections, clarification, and background. Rabin's view was that even though "the poor student may have found these interruptions disconcerting and intimidating, everyone else thought it was illuminating and a very important part of the seminar. When Tarski stood there in the middle of someone's presentation discussing the right way to say something, it was not always a criticism of the student. He wanted to find the right way for himself too."

At the same time, in the same seminar, Donald Monk, who finished his Ph.D. with Tarski that year (and would later work with him on *Cylindric Algebras*), had a different reaction that was shared by many others. He considered Tarski to be "extremely harsh on some people and exceedingly demanding." However, in the next breath Monk said, "I wish Tarski had made more demands on *me*. I would have preferred more feedback and suggestions. In my case, he was always sympathetic."[7]

Students Who Took Other Routes

Some students who did well in courses and seminars with Tarski had the opportunity to work with him but ended up choosing someone else as their Ph.D. advisor. Peter Hinman, who began his graduate work in 1959, and his friend Haragauri Gupta took courses on the foundations of set theory and foundations of geometry from Tarski in the early 1960s. Both stood out and Tarski, pleased with their problem-solving abilities, proposed to supervise their dissertation work. This they took as a great compliment. Gupta accepted right away, but Hinman was wary for a number of reasons:

> To me, it didn't look like a good bandwagon to hitch up to because a lot of the students hanging around him had been there forever. And those all night sessions! On the other hand, I was afraid to turn him down because I knew he had a temper and I was afraid if I made him angry he might make problems for me. It didn't turn out that way, though. Tarski never pushed the issue of my becoming his student.[8]

Instead, Hinman took a course in recursion theory from John Addison, and finding the subject and the teacher appealing, he soon became Addison's advisee. In addition, he went to Poland for a year, in 1964/65, and spent a good deal of time discussing logic with Andrzej Mostowski – as well as many hours explaining the free-speech movement in Berkeley, because Mostowski was reading about it in *Time* magazine. Upon his return to Berkeley he finished his Ph.D. within a year but not sooner than his friend Gupta, who had succeeded in finishing with Tarski in 1965. From Hinman's point of view, though, he had chosen the right path because, as he said, "I never got caught up in having to do a lot of things for him that had nothing to do with me."

One of the students in the group Hinman categorized as "hanging around Tarski forever" was Thomas Frayne, who had asked Tarski to be his advisor in the mid-1950s and was accepted as a student and research assistant – but there was a catch. The research assistantship Tarski offered involved translating some of his papers that had originally been written in French and German. "To be honest," Tarski said, "I have to warn you that it will be a big job and the work would not count towards your Ph.D. degree," Frayne recalled. "He said I should think it over before deciding."

In spite of the warning, Frayne felt honored that Tarski would select him for the task and accepted. All too soon, however, he realized that "I was in way over my head. Foolishly, though, I did not seek advice on what to do but just plowed on. Most of my meetings with Tarski were about translations and I was making no progress on my own work."

Needing money, Frayne accepted a job teaching at the University of San Francisco; he continued to translate as he had agreed, but "I was doing absolutely nothing on my thesis. Still, my main complaint about Tarski was that he was hard to reach. I was an early riser and Tarski slept until noon. I got sleepy at 10 P.M. and he got going strong at 2 A.M."

Their meetings at Tarski's house were very hard on Frayne. By the time he got home he was too exhausted to consolidate the many ideas Tarski had given him and by the next day they were gone, vanished before he could reconstruct them. "I have always blamed this mismatch in our sleep patterns for my difficulties, rather than any fault in Tarski or myself."

Frayne's assessment seems too kind. It was far from the first or last time that Tarski was inconsiderate of his students' needs; and it's not actually true that Frayne did "absolutely nothing" on his thesis. By 1958 he had obtained some results, in collaboration with Scott and Tarski, on a generalization of the ultraproduct construction, called reduced products. A few years later, he published a paper with Morel and Scott, which has since come to be regarded as a classic, on the properties and applications of reduced products; Frayne's contribution to that could conceivably have been part of a thesis. Begining with the 1960s, reduced products and especially ultraproducts have served extensively as a tool in model theory and set theory, among other things allowing formerly metamathematical arguments to be replaced by more purely mathematical ones. In particular, with these methods Jerome Keisler and Tarski obtained many new mathematical results about relations between unusually large transfinite cardinal numbers, the first of which had been pioneered by William Hanf in his dissertation using metamathematical methods.

In retrospect Frayne felt that he should have switched advisors – the "mismatch" was reason enough – but he couldn't bring himself to do so; as a result, he never finished his Ph.D. Happily, he ended up in a better situation for that reason. After taking a year off from his teaching job at the University of San Francisco to make a last try at finishing his thesis, he gave up for good and in 1969 accepted a computer programming position at IBM, where he found his niche. "If I had finished the Ph.D.," he later said, "I think I would have continued teaching and I might have been a mediocre researcher. On the other hand as a programmer at IBM, I was a top performer and very satisfied with my work. So my failure turned out to be an opportunity to switch to a satisfying, successful new career."

In an act of generosity that was a demonstration of his success on his own terms, after Tarski's death, Frayne and his wife were the first major contributors to an endowment fund supporting the annual Tarski Lectures at UC Berkeley, which have been presented in his memory since 1989.[9]

Jerome Keisler

Another student who, like Haim Gaifman, breezed through to a Ph.D. early in the sixties was Jerome Keisler; like Donald Monk, Keisler had no problems with Tarski's toughness. He was very much in tune with Tarski's systematic approach to problems that still left wide scope for his own ingenuity and creativity. As an undergraduate at the California Institute of Technology, Keisler had already begun to do serious research in logic and had published a paper on model theory. From C. C. Chang, who was at nearby UCLA, he heard details of Tarski's work and life. Chang loved telling stories about Berkeley and went into legends of the late-night sessions, the parties and the drinking, and the traumas of the seminar. He also spoke of how Tarski had changed his name because of prejudice against Jews in Poland. This struck a deep chord in Keisler, and when he met Tarski he immediately liked him "because he looked a lot like my grandfather who had emigrated from Poland in 1906 because of the pogroms."

"So," Keisler said, "since my grandfather was a mild-mannered person and I loved him, this resemblance made me feel more comfortable and less in awe of Tarski than I otherwise would have been, even though I knew by reputation that Tarski was the opposite of mild mannnered; and my basic impression of him did not change much over the years. I was aware of the difficulties others had with him but he always treated me well."[10]

Small wonder Keisler was treated well. Effortlessly, or so it seemed to others, he worked independently and steadily on original problems in model theory and set theory, which he solved one after another... after another. Two years after finishing his master's degree in 1961, he completed

H. Jerome Keisler at the Tarski Symposium, 1971.

his doctoral dissertation on "Ultraproducts and Elementary Classes" and was awarded a Miller postdoctoral fellowship at Berkeley immediately afterward. Among outstanding students he was *exceptionally* outstanding, but being mild-mannered and quiet, like his grandfather, there was nothing boastful or overtly competitive about him. He had the impression that Tarski thought he was "square," but that was all right with him. He regarded his teacher as a mentor and later a colleague, but they never developed the close if sometimes contentious "father–son" relationship that other students experienced, even though he spent many a late night at Tarski's home and did major work with him.

The Issue of Jewish Identity

It's amusing that Keisler's first impression and his warm feelings for Tarski were colored by the fact that he identified him with his "very Jewish" grandfather because Tarski always made the point of identifying himself as Polish rather than Jewish, in spite of the persecution that had been visited upon him and his family. Yet his attitude about "what he was" is murky, and what he said may not always have been what he felt. If someone Jewish straightforwardly made the point of their being affiliated, Tarski usually bristled. On the other hand, others besides Keisler felt a *Landsmann* affinity and, on some level, Tarski felt the same. Azriel Levy, an assistant professor visiting from Jerusalem in 1959/60, said, "Maybe part of the reason I felt close to him was that both of us were Jewish; I regarded him as such in spite of his conversion." With Levy, Tarski talked about "Jewish matters," particularly regarding names. "He told me that Sierpiński, his professor in Warsaw, was a Karaite [an ancient Jewish sect] and that his ancestors had become Christian and adopted the names of the months." (*Sierpień* is the word for August.)

Similarly, with Michael Rabin, Tarski spoke in an intellectually detached way about the possibility that he could be related to the Satmar rabbis, a Hasidic dynasty, all of whom are named Teitelbaum. (The name of the sect, whose origins lie in the eighteenth century, comes from the city of Satu Mare bordering Hungary in Romania.)[11] Rabin was never quite sure whether Tarski was serious about this supposed connection. In addition, he found it odd that, in all their conversation over a period of many years, this was Tarski's only allusion to his Jewishness. "He never spoke about the Holocaust," Rabin said. "He never mentioned that his own family was annihilated. He was in Israel at least twice and never made any statement about the revival of the Jewish people in Israel. Nothing about the Jewish State. Since he never said anything, I felt too shy to ask him what he thought. If I had been older or felt more comfortable, I might have, but not then."[12]

A decade later, Tarski did discuss his views about Israel with Menachem Magidor, a postdoctoral student in Berkeley in the early 1970s who twenty-five years later would become president of the Hebrew University of Jerusalem. The hospitable Tarskis invited Magidor and his wife to dinner and, during the evening, conversation turned to politics, Israel, and the 1973 Yom Kippur War. To Magidor, Tarski said that at first he thought Israel had no right to form a Jewish state based upon historical rights or any kind of continuity. "But" he said, "now I do believe that people born in a particular place have a right to live there. Once you are born there you have a right to national expression." He had nothing against a Jewish state, if you were born there, but he wouldn't define it in a religious sense. Then, in the context of religion, Tarski explained the reasons for his own conversion to Catholicism in the 1920s:

You know, I was born Jewish but I didn't stay Jewish. I considered myself, at that point, to be Polish, culturally and nationally; and I didn't care about religion at all, so it had nothing to do with religious belief. But there was the fact that 95% of Poles were Catholic so I thought it would be crazy to separate myself from that group.[13]

In this explanation Tarski omitted mentioning the practical, economic reasons that were relevant at the time of his name change and conversion from Jew to Pole. He did, however, admit that his family was not happy with that act.

In spite of Tarski's insistence on his Polishness, Rabin and others concluded that Tarski had a "warm spot" for Jews, that he considered Jews to be especially gifted as a "race," and – having it both ways – included himself in the gifted-by-race category. Tarski also voiced the opinion that the Chinese were superintelligent. On the flip side of the coin, to the dismay of his liberal colleagues, he was interested in research into the genetic factors accounting for differences in IQ based upon race; this research was generally scorned by the academic community because of its faulty protocols and because it fed into theories of the eugenics movement that advocated sterilization of people with low intelligence scores. It is entirely possible that Tarski was acting as devil's advocate in this case, a role he was happy to play simply to provoke argument, but it is also true that he was prone to racial stereotyping.

Poetry, Poles, and More Warm Spots:
Tarski and Czesław Miłosz

One more high point of the early sixties in which Tarski played a role on a matter having very little to do with logic was the appointment of Czesław Miłosz, the Polish-Lithuanian poet, to the Slavic Languages department of UC Berkeley. This was twenty years before Miłosz was awarded the Nobel Prize in literature and before that possibility had crossed anyone's mind. Tarski did not know him personally, but he knew of his poetry and a good deal more about him. Miłosz later said, "I learned Tarski was instrumental in bringing me to Berkeley. I don't know exactly how, but he was consulted and I heard that he backed me." Given Tarski's reputation for getting whom and what he wanted, this "backing" was certainly done with vigor. What was different in this instance is that Tarski used his clout outside of his usual sphere of influence in the mathematics and philosophy departments and urged the dean to make the appointment.

Miłosz, ten years years younger than Tarski, had lived through World War II in Warsaw and afterward supported the postwar communist regime in Poland. As a prominent member of the intelligentsia, he eventually became cultural attaché to France and lived in Paris. In 1951, finding his position morally untenable, he resigned and defected to France, remaining in Paris as a refugee. Tarski knew of all this through his regular reading of *Kultura*, the magazine begun in 1947 by Polish exiles in France. When Miłosz was invited to Berkeley several years later to visit, Tarski learned more about his academic qualifications from his colleague Frank Whitfield, who specialized in Slavic linguistics. Whitfield's wife Celina was Polish, and she and Maria Tarski had become very close friends; their common Polish background was key. The Whitfields and the Tarskis dined together once a week, gave each other plants, and talked about their gardens. They went to movies and other entertainments as a foursome, and Alfred, according to Celina, was always pushing to go somewhere "racy."

Quite naturally, when Miłosz first came to Berkeley to visit in 1959 prior to joining the faculty, they all met at the Whitfields' for dinner. As it happened, Czesław was good friends with the philosopher Jeanne Hersch, Alfred's distant cousin from a branch of the Teitelbaum family that had moved to Geneva in the early 1900s. He was also versed in the avant-garde circle of poets, playwrights, and artists in Warsaw whom Tarski knew, including the Whitmanesque poet Julian Tuwim, Tarski's favorite from the *Skamander* circle; Miłosz himself belonged to the next generation of poets. Apart from logic, there is probably nothing Tarski liked to talk about more than politics, poetry, and plants; their conversations must have been choice and lacked only a Boswell to record them.[14]

The following year, Miłosz and his family made a permanent move to Berkeley. By then Alfred and Maria were living apart, but each of them hosted the newcomers in their usual manner. Maria invited them to lunch and, as Miłosz later wrote, "Tarski drove us around to show us the area. The countryside to the east of the Berkeley hills – chestnut orchards in the valley; higher up, straw-colored slopes punctuated by black oaks throughout most of the year."

Tarski went out of his way to be extremely friendly to the Miłoszes: "He showered us with most moving tokens of friendship." This kind of generosity toward visitors or people new to California was Alfred at his best, and never was it more appreciated. Tarski felt amply rewarded because of their unspoken understanding of what it meant to be Polish. Their friendship lasted until Tarski's death.

In addition to the Whitfields, Alfred and Czesław had Leszek Kolakowski, the Polish philosopher who also spent some time in Berkeley, as a friend in common; but much of their conversation was about the intellectual and artistic circles in Warsaw and the avant-garde clique surrounding Stanisław "Witkacy" Witkiewicz in Zakopane in the 1920s and 1930s. Miłosz was very impressed by the Witkacy portraits that hung in the Michigan Avenue house, and he relished the first-hand tales Tarski told of Witkacy's idiosyncratic behavior and sense of humor. One of those stories was about the lewd annotations connecting metalanguage with oral sex that Witkacy wrote on his copy of Tarski's article "Pojęcie Prawdy w Językach Nauk Dedukcyjnych" [The Concept of Truth in the Languages of Deductive Sciences], about which every Polish logician, especially Tarski, tittered appreciatively (in Polish the word *język* may be used interchangeably for 'language' and 'tongue').[15] Another story recalled by Miłosz concerned what happened when Tarski and Witkacy went to a party at the German Embassy in Warsaw. There was a moment after the death of Piłsudski when the Polish government was making diplomatic overtures to Germany – this was after 1935, and Hitler was already in power. Dignitaries, intellectuals, and cultural figures were invited to the party; Tarski as a leading logician and Witkacy as a well-known artist were included. Once there, Witkacy very quickly became angry at the nature of the situation and said, "Listen Alfred, either I smash somebody's snout or I have to go to another room and take cocaine"; Alfred's reply: "Take cocaine."[16]

A New Woman

While Tarski's professional life was at a peak, his personal and emotional life was suffering. He was lonely in the big house on Michigan Avenue. Dorothy Wolfe had moved out, and though Wanda Szmielew was visiting for the academic year 1959/60, she did not move in. Maria was living on Regent Street and he saw her regularly there, but the dinners and parties he loved to host were a thing of the past. Both of his children were long gone, and even though they had never been close, their absence added to his feeling of emptiness. No matter how deeply engaged he was in work with his many students, visitors, and colleagues, and no matter how much he traveled or how many conferences he was involved in organizing, something was missing. He needed at least one woman as confidante and admirer, if not lover. Ever the romantic, he needed passion, and he thought he had found it when Verena Huber-Dyson came to Berkeley for postdoctoral work in 1961.

Tarski remembered her well from their 1957 meeting in Cornell, when she had been Georg Kreisel's striking and intriguing companion. After the Cornell conference, Huber-Dyson spent most of the following year in Europe: first in Reading, England, with Kreisel and then by herself in Switzerland. Encouraged by Kreisel, in the summer of 1958 she and Freeman Dyson divorced and she accompanied Kreisel to Stanford University, where he was a visiting professor for the academic year 1958/59. At Stanford she met Kreisel's colleagues – Feferman, Tait, and Suppes – and their spouses. Kreisel took perverse pleasure in introducing her at social gatherings as "My wife, Mrs. Dyson," but in fact they were not married and at the end of that year they separated for good.[17] Then she began serious efforts to rebuild her career in mathematics. In 1959 she took a teaching position at San Jose State University in order to support herself and her oldest daughter, who was living with her in Palo Alto; her two younger children remained on the East Coast with their father. The following year, Huber-Dyson attended the LMPS conference at Stanford and there Dana Scott mentioned (with Tarski's approval, of course) that it might be possible to arrange a visiting research appointment at Berkeley for her. Once offered the position, she eagerly accepted. Her friend Roger Lyndon, a mathematician whose interests were in algebraic and logical aspects of group theory, knew Tarski well and assured her it was a wonderful decision; Tarski would be brimming with ideas for her.

Lyndon's recommendation was right on the mark. Settled in Berkeley, Verena knocked on Alfred's office door, told him of her interest in group theory, and asked what she should work on.

As soon as I uttered the word "group" he stood up and exclaimed, "I've got just the right problem for you." Actually there were two: the decision problem for the elementary theory of free groups and the question of whether any two non-commutative free groups are elementarily equivalent.

She had no idea then that he had suggested these very difficult problems to many others before her, including Robert Vaught, and that many years later people would still be struggling with them.

Huber-Dyson plunged headlong into the work. She went to courses and seminars and joined the "logic family" – the clan – for the postseminar dinners at Spenger's Fish Grotto, and almost before she turned around, she was dining tête-à-tête with Alfred at Trader Vic's, one of his favorite restaurants. By October he was inviting her to Michigan Avenue for a *digestif* in the cozy breakfast nook with the view of San Francisco Bay while they discussed the problems in group theory she was working on. After a few drinks, he encouraged her to tell him her life story. "I believe, in my natural reticence, I kept to a rather slim sketch of essential facts," she said. "Still, it became clear that he was courting me. I tried to curb his advances and restrict our encounters to mathematical concerns ... [but] it was never possible to have any serious professional discussion with Alfred any place other than in his house, and with a bottle of vodka fixed up with juniper berries from his garden. After I told him about my past, he started probing my views of morals and codes of behavior. I told him of my belief in personal freedom, which included free love."

Teasing, Alfred asked Verena if she meant it. "Aren't you too much of a well-bred bourgeois Swiss girl for that?" She answered forcefully, she thought: "My concept of freedom extends to the freedom *not* to do what one does not want to do." She tried to remain polite, which was not so easy given the particular argument.

Well-bred, graceful, and cosmopolitan, Verena had been married twice and had three children; she was certainly not innocent. But contrary to outward appearances, at thirty-eight she was as insecure, vulnerable, and confused as a much younger student might have been. By her own account, "I was quite up in the clouds, my Ph.D. and various experiences and adventures notwithstanding ... and I was still a bewildered and uncompromising idealist in search of the real thing in terms of mathematics."[18]

Alfred found her enormously appealing, personally charming, physically attractive, and mathematically talented but clearly in need of direction. Playing Pygmalion was his natural bent, a role he loved. He enjoyed teaching her how to approach a problem and how to write about it. He wanted her to be a success – his success – and he wanted her to love him.

Verena said no as politely as she could. She wanted Alfred as a teacher and a friend, not a lover. As she put it, "I wasn't interested in him *that* way" – in fact, there was someone else she was interested in – but for good reasons she did not want to terminate their relationship. As a mentor he was extremely important to her because, ten years after her first postdoctoral fellowship at the Institute for Advanced Study, she was still struggling to establish herself as a mathematician. She truly liked him very much and enjoyed his company, his brilliance in so many areas, and his stimulating advice. With unusual insight, she also perceived his vulnerability and *almost* loved him for that. She, who was tall, felt he was insecure because he was short. Though most others labeled him Napoleonic and described him as utterly self-confident, she saw him as "an outsider, like me, who never totally adapted to American culture. I recognized the European disposition in him and identified with it."

Tenacious in the face of her refusal, Alfred continued his seduction, pleading and pressing, saying "just once, no strings attached, no expectations if you don't like it" and Verena, running out of ripostes, gave in.

I took the easy way out, easier than arguing – or walking out the door to my car. Why did I not do that? Or did I try and fail? I do not remember. But I do remember that the next day I woke up with a Tarski-vodka hangover calling myself a fool for not heeding my misgivings, because in spite of all his protestations, I knew this was not going to remain a simple, casual, one-time event.... I realized I had done something that I had not wanted to do, something that was going to haunt me.[19]

At the very moment that Verena was ruing the turn their relationship had taken, Alfred wrote, "This is to tell you, Verena, that I enjoyed last night as thoroughly as a human being can enjoy anything in this life. Thank you."[20]

Searching for a way to extricate herself from this morass, Verena replied:

Dear Alfred,
 It is not easy for me to write this letter.... I have been feeling ill and miserable ever since yesterday morning. Please understand that this has nothing to do with you personally. I very much appreciate your kindness. But I was a great fool, I ought to know myself better.... I have done something to myself which I should not have done, and I feel it was not very decent towards you either. Please forgive me. And I hope this letter will not hurt you ....[21]

Their exchange of letters settled nothing, and they continued as before. She submitted her papers to him and he helped her rewrite them sentence by sentence. She compared him to her exacting mother – who used to sit with her when she did her high-school homework to be sure there were no mistakes in grammar or spelling – and even as she resented it, she appreciated Alfred's "fatherly" guidance. He liked being a tutor but he wanted a sexual relationship to go with it. Although Verena protested that she did not, at times she acquiesced, running out of ways to resist. Once, when she refused and he asked her to be specific as to why she found him physically unappealing, she dropped her guard and said that she didn't like his hands and didn't like him to touch her. Furious, he hit her.

Other times, when they argued, Alfred shook his fist and sputtered, "You make me so angry, I am going to make sure you will not be mentioned in my biography!" To herself she said, "Alfred, you just made sure that I *will* be."[22]

Needing to escape the constant pressure of Tarski's pleading, in the spring of 1963 Huber-Dyson went east for job interviews. When she returned there was an incident between them that Tarski referred to as "the stormy Monday," after which Verena wrote him a letter counseling "wisdom in place of passion."

Alfred answered with a three-page letter that began,

I have done a thorough, ruthless soul searching since I saw you .... I blame myself as bitterly as I can. How different everything would look if during those days I had mastered more inner strength, managed to still the storm which was raging in me and responded with sympathy, understanding and love to your desires.... There is something terrifying in the knowledge that the time process is not reversible.[23]

He vowed that he could provide what she needed and said, "Father's love does not exhaust what I feel for you, but it has been a most essential element of it from the very beginning. The realization that I can give this love to you ... and you are ready to take it has always been for me a source of deep satisfaction."

But about wisdom in place of passion, at the end of his letter Alfred said, "I reject your suggestion, emphatically. I do not wish that my ability to feel strongly, to love and to suffer, die out before my physical death. Period!... if anything, this one is certain: I love you, Verena." The letter ended with a postscript: "I hope the azalea and the roses were delivered at the proper time and that you have put the roses in the ground."

Verena wrote back saying she was moved by his letter, that he should not blame himself, and that by wisdom she did not mean resignation but rather "a recognition of the impossible as impossible." Without changing her position, she said she felt warmly toward him but that he made it difficult for her to offer the feelings she had because he always wanted more. The letter was intended to make Alfred feel better. It did, and he came back with a new angle.

While chastising her for "not having played the game right," Alfred proposed marriage, saying it would be a good career move. According to Verena, "he said that if I married him I could be his scientific helpmate, meet all the best mathematicians, teach as little or as much as I wished .... But I simply kept repeating that such a move was incompatible with my conscience because I did not love him 'that way'."[24]

As with Anne Davis, the question arises: Was the proposal Alfred's way of keeping Verena close to him? Although by this time Maria was living on Regent Street, they were still married and she remained an important part of his life. As Verena astutely observed, Maria kept intact the old world in which at core he was truly at home, notwithstanding his "I am an American" proclamations. To divorce, to break with her formally, would have severed the main artery to his past. The question remained moot because in August of 1963, partly to escape from Tarski and partly to be closer to Princeton where her younger children lived, Verena left Berkeley for a position at Adelphi University in New York. She remained ambivalent about her feelings toward Alfred for having, as she saw it, driven her from the San Francisco Bay area, a region she loved and where there were other job opportunities, including a job at San Francisco State University that he had arranged for her. In her view, even the other side of the bay was not far enough away. "I stayed aloof for a long time," she said, "afraid of the old pattern of persecution surging up again."

Several years after Maria returned to Michigan Avenue to live with Alfred again, an easier friendship developed. Verena visited Berkeley in the summers and became close to Maria and Alfred as a couple. Maria put her at ease, and she relaxed and felt at home in their joint company. Their common European background served as a bond that was of great importance to Verena. After changing jobs a few times, she eventually settled in Canada at Calgary University where she taught logic for many years. In 1982, when Alfred Tarski was awarded an honorary degree from that university, there is no question that Professor Verena Huber-Dyson's nomination and campaign for it was crucial in his having been selected.

INTERLUDE V

Model Theory and the 1963 Symposium

What Is It?

THE WORD 'model', in its usual scientific sense, is quite different from the sense in which it is used in logic. For example, the Newtonian model of the physics of moving bodies employs the concepts of mass and force, velocity and acceleration, and posits relationships between those concepts in the form of Newton's laws of mechanics as they apply to idealized bodies of matter. A more up-to-date example is provided by the Watson–Crick helix model of DNA and its role in cell replication and other biological processes. Some scientific models are very speculative, such as cosmological ones that describe the evolution of the universe from the Big Bang to its final collapse or, alternatively, its endless expansion, or models that hypothesize continuously bifurcating universes. Roughly speaking, then, scientific models are theories that represent, in idealized form, various kinds of objects in the world and their (possibly dynamic) interrelationships in order to explain actual phenomena.

For the logician, models are the realizations of axiom systems that are given by interpreting their basic notions in one way or another. One of the simplest examples of an axiom system is that for an ordering relation $x < y$. The axioms are *transitivity* (for any x, y, and z, if $x < y$ and $y < z$ then $x < z$), *asymmetry* (for any x and y, not both $x < y$ and $y < x$), and *comparability* (for any x and y, if $x \neq y$ then either $x < y$ or $y < x$). Examples of models of these axioms are given by the integers and by the real numbers, both with their usual ordering relations. The real numbers also form a model of the statement of *density* (for any x and y, if $x < y$ then there is a z with $x < z$ and $z < y$). That statement is false for the integers, since (for example) there is no z between 1 and 2. Considered as a possible axiom, the assumption of density is both *consistent with* and *independent of* the basic axioms for an ordering relation.

More complicated models arise in algebra when one considers such operations as addition and multiplication along with the ordering relation. For example, the statement that every nonzero element x has an inverse $y = 1/x$ (i.e., $x \cdot y = 1$) is true in the algebra of real numbers but false in the integers. On the other hand, the statement that if $x \neq 0$ then its square is positive (i.e., $0 < x \cdot x$) is true in both domains.

Geometry provides a wealth of examples of models that can often be more easily visualized than those for algebraic axiom systems. The basic undefined notions in one form of treating plane geometry are "point", "straight line", the incidence relation ("point p lies on line l"), and the length operation ("the length of the line segment from point p to point q"). The standard model for Euclidean plane geometry is given in terms of a coordinate system, and it takes "points" to be pairs (x, y) of real numbers and "lines" to be the sets of points satisfying a (nondegenerate) linear equation $ax + by = c$; length is calculated by Pythagoras' formula for the hypotenuse of a right triangle in terms of its sides. Non-standard models can be obtained most simply from the standard ones by replacing the real numbers (used for the coordinates of points) by some other algebraic number system.

More interesting are models of non-Euclidean plane geometry – that is, models providing interpretations of the notions of "point", "straight line", the incidence relation, and "length" in such a way as to make each (or most) of the postulates of Euclidean plane geometry true, except for the notorious fifth postulate, and to make that one false. (Euclid used the word "postulate" for what is now more commonly referred to as an axiom.) In the 1830s the Hungarian mathematician János Bolyai and the Russian Nikolai Lobachevsky independently explored the development of geometry with the fifth postulate negated: permitting more than one parallel to a given line through a given point not on that line. Their work made it plausible for that form of non-Euclidean geometry to be consistent. Later in the nineteenth century, various models of non-Euclidean geometry were produced; one of the most perspicuous is due to Eugenio Beltrami in 1866. Beltrami's interpretation makes use of a *disk*, by which is meant the set of all points inside but not on the circumference of a circle in the plane. By "point" in Beltrami's model is meant a point of the disk, and by "line" is meant a chord – that is, the intersection of a straight line in the usual sense with the disk. The incidence relation is just the relation of membership of a point to a line, as in the standard interpretation. Only the meaning of "length" has to be changed, so that if p and q are points on a line (chord) then the length of the line segment from p to q approaches infinity as one of the points, say p, is kept fixed while the other one, q, approaches the circumference of the disk. The formula for this, which is a bit complicated, was supplied by Felix Klein a couple of years after Beltrami's work. Looking at a picture of a line in this model and a point not on this line, it is easy to see that many lines can be drawn through the point which do not intersect the given line on the disk.

A model of another form of non-Euclidean geometry interprets "point" to be a point on a sphere and "straight line" to be a great circle (i.e., by definition, the intersection of the surface of a sphere with a plane passing through its center). In this model, any two "lines" intersect, so there are *no* parallels to a given "line" through a given "point"; this model also falsifies another of Euclid's postulates.

These examples suggest that there is extraordinarily wide latitude when interpreting the basic concepts of geometry so as to make various combinations of the usual axioms true and others false. To make such freedom of choice vivid, David Hilbert famously said that instead of points, lines, and planes one could consider an interpretation of geometry in terms of "tables, chairs and beer mugs."

The general definition that logicians give for the notion of model is rather technical; what follows here is a fairly nontechnical explanation. Required at the outset is a *formal language* L that has one or more basic *kinds* of *variables* intended to range over the different kinds of objects in an interpretation; there may be some *constant symbols* of specified kinds and some *operation symbols* for building terms of various kinds from variables and constants. Finally, L is required to have one or more *relation symbols* to express relations between terms of specified kinds, usually including the *equality relations* for each such. By an *interpretation* of L is meant a *structure* consisting of a domain of objects for each basic kind of variable as well as an interpretation of the relation, operation, and constant symbols of L by actual relations between, operations on, and members of the domain of objects of the appropriate kind; equality relation symbols are interpreted to be the corresponding identity relations. For example, if L is the language of plane geometry as discussed previously then there are three basic kinds of variables corresponding to points, lines, and numbers; the incidence relation is a relation between points and lines and the length operation applies to pairs of points to yield a number.

The *(well-formed) formulas* of L are generated from what are called *atomic formulas,* which express relations between terms of appropriate kinds by closing under various *logical operations* such as negation, conjunction, universal quantification, and so on; the logical operations chosen constitute part of the specification of L. By a *sentence* of L is meant a formula without free variables. By an *axiom system* in L is meant a set of sentences of L.

The basic relation between a sentence S of a formal language and a structure M that interprets L is that of S being true in M. The applicable concept of truth here is that of truth in a structure described toward the end of Interlude III as a modification of Tarski's definition of truth in his famous *Wahrheitsbegriff* paper of 1935. A structure is said to be a *model of an axiom system* if each of its axioms is true in the structure. For example, if the system consists of the axioms of Euclidean geometry, then a model of it is simply one (such as the "standard" model) in which all of the Euclidean axioms are true. A model for non-Euclidean geometry, on the other hand, is one in which Euclid's parallel postulate is not true and various of the other postulates *are* true. Model theory is the subject concerned with the question: For a given language L and axiom system in L, which structures M, if any, are models of the system? The qualification "if any" is needed because an *inconsistent* axiom system has *no models*; put in contrapositive terms, one way to show that an axiom system is consistent is to produce a model for it. This is why the Beltrami–Klein model established the consistency of the non-Euclidean geometry of Bolyai and Lobachevsky.

It is unfortunate common terminology in logic these days to use the word "theory" for an axiom system, so that one speaks of models of theories. The primary examples of structures dealt with in model theory come from mathematics, and in that respect, *model theory* is a part of metamathematics; it is an *informal* mathematical theory whose subject matter is *formal theories and their models.*

Tarski and Model Theory

Tarski did not create this field, but as Robert Vaught later wrote in a survey of it for the occasion of his seventieth birthday celebration, his influence was decisive.[1] Tarski laid the conceptual foundations of the subject in a series of papers beginning in 1930, of which the high point was his paper on the concept of truth for formalized languages. Prior to 1930, the main results of a general character were few and far between. The first was the Löwenheim–Skolem theorem, which states that if a countable set of sentences in a first-order language has a model then it has a countable model. Next, Kurt Gödel, in his 1929 doctoral dissertation, established the completeness of the axioms and rules for the *classical first-order predicate* (or *quantificational*) *calculus*. That is the language in which formulas are generated from atomic formulas by the propositional operations expressed by "and", "or", "not", and "if ... then ..." together with the quantifiers "all" and "some" – interpreted as applying to arbitrary elements of a structure. The *completeness theorem* states that if a sentence of this language is true in every possible model then it is derivable by the usual axioms and rules of first-order logic, including the classical law of the excluded middle. More generally: If a sentence holds in every model of a given theory then it is logically derivable from that theory.[2]

Before taking up its general conceptual foundations in the 1930s, Tarski had pursued applications of model theory in the seminar that he conducted at the University of Warsaw in the years 1926–1928. As described in Interlude II, the most important application coming out of that period was Tarski's proof of the completeness and decidability of the first-order theory for the algebra of real numbers; he drew the same consequences for Euclidean geometry using the translation of geometry into algebra via a Cartesian coordinate system (as in the standard models for geometry).

In the history of model theory, the year 1945 is a natural dividing point: that was when Tarski started building his school in logic at Berkeley and when the great expansion of the subject began. In the survey of model theory after 1945 prepared for Tarski's seventieth birthday celebration, C. C. Chang wrote:

I shall not make a point of mentioning Tarski's name at each place in this paper where his influence is either directly or indirectly present. This is because his influence in model theory is felt everywhere .... It suffices to say at the outset that the majority of the work done in model theory in this period [1945–1971] is either due to Tarski himself, or to his close colleagues, or to his students, or his students' students.[3]

The first high points were the long-delayed publication of Tarski's decision procedure for the algebra of real numbers in 1948; then Leon

Henkin's new, simplified proof of the completeness theorem for first-order logic a year later; and his extensions of that to languages with uncountably many symbols for applications to algebra in the early 1950s. (Unbeknownst to Henkin, such extensions had been dealt with before the war by the Russian logician Anatolii I. Mal'cev.)[4]

The name for this emerging field seems to have been first used in print in several of Tarski's articles of the mid-1950s entitled "Contributions to the Theory of Models."[5] There Tarski established another result that was to be paradigmatic for further work in the pure (i.e., general) part of the subject by others; he characterized those sentences which, if true in a structure, are true in every substructure – namely, as being exactly those sentences expressible in purely universal form. (This result was established independently by Jerzy Łoś.) A few years later, Tarski and Vaught defined and established basic properties of the relation of elementary equivalence between two structures, which holds when they make true exactly the same first-order sentences; this and an allied notion of elementary substructure were fundamental for all further work in the subject.[6] Other contributions to model theory were being made at that time by Tarski's students Chang, Feferman, Montague, and Vaught. Independently of the Berkeley school and working more or less in isolation, Abraham Robinson was pursuing both pure and applied model theory in a series of well-recognized publications.[7] From the early 1950s on the subject gathered increasing steam, so that by the time of the Cornell conference in 1957 it was one of the major subjects dealt with there.

Infinitary Model Theory

In that same year, Tarski together with Dana Scott took a significant step in the enlargement of the subject by permitting the operations of conjunction and disjunction to be applicable to infinitely many sentences at a time.[8] Following that, a main topic in Berkeley came to be the study of languages with infinitely long expressions, allowing infinitely long quantifier sequences in addition to infinitely long conjunctions and disjunctions. Fundamental contributions to the study of such languages and their model theory were made by Carol Karp, a student of Henkin's, and by Tarski's students William Hanf and Jerome Keisler.

An exceptionally fast-developing synergy took place between Hanf, Keisler, and Tarski at the turn of the 1960s while the former were still graduate students. Tarski had proposed to Hanf that he look into the problem of extending the *compactness theorem* for ordinary first-order logic to certain infinitary languages. The ordinary compactness theorem for first-order logic tells us that, if every finite subset of a set of sentences has a model, then the whole set has a model. This is a consequence of Gödel's general completeness theorem for first-order logic, the proof of which was carried out by metamathematical means – that is, it involved the syntactic notion of sentences of a formal language and the semantic notion of truth of a sentence in a model. In the mid-1950s, the versatile Polish logician Jerzy Łoś (later to visit Berkeley) introduced a novel construction allowing the combination of any set of structures into a single structure, called their *ultraproduct*; Łoś's main theorem for these showed how the determination of which first-order sentences are true in an ultraproduct is reduced to which sentences are true in its factors. Toward the end of that decade, Tarski together with Anne Morel and Dana Scott used the ultraproduct construction to give a neat mathematical proof of the compactness theorem for first-order logic; in so doing, they ushered in extensive applications of that construction in model theory.[9]

The compactness problem that Tarski had suggested to Hanf concerned languages L_m with infinitely long expressions associated with any prescribed infinite cardinal number m; these languages allow conjunctions and disjunctions of infinitely long sequences of formulas and infinitely long quantifier sequences, but only for those sequences of length less than m. When m is the least infinite cardinal number – that is, the cardinal number of countably infinite sets – this language is simply another version of first-order logic because all the conjunctions, disjunctions, and quantifier sequences involved are finite. The statement of the compactness theorem has a natural generalization to each language associated with an infinite cardinal number m, and by the compactness theorem for first-order logic, it is true for the smallest infinite cardinal for the reason just given. The question that Tarski raised was whether there is any *uncountable* cardinal number m such that the compactness theorem holds for the language L_m. Using metamathematical arguments, Hanf showed that if any such language has the compactness property then the cardinal number m must be far larger than all "accessible" cardinals dealt with in standard systems of set theory – and, indeed, far larger than many "inaccessible" cardinals, whose existence does not follow from those systems.[10] Keisler then showed that Hanf's result about the extended compactness problem and another previously open result about Boolean algebras could be established by purely mathematical means using ultraproducts. Tarski quickly realized that both the metamathematical and mathematical methods were applicable to a wide variety of open problems involving infinite cardinal numbers that he had previously considered in his two papers with Paul Erdös.

This development progressed so rapidly that Hanf, Keisler, and Tarski all spoke about their work at the Stanford Logic and Methodology conference of 1960. Capitalizing on the methods that had been opened up in this way, within a few years Keisler and Tarski moved on quickly to establish a whole slew of results sensitive to the size of very large infinite cardinal numbers in a wide range of mathematical areas: set theory, Boolean algebra, measure theory, group theory, and functional analysis. This culminated in a long joint paper in which the proofs were carried out by purely mathematical methods involving forms of the ultraproduct construction.[11] Of the relation between mathematical and metamathematical arguments, Keisler and Tarski wrote in the introduction to that paper:

> The results we have ... concerning large classes of inaccessible cardinals were originally obtained with the essential help of metamathematical (model-theoretical) methods. These methods still provide the intuitively and deductively simplest approach [to] the topic in its full generality. In our opinion this circumstance provides new and significant evidence of the power of metamathematics as a tool in purely mathematical research .... Nevertheless we have decided to undertake in this paper an exhaustive purely mathematical treatment of the whole topic avoiding any use of metamathematical notions and methods.

The reason given for their choice is that

> we have been motivated by the realization of the practical fact that the knowledge of metamathematics is not sufficiently widespread and may be defective among mathematicians who would otherwise be intensely interested in the topics discussed, and to a certain extent also by some (irrational) inclination toward puritanism in methods.[12]

The "puritanism" referred to is the idea that purely mathematical results should have purely mathematical proofs. This "(irrational) inclination" is omnipresent in much of Tarski's work, where he strove constantly for this kind of purity of method – even when prior metamathematical arguments had established their fecundity and comparative simplicity – hoping thereby to reach a wider mathematical audience beyond that of his own constituency among logicians.

The 1963 Symposium on the Theory of Models and the New Kids on the Block

Concurrent with the development of the model theory of infinitary languages, the subject flowered in other directions in the 1950s and early 1960s, among them: languages with generalized quantifiers (e.g., those that assert the existence of uncountably many objects having a given property); nonclassical logics (e.g., many-valued and modal logics); and applications extending beyond algebra to number theory, geometry, analysis, set theory, and even to certain empirical sciences. Of special interest was Abraham Robinson's development of non-standard models of analysis that used proper extensions of the real number system satisfying all its axioms; in these models, certain positive nonzero elements are smaller than all positive real numbers and thus may be regarded as infinitesimal quantities.[13] For the first time since Leibniz, one could now develop a rigorous theory of infinitesimals having the same properties, in a suitable sense, as the standard real numbers.

Thus the time was ripe in 1963 for a conference in Berkeley devoted entirely to model theory.[14] Spearheaded by Tarski and organized with the primary assistance of Leon Henkin, Dana Scott, and their recently appointed colleague John Addison, more than 150 participants from seventeen countries attended. A special effort was made to draw participants from behind the Iron Curtain, but the tense political standoff between East and West meant that few were able to come except for several Poles and one Czech. The conference itself was dedicated to Thoralf Skolem, one of the founders of the subject, who had died earlier that year at the age of seventy-six. Tarski's old friend and colleague Evert Beth was an invited speaker but could not come because of illness. Sadly, he died a year after the meeting at the age of fifty-six.

Among those who had made waves at the 1957 Cornell meeting, Georg Kreisel and Abraham Robinson were back as invited speakers – the former with the use of model theory in generalizations of recursive function theory, the latter with applications of model theory to analysis. In Berkeley, Robinson spoke about further advances in non-standard analysis making use of his powerful new techniques. But the greatest excitement at the meeting was generated by the lectures of two relative newcomers. One was the young genius Saul Kripke, who had made a splash with his publication four years earlier (at the age of nineteen) of a completeness proof of propositional modal logic, the logic of possibility and necessity. For that he introduced a model-theoretic representation of the idea of "possible worlds" in a form that later came to be called Kripke models and were applied extensively in a variety of modal, intuitionistic, and related logics. Kripke subsequently became noted as a philosopher as well as a logician.

The other newcomer was Paul J. Cohen, a mathematician at Stanford University who in the spring of 1963 settled two long-outstanding problems in the metamathematics of set theory that had been challenging logicians since 1939, when Gödel published his proofs of the consistency of the Axiom of Choice and the Continuum Hypothesis with the usually accepted axioms for set theory by use of a model of set theory in what is called the constructible sets. Cohen had made an early mark in mathematical analysis and subsequently pursued important problems in analysis and other areas. At Stanford, in contact with Kreisel and Feferman, his interests turned more specifically to logic and he proceeded to attack the problems left open by Gödel's set-theoretical work. By introducing a novel technique – called the method of *forcing* – for the construction of unusual models of set theory, Cohen succeeded in establishing the independence of the Axiom of Choice and the Continuum Hypothesis from the accepted axioms. His lecture at the Berkeley model theory conference, entitled "Independence Results in Set Theory" (which happened to be delivered on Independence Day, 4 July 1963), was one of the high points of the meeting.[15] Cohen's remarkable achievement was inevitably compared with the proof of independence of the parallel postulate from the other axioms of Euclidean geometry, which also proceeded by the construction of unusual models, and there was talk of an emerging non-Cantorian set theory analogous to the development of non-Euclidean geometry. Whatever the value of the comparison, it was justified in one sense by the immediate flood of independence results in set theory and related systems that were unleashed by Cohen's method of forcing.

Already at the Berkeley meeting, both Solomon Feferman and Robert Solovay were able to announce further results using extensions of Cohen's method of forcing. Solovay, with a recent Ph.D. in topology, would join the UC faculty of mathematics two years later; like Cohen he had wide interests but turned to concentrate on set theory, to which he soon made major contributions. As explained in Interlude I, one of his most important results was the construction of a model of the axioms of set theory with a weakened version of the Axiom of Choice in which, among other things, the Banach–Tarski decomposition theorem is false; this shows that the full Axiom of Choice is needed to derive the Banach–Tarski "paradox".[16] Later, Solovay and Scott also showed how to construct generalized models of set theory in which the "truth values" of statements are taken to be elements of a suitable Boolean algebra instead of just 1 (for true) and 0 (for false); these so-called Boolean-valued models provide an alternative method to that of forcing in order to obtain a wide variety of consistency and independence results in set theory.

In all respects, the Berkeley Model Theory conference was a great success. It made plain the enormous progress in the field since its emergence as a subject in its own right after 1945, and it set the stage for still greater progress to come – running from number theory and algebra to analysis and set theory, and from pure mathematics to computer science and linguistics. Tarski could feel well satisfied with having helped to inspire so much of the work in model theory and for bringing together so many of the people who were carrying it on.

12

Around the World

TARSKI'S TASTE FOR TRAVEL continued, unabated, until the very end of his life. He loved it partly for the pleasures of the exotic experience of new places and new people, partly as a missionary spreading his gospel of the centrality of logic, and partly to enhance his own image. He liked juggling the possibility of one trip against another and was not above choosing his destination for personal rather than professional reasons, even though almost every trip he took was connected to a conference or an independent invitation to lecture and was funded by a grant of some kind. In January 1963 he wrote to Paulette Février, who had invited him to Paris: "I wish I could go ... in the near future but the prospects are not too good [the Model Theory symposium in Berkeley was scheduled for June and July]. But I am thinking of going to Mexico by the end of August – officially to attend the International Philosophical Congress (but actually to see again Mexico City which I love) ... the scientific level of this undertaking will probably be very low, but who cares? My stay in Mexico a few years ago was a source of some of the strongest impressions in my life."

In a postscript he added his standard complaint, "I am working harder than ever." Yet busy and driven as he was, Tarski seemed to thrive on trips. Yes, they took precious time, but on the other hand his scientific mind was at work wherever he was; the flash of insight, the last piece of a puzzle could be found anywhere. He could shut out his surroundings and carry through an argument on a plane, on a train, on a walk, or, as he repeatedly said, "in the dentist's chair just as the tooth came out." Nor was jet lag a hindrance. He was accustomed to long hours without sleep and used his Kola Astier to keep alert if he found himself drowsy.

Ina Tarski, late 1950s.

Jan Tarski, 1956.

Jan and Ina

In the summer of 1964, in the course of a trip abroad to Warsaw and Jerusalem, a rupture occurred between Tarski and his son and a reconciliation took place with his daughter. Tarski and his children had never been at ease with one another. Once the brief honeymoon of their 1946 reunion in Berkeley was over, both of the Tarski children felt they rarely had their father's approval. As young children, as teenagers, and as college students, they tiptoed around his ego and generally avoided confrontation. But in 1964, in Warsaw, a fifteen-year-old wound was opened, and Jan revealed his pain and anger.

Jan, by then thirty years old with a Ph.D. in physics, had come from Moscow – where he had been a visitor at the Steklov Mathematical Institute – to Warsaw to visit family and friends. Coincidentally, Tarski was also there for a meeting. After receiving his Ph.D., Jan had spent two years as a visiting fellow at the Institute for Advanced Study in Princeton and two more years at research positions in New York and London. He was to continue his career as a visiting scientist and scholar in France, Germany, and Italy for the next twenty years, but never did he hold a permanent position in any one place. This unsettled existence displeased his parents, especially his father. Jan could hardly set foot in the house before there would be an argument about it; but what led to the explosion in Warsaw was personal, not professional, behavior.

Jan and Alfred had arranged to meet at the classic old Europejski Hotel and Jan, who was first to arrive, sat down to wait in the lobby. As his father came into the hotel, so did Wanda Szmielew, and Jan – aggrieved to see that Alfred had included her at the same time – did not greet her. "She walked up and I didn't budge out of my chair," Jan later said; "I had assumed that my father and I were going to be having a private 'family-type' meeting. The day before, Jerzy Łoś had asked if he might see my father and he had said it wasn't possible because we were to have a family meeting. Then Wanda appeared. I did not consider her to be a part of the family and I did not speak to her at all."[1]

Still carrying a grudge from 1949 when Wanda had lived in the house on Cragmont Avenue and created such turmoil, Jan chose this moment to openly slight his father's friend. As a teenager he had become aware of his father's infidelity and his mother's humiliation and now, fifteen years later, his hostile silence made it clear how he felt about his father's relationship with this brilliant, sensual woman. Tarski, not generally sensitive to the feelings of his children, may never have appreciated the depth of his son's anger, but in any case he completely lost his temper at what he considered to be unacceptable rudeness. That was the end of their meeting. He was so angry that two days later, when they chanced to meet in the airport departure lounge, he turned away and refused to speak to

Jan and maintained this silence for six more years, even on the rare occasions when his son came back to Berkeley. Finally, in 1970 they had a rapprochement of sorts at a meeting of the International Congress of Mathematicians at Nice, on the Côte d'Azur.

What was it like to be Tarski's son? Jan's answer was: "Well, hah! Not easy." Not easy to measure up to his expectations, to be subject to his constant criticism and his conviction that he was always right. It was not the same as being a graduate student and having "Papa Tarski" as your *Doktor-Vater* controlling your work; it was a lifelong relationship. Yet Tarski did wish his son to do well; he wanted to be proud of him and was disturbed at his failure to secure a tenured academic position. Jan, too, wanted that for himself but somehow it never happened. Since the job market at that time was generally favorable for a physicist with a Ph.D. from Berkeley, a two-year fellowship at the Institute for Advanced Study, and other postdoctoral appointments at important research institutes, perhaps something personal was at play. Jan was no happier about his situation than Tarski, but he was unable to change it. Whatever the reason, the effect was a continuation of their lifelong pattern in which he could never satisfy his father.

While father–son relations were at their worst, father and daughter were in rare accord. Ina and Andrzej Ehrenfeucht, who had been living together in Warsaw for several years, decided to get married officially, and Tarski was quite happy about that. "It was one of the few times my father was openly pleased by something I did," Ina recalled. "He said he thought it was a good choice." Indeed, Tarski liked Ehrenfeucht very much for his personal appeal as well as his intellectual capabilities; to top it off, he was the student of his first student.

For Ina and Andrzej to get married legally turned out to be farcically difficult. In Warsaw they were shunted from one agency to another, beginning with the American Embassy, which at first seemed most promising. They were ultimately told that it was impossible for them to marry in Poland, allegedly because the laws were being changed. Not unduly troubled, the couple decided they'd try another country.

Ehrenfeucht was an invited speaker at the International Congress of Logic and Methodology in Jerusalem in August 1964, and Tarski – hearing of the marriage plans and the difficulties they were having – offered to pay Ina's expenses so that she could come to Jerusalem, too. This was unusual generosity on Tarski's part, and perhaps he was making up for his lack of financial help when Ina first announced she was going to Poland to study. On their way to Israel, Ina and Andrzej stopped in Zurich thinking they would marry there, but Switzerland had declared a moratorium on all marriages between Americans and foreigners because too many American soldiers stationed in Germany were bringing their German fiancées to Switzerland for that purpose (such marriages were not permitted in Germany). Then it seemed to Ina and Andrzej that Jerusalem would be the ideal spot for the wedding, especially since Alfred and many friends and colleagues would be present. But in Israel they needed permission from the chief rabbi and they needed to belong to a congregation, which of course they did not. Inquiring how nonreligious Israelis managed to marry, they were told that Cyprus was the country of choice; but because a civil war was then raging in Cyprus, their visas were denied.

The young couple's travails became a much-discussed side issue during the Jerusalem conference, where many suggestions were offered for how they could tie the knot. They went to various embassies in Jerusalem: the Italians informed them they would have to visit a priest; France required three months' residence. Nothing worked; the conference ended and the attendees, including Tarski, left. Ina was ready to give up; they had already been living together, she said, so why was it important to have that piece of paper? But in fact it was important for Andrzej, who wished to work in the United States, and being married to an American citizen would make his green-card situation very much simpler. Finally at the British Embassy they were informed that after a short stay in England it would be possible to marry; thus they became a "legal couple" in London and everyone heaved a sigh of relief. Not long afterwards, the newly wed Ehrenfeuchts began their life in the United States. They came first to Berkeley and moved into Maria's house on Regent Street.[2]

∼ ∼ ∼

By 1965, after much courting, Alfred convinced Maria to move back to Michigan Avenue and re-establish their life together. Ina's marriage to Andrzej may have played a part in reuniting them, or perhaps Maria took pity on Alfred because he was lonesome. Haim Gaifman remembered Tarski calling him to say "Maria is back with me!" – he was very excited and happy.

For a time Ina and Andrzej continued to live in the Regent Street house, but after they left the house was sold. Andrzej taught at Stanford University for two years, then at the University of Southern California, and finally he and Ina moved permanently to the University of Colorado. They had three children – Anya, Renya, and Michael – and as a family they made regular visits to Berkeley. Tarski liked his grandchildren and liked to tell stories about how charming and clever they were, but he wanted them to be quiet when they were in his house. He thought they were cute, but according to Ina, "he never really related to children."[3]

Maria, on the other hand, fully enjoyed her grandchildren and at least once a year went to visit the Ehrenfeuchts in Boulder, often over Tarski's objection: he didn't like her to leave him alone because he was hopeless about attending to details in the house. When she went, she arranged for her sister Agnieszka (by then in the U.S.) to come cook and clean for Alfred. On one occasion, three days after she left, Alfred called Maria and told her "something smells in the refrigerator." Maria called her sister and asked her to clean it thoroughly but Alfred called again to say the smell was still there and that Maria had to come home. Resentfully, she left Boulder much earlier than planned, returned to Michigan Avenue, and called Agnieszka. Then the three of them enacted the following scene: Alfred was seated on a chair in the middle of the kitchen. Maria and Agnieszka took one item after another out of the refrigerator, offering each item for him to smell. He said, "No, it's not that, no not this" until finally a tiny piece of old cheese was discovered at the very bottom. Both Alfred and Maria told this story to Verena Huber-Dyson. His emphasis was on the fact that Maria left him alone to fend for himself; her version was how outrageous it was of him to ask her to return home.

Honors

In the mid-sixties, coincident with the trips connected to conferences and invitations to lecture, a number of high honors were awarded to Tarski. In 1965 he was elected to membership in the National Academy of Sciences of the USA. In the same year, he was named as a foreign member of the Royal Netherlands Academy of Sciences. And, in 1966, Tarski was named Corresponding Fellow of the British Academy. It is usually the case that many fewer foreign or "corresponding" members are named to such an academy than ordinary members from among its nationals, so the distinction bestowed is that much greater.

Also in 1966, Tarski received the Millennium Award from the Alfred Jurzykowski Foundation in New York. The "Millennium" refers to the conversion to Christianity of the nascent Polish kingdom under its first dynasty in 966. This award was indeed a high honor in Polish-American circles – and more than a touch ironic in view of Alfred's Jewish origins, his own conversion to Catholicism, and his declared atheism. The donor, Alfred Jurzykowski, a very successful Polish-American businessman, used his fortune to establish the foundation carrying his name and to provide major financial support to the Polish Institute of Arts and Sciences in America (PIASA). As described in the history of PIASA,

> the Institute was established in the midst of World War II when Poland was under the Nazi occupation and the Polish nation was subject to enslavement and genocide. As it became clear that the Nazis intended to destroy the Polish national culture, several outstanding Polish scholars who managed to escape to the West, decided to create an institution that would ensure the continuity of Poland's cultural development on free American soil.[4]

Besides Tarski, the Polish Institute counts among its past and present prominent members Czesław Miłosz, Jerzy Neyman, and Stanisław Ulam.

In England Tarski received yet another high honor when he was invited, for a second time, to give the Shearman lectures at the University of London. This series of talks were endowed in 1938 in memory of A. T. Shearman, a lecturer in philosophy at University College who had worked in logic early in the twentieth century. The terms of the bequest were that the lectures be in the general field of methodology or symbolic logic, but in practice they have ranged more widely through logic, the philosophy of mathematics, the philosophy of science, linguistics, and even farther afield. Given every two or three years, the series started with Bertrand Russell in 1945; Tarski's first turn was in 1950, and by the time of his second round in 1966 the list of distinguished speakers included L. E. J. Brouwer, Erwin Schrödinger, Karl Popper, Alonzo Church, Abraham Robinson, and Noam Chomsky. Tarski was the only one honored with a second invitation.[5]

The *Herrenvolk* Argument

Tarski stayed in London for six weeks to give not only the Shearman lectures but also several talks at Bedford College. The British philosophers who heard Tarski speak and met him personally at parties given in his honor retained vivid memories of Tarski's presence and his engagement on many topics besides logic. At a dinner party hosted by Richard Wollheim, chair of the philosophy department at London College, Hidé Ishiguro – a lively young Japanese philosopher who was living and teaching in London – had a memorable encounter with him. She was exactly the kind of woman Tarski found exciting to meet and in the course of the evening he inevitably gravitated toward her. Thirty years later, Ishiguro was still perplexed when she recalled their conversation, which began with a discussion of Japanese movies:

Tarski told me how much he liked the film *The Forty-Seven Ronin* in which a lord is tricked by his political enemy, loses his temper and draws his sword at the Shogun's palace and is therefore ordered to commit suicide. That was the legal punishment for drawing a sword in the palace. The lord's forty-seven retainers [ronin] avenge their master's death by killing his enemy but then *they* are required to commit suicide themselves. Tarski considered this admirable behavior.

I said to Tarski that people who want to be faithful servants and heroes can behave as they wish but shouldn't expect others to respect them if they kill someone.

Tarski said he always thought that both Poles and Japanese knew what it was to live with an arrow in their heart.

I said, "I would prefer a soldier Schweik to a hero who acts with an arrow in his heart, believing himself to be noble."[6]

Tarski said "I am amazed to talk to a Japanese in London who defends the shrewd Schweik," to which I said, "At least Schweik doesn't believe that he or anyone else is a hero." And Tarski replied: "The difference between the Poles and the Czechs is that the Czechs lost all their *Herrenvolk* in the battle of the White Mountain and that [afterwards] they were non-heroic Mediterraneans."

I had never heard anyone utter the word *Herrenvolk* in my life; I had only read about the expression as used by the Nazis to mean the master race. As I knew that Tarski had to leave Warsaw because of the Nazis, I was dismayed by his Polish nationalistic tone, so I said, "Julius Caesar was a Mediterranean and he seemed aggressive enough." I was ashamed later to have become angry at a guest who had given a most clear and wonderful lecture and was friendly to us all. But I was surprised at someone who was a victim of the Nazis praising the character of the *Herrenvolk*. I am a great admirer of his work but I was puzzled and shocked.[7]

While reporting the contretemps, Ishiguro was worried that it "may not show a nice side of Tarski"; but nice or not, the sparring was characteristic. However, in using the term *Herrenvolk* it is most unlikely that he would have been praising the Nazis. Given the context, he was probably thinking of an earlier use of the word. In the battle of the White Mountain (1620) to which he refers, the brave Protestant nobles – the *Herrenvolk* – of Bohemia were crushed by the Hapsburg forces and lost their estates and their rights, and Catholicism was reimposed upon the region. Nonetheless he did mean to compare the modern Czechs unfavorably with the brave Poles with whom he naturally identified.

Hans Sluga, then an assistant professor teaching logic and working on the philosophy of logic at University College, was frequently Tarski's guide, troubleshooter, and companion during the latter's London stay; he also recalled many detailed discussions with Tarski on the same subject.[8] "To my stunned surprise, I realized he was a rabid Polish nationalist who had all the prejudices that Poles have about Czechs," Sluga said. "He thought they were political cowards and [worse yet] totally unoriginal thinkers. When I asked, 'What about Czech movies?' he said, scornfully, 'Czech movies?' 'What about Kafka?' I asked. He said, 'Kafka, who is he? He is a German Jew.' In other words, nothing and nobody Czech was good."

As it happened, Sluga's family came from a part of Silesia that had been ceded by Germany to Poland after World War I, but their identification was always German; another part of Silesia was absorbed into Czechoslovakia. Tarski was very interested in that complicated history and wanted to know all the details. The Polish connection created a special bond between them in addition to their common logical and philosophical interests. Sluga reacted to Tarski through his own personal lens. He found him "Napoleonic, like my father who even looked like Tarski and was about the same height. He was sturdy and strong – a vibrant personality. It was a bit of a shock to me," Sluga said, "to have to go through all that again. I grew up knowing that these little men often succeeded by being very assertive because my father was like that." Sluga's reaction was in striking contrast to Jerome Keisler's being reminded, upon first meeting Tarski, of his own gentle Jewish grandfather.

During the six London weeks, Sluga spent many days and nights with Tarski. He borrowed books for him from the library and arranged matters at his hotel; they went to dinner together and went out drinking, and he was asked to provide other entertainment. "He was always interested in young ladies," Sluga said. "This was a major obsession of his. Where could he find one? Not that we ever found any appropriate young ladies, so nothing came of it, as far as I know." But Tarski did not limit his requests to Sluga. He asked others. According to the philosopher David Miller, who was Karl Popper's research assistant at the time, "Tarski asked Imre Lakatos, another colleague, if he could find him a female dining companion. Lakatos suggested he might like to take his current girlfriend to dinner. She was happy to have dinner with Tarski but for reasons which are obvious needed somewhere to take him afterwards." The "somewhere" turned out to be a party given by Miller at which Tarski "sat in an armchair and held forth on the topic of logic and insanity. He sensed a strong correlation between the two and cited the usual examples of logicians who had had psychiatric problems of one sort or another. At the end of his survey, we all had the feeling that ... there was at least one distinguished logician who was perfectly sane."[9] This was not the first or the last time Tarski would make this point about himself.

In London, Tarski gave two lectures – "Truth and Proof" and "What Are Logical Notions?" – which were to become famous and which he was to repeat with variations many times thereafter. On this occasion, after the first lecture a woman in the audience rose during the question period and asked: "That is all very well, Professor Tarski, but what has logic done to solve the problem of the Vietnam War?" Tarski was silent. David Miller said it was the only time he ever saw Tarski rattled; he had no adequate response. Fortunately, "quick-witted Bernard Williams, who chaired the lecture, answered the woman's question without really answering it."[10]

"Truth and Proof" in the *Scientific American*

The posters announcing Tarski's Shearman lectures in London remained in place long after the lectures were over. In 1968 Dennis Flanagan, then editor of the *Scientific American,* was visiting University College and saw the announcement for Tarski's lecture on "Truth and Proof." Thinking this would make an excellent article, he wrote Tarski asking if he could publish the lecture and offered the standard fee of $500. Tarski agreed and wrote up the lecture in a style he thought suitable for that magazine.

Joseph Wisnovsky, who coincidentally had been an English major at UC Berkeley, had his first job at the *Scientific American* just then, and he was assigned to edit Tarski's paper. He had never met Tarski but he knew of his reputation. Usually the *Scientific American* articles are heavily edited; many are rewritten and some even ghostwritten, but Wisnovsky knew better than to tamper with Tarski's work and did not – except for his usage of 'which' and 'that'. It seemed to him that Tarski did a 180-degree reversal of these words, so he changed every 'which' to 'that' and every 'that' to 'which' and sent the proofs to Tarski, who changed everything back to the way it had been. Wisnovsky got out Fowler's *Dictionary of Modern English Usage,* the house bible, and called Tarski on the telephone. "I asked if I could read him the relevant passage on 'that' and 'which' and he said, 'yes'. It goes on for pages, but he listened very patiently until I finished. Then he said, 'Well, you see, *that* is Fowler. *I* am Tarski.' The minute he said that I caved in. I felt cut off at the knees and I gave up trying to make any changes at all."[11]

"Truth and Proof," the only article Tarski wrote for a popular journal, was published exactly as Tarski wrote it in the June 1969 issue of the *Scientific American*; many remember it as the first and perhaps the only thing of Tarski's that they read.[12] It has since been translated into Polish, Russian, Rumanian, Spanish, and German.

"If *I* tell them to take you, they will take you"

On the professional level, Tarski was so impressed with Hans Sluga's acuity and breadth of knowledge that, at the end of his stay in London, he asked him if he would like to join the faculty in philosophy at the University of California in Berkeley. Sluga, enormously pleased but incredulous at Tarski's invitation, asked: "You mean it could happen – just like that?"

"If *I* tell them to take you, they will take you!" Tarski answered. Perhaps it wasn't quite so simple, but it was that attitude that inspired some to say that, in the matter of academic appointments, "Tarski behaved like a gangster."[13]

Actually, Tarski had to do some serious campaigning on Sluga's behalf. He began by convincing his colleagues Benson Mates and William Craig of Sluga's potential value to their department and more specifically to the logic and methodology program. These men argued so persuasively in Sluga's favor that it was assumed they knew him and his work well. After a preliminary visit in 1968/69, Sluga was invited to join the department and he did so, two years later, after satisfying his commitments at the University of London. He became a permanent member of the philosophy department in 1971 and an active teacher in the logic and methodology program, but to Tarski's ultimate disappointment he became less interested in work on Frege and logic in general and more interested in the "continental philosophy" of Friedrich Nietzsche and Martin Heidegger. According to Sluga, "Tarski thought that was a terrible mistake. He thought I had lost my way and he let me know I had disappointed him. He thought he had hired someone who was going to do logic and now I was wasting my time and not fullfilling my obligation to him by working on this other stuff. I thought that wasn't quite right of him. After all, it was part of my development." In fact, Sluga also continued to be interested in the intersection between logic and the work of Frege, Russell, and Wittgenstein.[14]

During Sluga's first visit to Berkeley, Karl-Heinz Diener was Tarski's research assistant; he later recalled how amused and delighted Tarski was to have two assistants, both named "servant" – *Diener* in German and *Sluga* in Polish. As far as Sluga was concerned, those days were over when he turned to Nietzsche and Heidegger.[15]

Trip to the Soviet Union

Fired by his success in London in 1966 and not at all fatigued, Tarski attended conferences in Paris and Hanover and returned to the United States briefly; then, with Maria – with whom he was now definitely reunited – he went to Moscow in August to attend the International Congress of Mathematicians (ICM). Meetings of the ICM had been held every four years since the end of the nineteenth century, except for a hiatus during the Second World War. They had taken place in various major cities of Western Europe and North America, so the choice of Moscow for 1966 was geographically unusual. But it was the background and political circumstances of the meeting in the Soviet Union that made it truly extraordinary. Long before the advent of the Soviets, Russia had a tradition of mathematical research going back to the establishment of the St. Petersburg Academy of Sciences by Peter the Great in the 1700s. At first, mathematicians from Europe were appointed to the academy, including the famous Swiss Daniel Bernoulli and Leonhard Euler. Russian mathematicians came into their own in the nineteenth century, and their contributions soon became internationally recognized. The tradition of mathematical excellence was continued after the Russian Revolution as part of a general recognition by the Soviets of the importance of the sciences and also in their drive toward modernization. Soviet mathematicians worked and were up to date in all the leading areas of the subject and had considerable exchange with colleagues in the West until the Second World War. Although science and mathematics continued to be vigorously pursued after the war as soon as conditions permitted, the atmosphere changed with the onset of the Cold War. In 1950, the secretary of the ICM meeting in Cambridge, Massachusetts, wrote that "mathematicians from behind the Iron Curtain were uniformly prevented from attending the meeting." The Soviet explanation, sent by cable from S. Vavilov, president of the USSR Academy of Sciences, was: "Soviet mathematicians being very much occupied with their regular work unable to attend Congress."[16]

After Stalin's death in 1953 there was some relaxation in Soviet policies, which gave rise to hopes for a rapprochement with the West. A few of the top Soviet mathematicians were allowed to attend the 1954 congress in Amsterdam and the next one in Edinburgh. Meanwhile, the Cold War continued in varying degrees under a succession of leaders, including the erratic Nikita Khrushchev from 1958 to 1964. The greatest moment of tension during that period was the 1962 standoff between John F. Kennedy and Khrushchev over the Soviet installation of missiles in Cuba, which brought the two countries within a hair's breadth of nuclear war. After that crisis was averted, Leonid Brezhnev took over the reins and peaceful coexistence with the West was again pursued as a foreign policy, although there were still major confrontations – the main one concerned the steadily increasing U.S. participation in the Vietnam War on the side of the South Vietnamese while the Soviets were providing support to the communists in the North.

It was in the general atmosphere of détente that the USSR succeeded in its bid to hold the 1966 meeting of the International Congress of Mathematicians in Moscow. Understandably, Moscow as the site of the meeting attracted a great deal of interest from mathematicians worldwide, eager to meet personally with their colleagues – who, for the most part, they only knew through their work – and to see with their own eyes what kind of society the communists had wrought. More than four thousand researchers attended, twice as many as at any previous meeting of the ICM; almost fifty countries were represented, and the number of special sections representing different areas of mathematics was also double that seen at any previous congress.[17] This reflected the remarkable advancement in mathematics worldwide following the end of the war, during which time the United States and the Soviet Union had emerged as leaders in both the number of mathematicians and the quality of research.

Mathematical logic had been pursued in the USSR since the 1920s, at first as a side topic by the noted probabilist Andrei N. Kolmogorov. Work in the subject flourished especially after the war when its senior leaders were Anatolii I. Mal'cev[18] in Novosibirsk, Andrei A. Markov in Leningrad and Moscow, and Petr S. Novikov in Moscow, all of whom had interests that spanned the field. Also, as in the West, a great new crop of younger logicians had started coming up in the 1950s, especially in these three principal centers, and the members of that generation, East and West, were eager to meet each other. At the event, there was an outpouring of good feeling and comradeship, and people got along famously.

Though logic as a part of mathematics in the USSR prospered, in philosophy its position was not secure when it seemed to conflict with Marxist ideology. When Tarski's *Introduction to Logic* was translated into Russian in 1948,[19] the preface criticized Tarski for holding "a confused idealistic viewpoint" and for being allied with the Vienna Circle, whose leader, Rudolf Carnap, was labeled "a reactionary bourgeois philosopher" (ironic, because in the United States he was accused of being a communist sympathizer). In the appendix Tarski was criticized further for adhering philosophically to the school of logical positivism, close to which are "such double-dyed reactionaries as Russell, Whitehead and Dewey."

In response to Tarski's statement in the 1941 English edition that "the future of logic, as well as of all theoretical science, depends essentially upon normalizing the political and social relations of mankind, and thus upon a factor which is beyond the control of professional scholars," Sofia Yanovskaya, author of the preface, wrote: "The attempts of reactionaries in the USA to make the progress of science serve the goals of imperialist aggression provide new evidence that an intensive struggle is inevitable in and around science, and that a scientist cannot just be an observer without risking becoming an accomplice of imperialism and reaction."[20] Yanovskaya was reportedly under great pressure during those times to pen such criticisms, with obligatory references to the writings of Lenin and Stalin.[21] But all these matters were treated as a thing of the past by the time Tarski came to the USSR in 1966, and though his book had been translated without his permission (this was standard Soviet practice at the time) he took pleasure in its being one of the first of many translations into other languages that followed its initial publication.

The opening ceremonies of the Moscow ICM took place on the afternoon of August 16 in the enormous Palace of Congresses built by the Soviets next to the Kremlin. Four Fields Medalists (for mathematicians, the Fields Medal is the equivalent of the Nobel Prize – without the money) were announced and their work described. Among them were Paul Cohen of Stanford University for his proofs of the independence of the Axiom of Choice and the Continuum Hypothesis, and the Berkeley topologist Stephen Smale. Smale had arrived late for the Congress and had not had a chance to register before coming to the opening ceremonies. Lacking proper identification, the door guards prevented him from entering the hall in spite of his protestations: "But, I'm supposed to be on stage receiving a medal!" Smale, a leader of the anti–Vietnam War movement in Berkeley, later created another sensation when he held a press conference on the steps of Moscow University criticizing both the United States and the USSR, the former for its involvement in the war and the latter for its lack of the basic means of protest.[22]

The schedule of the regular sessions that were held at the university, located in the Moscow hills, was divided between invited addresses and fifteen-minute contributed lectures; the latter took place in many simultaneous sessions. Logic was represented in one-hour lectures by Anatolii Mal'cev and the German proof theorist Kurt Schütte, and in the half-hour lectures by Paul Cohen, Yuri Ershov, Nikolai Shanin, and Robert Vaught; Ershov was the young shining light of the Mal'cev Novosibirsk school of model theory and algebra, and Shanin was vigorously carrying on Markov's approach to constructive mathematics in Leningrad. Tarski was not invited to talk because he had served as chair of the panel to select speakers in logic. Instead, like thousands of others in attendance, he was satisfied to listen, to meet the Soviet logicians, most for the first time, and (with Maria) to take in as much of the attractions of Moscow and its environs as the demanding schedule of meetings would allow.

Tarski had mixed feelings about being in the Soviet Union. He hated communism, and while the worst of Soviet practices had ended with the death of Stalin, the USSR was still a severely repressive regime, both internally and in its control of the satellite nations behind the Iron Curtain, including his beloved Poland. But he responded positively to individual Russians, who were warm and expansive. His meeting with Mal'cev, another pioneer in model theory, was particularly significant; at Mal'cev's invitation, arrangements were made for the Tarskis to visit Novosibirsk after the congress. In these individual face-to-face meetings, Tarski relished speaking Russian, which he had learned in school, and most Russians were surprised at his proficiency. In an odd way he felt at home or at least familiar with the culture. He was very pleased to tell everyone about the day a table had been reserved in his name at the hotel restaurant. When he arrived, the table was pointed out to him and he seated himself. Eventually (service was very slow) a waiter came over and Tarski began to order in Russian, at which point the waiter stopped him and asked him to leave, saying that the table was reserved for an American professor.[23]

Personal contact was enhanced further when a number of American logicians were invited by their Soviet friends to a dinner party at the exclusive Praha Restaurant. Among the Russians present were Mal'cev, Markov, Shanin, and Ershov; the American contingent included Tarski, Kleene, Curry, Chang, Feferman, Vaught, and their wives. The invitation was last-minute, which created difficulty in gaining entry for those who arrived unaccompanied by a Russian. Nevertheless, the dinner party itself was elaborate if not exactly formal. It took place in a large ballroom that had been divided in half by a drawn curtain. On the other side of the curtain (this one was not "iron"), a wedding party was in full swing with much drinking, loud singing, and dancing. By necessity this raised the noise level on the logic side, and the endless vodka-enhanced toasts that were shouted to logic, to brotherhood, and to international friendship and unity took on the spirit of the nuptial celebration.

One last extraordinary banquet on the closing day of the congress was like a Brueghel painting in modern dress. Not a sit-down dinner but a boisterous "stand-up" affair in the vast modern hall of the Kremlin, hundreds of large square tables groaned under platters of caviar, brown bread, rye bread, butter, smoked salmon, blinis, sour cream, piroshki, and endless supplies of vodka. Partaking of this bounty were thousands of voracious mathematicians from everywhere, moving from table to table to eat, drink, embrace friends old and new – who, in that atmosphere, felt like intimates. It was one of those moving moments of communal camaraderie and farewell that promises more than anyone can really hope for, but the promises are made all the same.

Soviet Georgia and Siberia

In the course of the ICM meeting, Leo Esakia, a young logician from Tbilisi, the capital of Soviet Georgia, gave a talk on topological interpretations of modal logic, a subject well known to Tarski because he and J. C. C. McKinsey had done pioneering work in that area in the 1940s. In his presentation, Esakia referred to their results several times, but not until he finished speaking did he learn that Tarski was in the audience. "This is the first time that I am in the presence of a living classic," he said with embarrassed pleasure when they were introduced. Soon afterward, Esakia said, "I dared to suggest that Alfred and Maria Tarski visit Georgia immediately after the congress."[24] Surprised that they accepted the invitation forthwith and even more surprised that the schedule for the improvised trip was arranged without the usual endless bureaucratic delays, Esakia and the Tarskis flew to Tbilisi. On arrival they were greeted by a delegation of the leading logicians and philosophers, whose warm embraces included male-to-male kisses, Georgian style, on both cheeks. "Later," Esakia said, "Tarski whispered to me that in America this might lead to far-reaching conclusions." According to Esakia, "Tarski was happy during his week-long stay. He hiked with ease and pleasure, was interested in everything, joked and took pleasure in drinking our Georgian wine and our cognac. We visited the old town, the history museum and Mtskehta, the ancient capital of Georgia. Pani Maria impressed me (and not only me) by her tactfulness, wisdom and warmth." At the university, Tarski gave a major lecture in Russian on the foundations of geometry that happened to coincide with the jubilee celebration of Tbilisi University, and for the local logic seminar he gave two informal talks based upon his Shearman lectures.

In addition, Tarski handed out reports and reprints of the work done by him and others in the Berkeley logic group as well as copies of his own bibliography. He described and, in effect, bragged about his excellent students, and he told about how closely he worked with them in long sessions at his house at night; according to Esakia, "he recommended us to try the doping pills which they take during those intense work periods to be in good scientific form." Tarski was, as usual, exaggerating a bit. Although he was a regular user of Kola Astier and other "uppers" and did urge others to try them, he made few if any converts among his students.

During one of their meetings, the Georgians asked Tarski for permission to translate his *Introduction to Logic* into the Georgian language, and he agreed. They presented him with copies of the ancient and modern Georgian alphabet, which pleased him greatly; he especially liked the first letter of the ancient alphabet. Five years later, the Georgian Misha Bezhanishvili ceremoniously delivered the Georgian edition to Tarski at the Logic, Methodology and Philosophy of Science meeting in Bucharest. Because of the exotic language and the beauty of the script, this was one of the translations of his book that pleased Tarski the most. The Georgians had also wanted to include translations of Tarski's Shearman lectures, but that idea was vetoed by the authorities in Moscow who said it was forbidden to publish in Georgian what had not previously been published in Russian.

The Tarskis had such a good time in Georgia that toward the end of their visit, in a warm burst of affection, Alfred surprised his hosts by proposing to return for his sabbatical year 1970/71. It's not clear how serious he was about this or whether he said it impulsively, knowing he could change his mind. The astonished Georgians, however, took it at face value and an official follow-up proposal, written by Esakia, was approved by the Georgian Academy of Sciences. The president of the academy himself brought the proposal to Moscow, where it died a quick death. Later Esakia said, "At that time we did not think (or wish to think) how insolent this looked in Moscow: that at some provincial place they decided to invite some American, a citizen of a hostile country. This would have been without precedent even in the capital and no détente can help here!"[25]

Novosibirsk, 2000 miles east of Moscow, was the next stop on the Tarskis' journey through the Soviet Union. The city was built in the late nineteenth century at a junction of the Trans-Siberian Railroad with the Ob River. Near the great natural resources of the region, it became a transportation hub and grew into the leading industrial center of Siberia. In the late 1950s, taking advantage of the location, the Soviets created a major new research and teaching center in nearby Akademgorodok that housed the Siberian branch of the Academy of Sciences and Novosibirsk State University.

Tarski had been invited to Novosibirsk by the Mathematics Institute of the Siberian Academy of Sciences, whose section in logic and algebra was headed by the expansive Anatolii Ivanovich Mal'cev, a huge bear of a man. A student of Kolmogorov's in Moscow during the 1930s, Mal'cev had done pioneering work on the applications of logic to algebra. His 1941 paper (in Russian) made use of a compactness theorem for first-order languages allowing uncountably many symbols, which anticipated much that was done after the war by Leon Henkin and Abraham Robinson. This work was largely unknown in the West until the 1950s, and only in 1971 would Mal'cev's papers become widely available with the publication of a volume of English translations by Tarski's student, Benjamin ("Pete") Wells.[26]

Given the lively interest by Western logicians and mathematicians in what was being done in the Soviet Union and vice versa, timely translation was an important issue. Although English was already becoming the scientific lingua franca and most Russian academics had studied it, journals in English were not readily accessible except in Moscow and Leningrad. Moreover, it was books that were being translated from English into Russian, not the journal articles that represented the cutting edge of research. In the other direction, some regular programs of translating both books and journal articles from Russian into English were begun in the 1950s, but the choices of what to translate depended on subjective estimates of importance and were to some extent hit or miss. In particular, the journal *Algebra i Logika* of reports from Mal'cev's seminar was not regularly translated into English until 1968, two years after the ICM meeting. Another public source of information about research was through journals devoted to reviews, but these were often a few years behind the times. The upshot of all this was that there was a high level of competition on all fronts between the Soviets and the West as to who would obtain significant

Tarski in Novosibirsk, 1966. Anatolii I. Mal'cev (to the right of Tarski), Yuri Ershov (to the left of Tarski), Boris Trakhtenbrot (with camera, first row), and Mikhail Taitslin (second from the right, first row).

results first. Given the vagaries and time delays of translations, there were a number of arguments on both sides about priority. A prominent example was the characterization of the first-order properties of p-adic fields published just a year before the Moscow ICM by James Ax jointly with Simon Kochen in the United States and independently by Yuri Ershov in the USSR. This was considered by many to be the most significant advance in the model theory of fields since Tarski's characterization of the first-order properties of the field of real numbers. Abraham Robinson called it the most spectacular application to date of logic to algebra. Inevitably, the question of who did it first was raised, and no answer ever settled the question to everyone's satisfaction.

As befitted his eminence, Tarski was treated royally in Novosibirsk. He gave several lectures in Russian and had private conversations with members of Mal'cev's seminar. Mal'cev gave a party for Tarski at his home to which all the members of the seminar were invited *except* Boris Trakhtenbrot and Mikhail Taitslin. Their absence was surprising since Trakhtenbrot, well known in the West as a leader in logic and theoretical computer science, was on the faculty of the Mathematical Institute and a professor at the university while Taitslin, who held a junior position, had been a Ph.D. student of Mal'cev's. Tarski didn't know what to make of this until he became aware that they were the only members of the seminar who were Jewish.[27]

Denied the opportunities for personal contact that the others had been given, Trakhtenbrot invited the Tarskis and the Taitslins for a social evening in his apartment. This was done at some risk, since ordinarily Soviet citizens were not permitted to invite foreigners to their residence without official approval, which in this case was not sought. As a guest of the Academy of Sciences, Tarski was put in an awkward position, but he decided to accept. In spite of his fluctuating notion of his own identity as a Jew or a Pole, the fact that colleagues were excluded from his company because they were Jewish was very upsetting – especially as it echoed what had happened to *him* in Poland thirty years earlier. He did not feel he could make an issue of this instance of anti-Semitism at the time, but he would have to deal with accusations of more egregious anti-Semitism on the part of leading Soviet logicians in the future.

Mal'cev died at the age of fifty-eight, just a year after Tarski's visit. Following his death, Yuri Ershov – then considered the most brilliant and enterprising young logician in the Soviet Union – took over the seminar in algebra and logic; he became director of the Mathematical Institute and eventually rose to the position of rector of the university. During this period, Ershov, along with others in the mathematical establishment, was accused of participating in anti-Semitic practices, though he steadfastly denied such accusations. These matters would come to a head during a visit that Ershov made to the United States in 1980, and then Tarski would have to confront the issue head-on.

Around the World

The idea of making a trip around the world had great appeal for Tarski, and once he knew that he would attend the meeting in Moscow he decided to include Japan in his itinerary. "I am very happy," he told his Soviet friends, "because I will travel home against the rotation of the earth and gain a day!" He had informed his Japanese colleagues of his plans, and timely arrangements were made for him to lecture in Tokyo and Nagoya. Also, on an impulse Tarski decided to squeeze in a brief stay in India, so Calcutta, Delhi, and Katmandu were added to his list of cities.

However, in India the last-minute efforts made by his student Haragauri Gupta failed to produce an invitation to lecture, and the Tarskis were without a personal host. This undoubtedly colored his view because he was accustomed to being treated as a celebrity and being fussed over by attentive academics. Poverty in the midst of grandeur saddened him, he later told Gupta, and the sight of undernourished children and beggars in the streets of Delhi and Calcutta depressed him. His impression of Katmandu was more favorable. Children seemed happier and healthier, and undoubtedly being in the vicinity of the Himalayas elevated his spirits.[28]

In Japan a few days later, Tarski did get the attention he expected. His host was Katuzi Ono, professor of mathematics at the University of Nagoya and later president of Shizuoka University for several years. In the 1930s, Ono was one of the fathers of mathematical logic in Japan; in addition to his work in that field he contributed to applied mathematics. His son Hiroakira Ono, then a student and later to become a noted logician as well, recalled helping his father lead Tarski on a sight-seeing tour of Tokyo, including the Imperial Palace. Of his lecture the next day at the department of mathematics, the younger Ono said, "What impressed me most was Tarski's energetic way of talking and his chain-smoking."[29]

13

Los Angeles and Berkeley

The Flint Professor and His Student Judith Ng

IN THE WINTER OF 1967, as part of a year's sabbatical leave from Berkeley, Tarski taught for a semester in the philosophy department of the University of California at Los Angeles and was honored with the title, Flint Professor. As such, he gave a public lecture as well as a colloquium, taught a course on undecidable theories, and participated in seminars. His student Judith Ng went to UCLA with him in order to continue her work. Ng, a warm, empathetic young woman, had been a Tarski graduate student since the early 1960s and was well on her way to becoming something like a surrogate daughter – not that she didn't have a family of her own. Only in retrospect did Ng understand the benefits and drawbacks of her situation:

For better or for worse he was a father figure for me. If I had it to do over again I would *never* have chosen him for a thesis advisor. I would choose someone I could keep at arms length. Even then, I had been thinking of asking Robert Vaught to be my advisor but I waited because at the time he had many students. Meanwhile I took a seminar with Tarski and Henkin and after class the whole group would follow Tarski to the Bear's Lair and while he had lunch we'd ask questions, one at a time. It was very exciting and when he asked me to be his research assistant I accepted. Then I sort of eased into being his student.[1]

As his research assistant in Berkeley, she followed Tarski's rhythm, working at his house until the crack of dawn, helping in the preparation of project reports and of papers and books for publication. After work, since she didn't have a car, Tarski would drive her home. Contrary to what usually happened, Alfred did not make a pass at Judith. Her

Judith Ng, Tarski's student, 1972.

assumption was that he felt students were "off limits" but obviously she was ignorant of past history. Perhaps her decorum mixed with generosity and her close friendship with Maria kept him at bay. Maria once said to her, "I like the way you behave. Other people try to act so sexy and you don't."

When the Tarskis drove to southern California along Highway 101, Ng rode with them. First Alfred was at the wheel of their Nash automobile but, after lunch, Maria drove. It was *her* uncertain driving, not Tarski's, that frightened Ng and made her resolve "to get my own driver's license so that the next time I could do my share." In other matters, though, Maria was faultless. Upon arrival in Los Angeles, they went directly to the apartment near UCLA that had been rented for the Tarskis. "Maria unpacked in half an hour," Ng said, "and then served tea." Ng had her own small apartment in a hilly area nearby, and Alfred and Maria often came to pick her up for dinner. She would wait outside for them and when she saw a car coming over the hill that looked as if it had no people in it, she knew it was them. She recalled that when she first came to the Tarski house in Berkeley, Maria opened the door, looked her up and down and said, "So tall?" – which amused Ng greatly since she thought of herself as an average-sized Chinese person.[2]

Teaching at UCLA offered Tarski a change of scene and a new audience as well as the opportunity to reconnect with old friends and former students. Rudolf Carnap was there as well as Richard Montague and C. C. Chang, who were now professors. Montague – his brilliant, organ-playing "problem child" – had been at UCLA for more than a decade. From the beginning Montague had adopted Tarski's stringent standards for students and, as his reputation grew, had become a force to deal with concerning curriculum and appointments in the philosophy department. David Kaplan, a student of both Carnap and Montague and, in a few years, also a philosophy professor at UCLA, recalled that early on

Montague talked about Tarski all the time and in his courses he was clearly imitating Tarski. Throughout his career he followed Tarski's pattern of insisting that things were done right. He wanted to get the idea across that a proof is *a proof*, not just some connected lines. It was hard on some people and it was good for others. But Montague did not have Tarski's majesty so he was less intimidating.[3]

Kaplan's view of Tarski in the flesh is an interesting projection of the intellectual onto the physical. He saw him as a "personification of elegant precision, well put together, not a touch of slovenliness, his physical appearance communicated nothing out of place, no gaps in the proof. I'm sure he got up every morning knowing exactly what tie he was going to wear and exactly what he was going to do. A well honed, self contained, compact little man." Because Montague was such an important person in his life, Kaplan said "I have enormous intellectual respect for Tarski and I always thought of him as an intellectual grandfather, because a huge amount of what I got from Richard, Richard got from Tarski."

Leisure Interests

As was his wont, Tarski made his presence felt to the students and the faculty in southern California in a very personal way. Donald Kalish, then chair of the philosophy department, commented on Tarski's collegiality, his willingess to talk to anyone, and his accessibility to any interested interlocutor. Kalish had experienced this openness years before during a sabbatical year he spent at Berkeley in the 1950s; now he saw that this laudable trait was still part of Tarski's character.[4] Among those benefitting from this quality was Hans Kamp, a doctoral student from the Netherlands studying with Montague. Kamp was inspired by the "extraordinary smoothness" of Tarski's lectures and his remarkable ability, noted by many others, to present, almost from the first lecture, unsolved problems for students to work on.

As Montague's protégé, Kamp became a member of Tarski's entourage and had many opportunities to spend time with him that semester. One of the things that impressed him most, partly because he hadn't expected it, was Tarski's extraordinarily encyclopedic knowledge of animals and plants: "It wasn't just love of knowledge, but a genuine love for the things known." Kamp saw this firsthand when they went to the Los Angeles Zoo, an excursion Tarski had requested Montague to arrange as part of his southern California entertainment. Since Montague abhorred zoos, he commissioned Kamp – who described himself as "pink with pleasure and deeply honored" – to be the escort.[5]

Wherever he was, Tarski loved zoos and was not above dragging anyone he could along with him to enjoy a visit. His excuse was his interest in primatology and genetics, which were among the subjects he listed under the rubric of "leisure interests" in a form he filled out for *Who's Who* (curiously he omitted botany and biology, about which he was also passionate, but did include poetry and visual arts). Judith Ng reported that she and Tarski shared an interest in evolution and in the study of primates, and that she contributed a number of books to his home library about primates and attempts to teach them language.[6] Frits Staal, an expert in Sanskrit, Indian logic, and philosophy, once went to the San Francisco Zoo with him. "Tarski insisted that we go to the zoo and that we look at the monkeys for a long time. I don't like monkeys," Staal said, "because they are so much like humans I find it embarrassing to watch them, but Tarski was mesmerized."[7] Of course it was precisely because they are like humans that Tarski found them so fascinating.

For further entertainment that semester at UCLA, the Tarskis went to parties at Richard Montague's house on Mulholland Drive. At one such occasion some of the other guests were taken aback to see Tarski smoking marijuana, a common recreational drug at student parties in those days.

Always ready to experiment and certainly not a novice at taking stimulants and intoxicants, Tarski joined right in. At these parties, as he had done at the Tarskis' years earlier, Montague played music – this time not on the piano but on the beautiful harpsichord he had purchased. (On Sundays he also played the organ in a local church, alternating with the regular organist.) Montague had other unsuspected capacities: through wise investments in Los Angeles real estate he had become rich. He purchased land and apartment buildings and made his mother and father the manager of these places while he kept his focus on his brilliant academic career. He was not ostentatious, said David Kaplan, tongue in cheek, because instead of a Rolls Royce he drove a Bentley with bumper pads that hung from its windows to protect the car from "dings" when it was parked in a garage.[8] This impressed his friends enormously, for it certainly was not a common sight to see a professor driving such a car. He also had a circle of friends in the homosexual community, but mostly he kept his sexual proclivities under wraps and this aspect of his social life did not hamper his academic success.

During Tarski's time at UCLA as well as before and after, he, Montague, and Scott were working on a book on the foundations of set theory – a book which, as it turned out, was never to see the light of day, only partly because Montague was quite busy on a number of other fronts. In the early 1960s, in addition to his teaching and research work in logic and set theory, he co-authored a logic textbook with Donald Kalish that proved to be very successful in the following years. Later in the 1960s, Montague turned to a new research direction on the model-theoretic semantics of parts of natural language based on strict principles of compositionality, an extension of the tradition of Tarski and Carnap. According to the linguist Barbara H. Partee, "Montague's idea that a natural language like English could be formally described using logicians' techniques was a radical one at the time. Most logicians believed that natural languages were not amenable to precise formalization, while most linguists doubted the appropriateness of logicians' approaches to the domain of natural language semantics."[9] At the time of Montague's work, the dominant paradigm in linguistics was generative syntax as developed by Noam Chomsky, but the relation of semantics to it was a matter of heated debate between two groups. In the 1970s a few linguists, led by Partee, argued that Montague's work "offered the potential to accommodate some of the best aspects of both of the warring approaches, with some advantages of its own." Since then, Montague grammar – as it has come to be known – has undergone considerable development and its broader extension to formal semantics has become one of the mainstream approaches to the semantics of natural language.[10]

In April 1967 Tarski left UCLA and was on the go again: first to give the DeLong Lectures at the University of Colorado, then some invited lectures in the philosophy department at Michigan State, followed by still more lectures for the mathematics department at the University of Michigan in Ann Arbor. Then he touched base for a while in Berkeley to give himself time to prepare for the next round of summer meetings. The first was a monthlong summer research institute on axiomatic set theory, back at UCLA. Organized by Paul Cohen, Abraham Robinson, and Dana Scott, it bore witness to the recent explosion of work in that subject – in part due to the progress of the Berkeley school on the role of large cardinal numbers and in part due to the impact of Cohen's powerful new method of "forcing," developed in 1963 to settle long-standing independence problems. Tarski's influence was manifested by the number of students and colleagues participating. For the UCLA meeting, Tarski's student John Doner lectured on their joint work in a more traditional part of the subject brought up to date: the arithmetic of ordinal numbers extended to higher operations beyond the familiar ones of addition, multiplication, and exponentiation.

"I think this young lady wants to talk to me"

Immediately following the set theory institute at UCLA, Tarski left for Amsterdam to take part in the International Congress for Logic, Methodology, and Philosophy of Science. At any conference reception Tarski played the gracious prima donna, the "star" surrounded by admirers wanting to talk to him or simply to be in his presence, listening. Cigarillo at his lips, drink in hand, he took on all comers. At the Amsterdam meeting he noticed an attractive young woman standing hesitantly in the background but obviously waiting for the right moment to approach. It was Anne Preller, a logic student from Marseille who had already heard a great deal about Tarski from her professor Roland Fraïssé, who knew Tarski as a result of his visit to Berkeley in the early 1950s. Preller retained a vivid memory of their first encounter:

He was dressed in a grey jacket, white shirt and tie, standing and gently rocking back and forth, facing the crowd around him. I wanted to present myself to him because I wanted to do part of my *thèse d'état* in Berkeley. I waited quite a while, watching the more courageous people talk to him. I tried to edge closer, and then he waved someone away, turned to me, his face flushed, and said, "I think this young lady also wants to talk to me."[11]

Preller was working on algebraic logic, a subject of the greatest interest to Tarski, but she was doing it in the framework of category theory, an approach that he knew little about because it did not appeal to him. Nevertheless, by the winter of 1968, with funding arranged by Tarski, Preller was in Berkeley and taking his class on universal algebra, in which he gave "the best lectures I've ever heard. He never used notes and always asked questions encouraging competition for who would find the best answer first." When *she* came up with some interesting answers, he suggested that she write a paper, and, in the Tarski manner, invited her to his house so he could help her. Preller's account of the late-night sessions with "a midnight coffee from Maria" is familiar:

I was taught by the great Professor Tarski how to write a paper: place the subject in its context, illustrate by examples, avoid repetitions, look for the precise word, replace this comma by a period, use this kind of font and this kind of red and blue underlining. I was so impressed by the details I sometimes forgot the more general advice.[12]

Tarski also taught Preller, who was raised in the proper European manner, about the lack of social formality in California. He had an "axiom" on first names that he recited to foreign visitors: "In the U.S. everyone uses first names, even the policeman who gives you a ticket." Then he encouraged her to buy a used car, telling her of the Polish visitors who bought one at the beginning of their stay and then sold it at a profit when they left. When later she sold hers at a loss, he sighed, "Ach, you are not Polish." As they became friends, he inevitably revealed his belief that nationality (or, loosely, race) determines behavior. She cringed when he told her one of his favorite jokes, a story with something to offend everyone:

What happens if a German in a restaurant finds a fly in his coffee? He is outraged, he calls the waiter and demands a new cup. What does a Frenchman do? He takes out the fly, shows it to the waiter, makes a big fuss, but eventually drinks the coffee. An Englishman takes out the fly, quietly disposes of it without saying anything, and drinks the coffee. A Chinaman takes out the fly and eats it. And a Jew? He takes out the fly and sells it to the Chinaman.

Preller laughed, but she found Tarski's penchant for stereotyping disturbing. She was German-born and had grown up in the dismal post–World War II era; unhappy in her native land, she had emigrated to France and become a French citizen. In Tarski's presence she always tried to be agreeable and pleasing; yet one day, when the topic of whether one could assign collective guilt to a nation came up, she insisted – her voice quavering with emotion – that there was nothing *specifically* German that accounted for the Holocaust. Usually Tarski was all too ready to argue any point, but in this case he said, "I see it bothers you to talk about this," and changed the subject.[13]

Inevitably, Alfred made a pass at Anne after inviting her to dinner at Trader Vic's (six years after his affair with Verena Huber-Dyson, he was still using the same modus operandi). The dinner ended with "a hand kiss a trifle too long, so I withdrew my hand somewhat abruptly," Anne said. She had been forewarned about Alfred's Don Juanish tendencies, so she was not surprised and knew how to handle the situation. Fortunately, she said, "almost as soon as I arrived in Berkeley, I met the man I would later marry. I would not have been smart enough to invent him, but since he was real, my defense against Tarski's advances was to talk about Larry – a lot. I also avoided further personal invitations but kept on asking for personal advice, which he was happy to give."

During one of her visits to Berkeley, Verena Huber-Dyson became acquainted with Preller and observed her interaction with Tarski. Verena marveled at her ability to handle Alfred with "skill and elegance.... Unlike me, she was able to have a charming flirtation without getting into anything serious or threatening." It was indeed unusual that Anne became a close friend and confidante without the attendant risks of a sexual liaison. Tarski told her how he felt about many of the women he had known, including Maria. As much as anything, it seems Alfred needed a woman with whom he could share his intimate thoughts about other women.

Preller returned to France in mid 1969, but their close friendship and professional relationship continued. In 1970, after she married Lawrence (later Laurent) Ellison, he and Maria visited her newly acquired country house in Dolgue, a hamlet near Montpellier in southern France, and in 1979 they stayed there for more than a month prior to his being awarded an honorary degree from the University of Marseille. Anne played a large part in Alfred's having been selected for the honor.

More Newcomers: Frits Staal and Ralph McKenzie

Professor Frits Staal, another Tarski "recruit," arrived in Berkeley in 1968. Tarski had met Staal, a student of Evert Beth, during an earlier visit to Amsterdam and liked him for his personal liveliness as well as for the wide range of his expertise in philosophy, South Asian languages, linguistics, and logic. When Staal indicated strong interest in an appointment at UC, Tarski wrote him a long letter explaining in detail how appointments were made and emphasizing the fact that – in the United States, as opposed to Europe – one did not have to wait until a chair was vacant to apply for a position. Staal was forever grateful to Tarski for being interested in him and for encouraging him to apply for a position without delay. Not only that, once Staal and his family were in Berkeley, Tarski with his characteristic insistence convinced the Staals to buy a "too expensive" house on San Luis Road, not far from his own home, "the only one we really liked." When Staal said, "But I can't afford it" Tarski replied, "Buy it anyway." Staal repeated that the price was more than they could afford but Tarski said, "If you have found the house you like, you *must* buy it." So the newcomers borrowed more money than they thought prudent, and in the end "it turned out to be the wisest investment we ever made in our lives."[14]

The Staals were frequent guests at Tarski's dinner parties, where conversation ranged over a wide variety of interesting subjects. More than once, Staal recalled Tarski saying that he did not like to talk much about philosophy because he thought it was like giving an "after dinner" speech – in other words, it was not rigorous. Although they never worked together directly, Staal, having been a student of mathematics as well as philosophy and linguistics, had greater comprehension about the nature of Tarski's work than vice versa; of course, Sanskrit grammar and Indian logic were esoteric topics by anyone's standards.

Ralph McKenzie – tall, blond, quiet, and a mountain climber – came to Berkeley in 1966 just after Tarski returned from his tour around the world. He had been Donald Monk's first doctoral student at the University of Colorado where Monk, fresh from his own exciting Ph.D. experience with Tarski, enthusiastically poured everything he had learned into his gifted student's eager ears.

Ralph McKenzie at the Tarski Symposium, 1971.

Inculcated with the notion that for a logician Berkeley was the center of the world, McKenzie wanted nothing more than "to put himself under Tarski's influence." By the time Monk finished his indoctrination, McKenzie's desire to be in Berkeley was so strong that he gave up a better offer he had already accepted elsewhere to come to Berkeley as an instructor. Fortunately, he knew immediately that he had done the right thing: his infatuation with Tarski and his subject was totally absorbing. "I was completely dazzled by him, he overwhelmed me," McKenzie said. "I was interested in every single thing he was doing. I was still just feeling my way and he was my mentor, leading me."[15]

They talked at night – all night – in Tarski's smoke-filled study, and that suited McKenzie perfectly, for he too was a smoker and a night owl, a rare bird among Tarski's followers. McKenzie and Tarski talked about a number of finite-basis problems for equational theories. "It was mostly him talking and me listening," McKenzie said, "but I worked hard on those problems and solved some of them." Later, when he was being considered for tenure, the chairman of the mathematics department said: "We hear you have all these results but you don't have any papers. You've got to have publications to get promoted." McKenzie said, "So I then wrote about ten or fifteen papers, all at once, and at least half of them were about problems Tarski had suggested. Amazingly, those same themes continued to pop up in my work for the next thirty years."

Surprisingly, in spite of their intellectual compatibility and the force of Tarski's inspiration, McKenzie, soft-spoken and introverted, felt their personal relationship was strange. "We had very important things in common: mathematics and our work habits, but I don't think he ever felt comfortable with me and I didn't feel close to him personally either, even though I spent more than a few nights with him. Then, when I went on leave and came back, everything was different. I had almost no contact with him for ten years. I really don't know what happened."

The best explanation for what happened (acknowledged or not) was that McKenzie managed to avoid becoming part of the team that Tarski recruited to help finish his long-term projects. While Donald Monk, Leon Henkin, and research assistants Donald Pigozzi and Steven Givant were working away on "Papa's" projects under "Papa's" direction, McKenzie the "grandchild" was going full steam ahead with his own ideas and his own graduate students. He felt guilty about it, though, because he thought he owed so much of his success to Tarski's influence. Only later would he feel that he had repaid his debt: first when he undertook the production aspects of the edition of Tarski's *Collected Works* and later, as Tarski's health began to fail in the 1980s, when he took over some of Tarski's last students and saw them through their doctorates.[16]

Donald Pigozzi

Donald Pigozzi, another student who worked with Tarski during most of the 1960s, began as an auditor, just sitting in on courses. Captivated by the lectures (like so many others), he decided that he too wanted to do logic with Tarski despite the snares that he knew might lie ahead. Insecure about being accepted as a Ph.D. candidate, he consulted Judith Ng. "Oh, he will take aleph-nought [infinitely many] students," she said, "it's finishing that's the hard part."

Indeed Pigozzi was "taken" and he took up the challenge, accepting the fact that he might spend at least four more years working under Tarski. But he had not understood what Thomas Frayne had discovered so painfully: that working "under Tarski" also meant working *for* him on material not necessarily connected to one's own projects. This became clear when he was caught up in editing Part I of *Cylindric Algebras*.[17] But

Donald Pigozzi, c. 1990.

Pigozzi was somehow spared the emotional trauma experienced by other students. He was able to say, "What I remember most about Alfred is his wonderful sense of humor. It never seemed to fail him." Although there is no question that Tarski was lively and interesting, not everyone would agree about his sense of humor.

Cylindric Algebras, written mostly by Leon Henkin and Donald Monk (except for Tarski's introductory chapter on universal algebra), was ostensibly ready for publication – down to the color coding for the typography – when it was sent to Tarski in the spring of 1964 for final editing. This was when Pigozzi became his research assistant and discovered firsthand what "final editing" meant to his teacher. As soon as Alfred saw how the book had been written, it was impossible for him not to rewrite it – even though he complained that he had been "sucked in" to the editing (the book had been Monk's idea). During the rewriting, he wasn't much interested in the proofs of theorems and, once he gained confidence in Pigozzi's ability, left that part to him. What he *was* interested in was the web of ideas. He would spend hours or whole nights composing statements of a few interrelated definitions and theorems and the remarks that connected them, making sure they conveyed the idea he was wanting to get across in exactly the right way. For Tarski, mathematics was always more than just a string of theorems; it was a work of art that formed an architectonic whole. In the end, the book was completely revised and doubled in size; Part I did not appear until 1971 and Part II in 1985, two years after Tarski's death.

During long, late-night sessions (like others before and after), Pigozzi sat struggling to stay awake, waiting for Tarski to formulate a remark. He was a sounding board, a stenographer, and an advisor on English idiom. "Tarski would take a long drag on his cigar and look off into space; the veins on his forehead turned blue," Pigozzi recalled. "I would get very anxious waiting for him to exhale and when he did I was even more frightened because I never saw any smoke expelled. Finally he would dictate a perfectly formed paragraph and ask me what I thought of it. We would discuss it and possibly reformulate it. Hours went by like this – often on a single remark."[18]

Nevertheless, Don and Alfred had an easy relationship – due, no doubt, to Pigozzi's capability and willingness to work hard. He laughed at Alfred's jokes and did what was asked of him. He was at Tarski's two or three times a week and sometimes on the weekend. During that period, they often went out for dinner and frequently Maria joined them. One of Alfred's favorite places was Paluso's, near Jack London Square in Oakland, where for ten dollars one could purchase a "gold card" that entitled you and your party to a substantial discount. "Tarski was not cheap," Pigozzi said, "but he was always concerned with getting the best possible value for his money. He also liked figuring out what he thought was the best route between his house and the freeway to Oakland or San Francisco, a route which involved winding around the hills of Berkeley rather than taking the more obvious straight shot." He appreciated any opportunity to enlighten his friends in this matter, and Judith Ng recalled an algorithm he devised for the return home: "*Links, dann links*... and take the uphill branch at any decision point. People who used his suggestion usually ended up totally lost."

Pigozzi and Tarski and Maria got along so well that late in the summer of 1969, at Alfred's suggestion, they drove across the country together in Don's blue Ford Maverick. With his thesis approved and the last details nearly finished, Pigozzi had gotten his first job at Indiana University and was moving to Bloomington. They were headed for Pennsylvania, where Alfred was to be the principal speaker at a meeting of the Conference Board of the Mathematical Sciences at Pennsylvania State University. The Tarskis never lost their enthusiasm for the road, no matter how many trips they made. Pigozzi recalled:

It was quite a sight, the three of us, with their luggage for a three week trip and all my worldly belongings packed into that little Maverick. Alfred and I shared the driving. He smoked carelessly, of course, and for years afterward

I took pleasure in showing off the scorched marks on the dashboard where Alfred had missed the ashtray. I called them "the Alfred Tarski Memorial Burns."

We drove through Nevada, Idaho and Wyoming to Yellowstone Park and the Badlands of South Dakota, to places Tarski had not seen before. He wanted to see buffalo on the range because Kolakowski had raved about them – and we did.

He was especially fussy about choosing a motel and always on the lookout for a good room at a reasonable price. He would never consider a Holiday Inn and was suspicious of all chain establishments, claiming "they charged hotel prices without providing hotel services." For the same reason, he avoided AAA recommendations. So we would drive around randomly looking for motels and then have to look at several before he found one he liked. This was a problem because we drove until seven at night. Maria and I always had a hard time persuading him to stop. He couldn't alter his usual work and sleep habits so we never got on the road until late morning.[19]

At the conference, Tarski gave ten lectures in five days on topics in general algebra and the metamathematics of algebra. Then the travelers got back into the Maverick, reloaded the suitcases, and drove to Columbus, Ohio, where the Tarskis caught their return flight to California. On the way to Ohio, Alfred asked Don to write up the lecture notes and Don said yes because he felt he couldn't refuse; but in fact, he never did. Perhaps it was because of the pressures upon him at the very beginning of his academic career or perhaps he finally felt he no longer had to do everything Tarski asked of him.

Benjamin ("Pete") Wells

Many of Tarski's students worked for as long as eight years before the happy moment when he blessed their Ph.D. theses with his signature and freed them to move on in life. But Benjamin Wells and Judith Ng were in a special category; with time out for other endeavors, they each spent closer to twenty years before their degrees were finally granted.

Wells began graduate work in 1962 and completed his thesis in 1982, when Tarski was already seriously ill; his Ph.D. was the last one Tarski signed officially. Their interaction over those long years is a variation on the "Tarski–student" theme, but parts of the story strike an unusual note and highlight Tarski's tolerance for the foibles of those he liked.

During his first year as a graduate student, not yet committed to a specific topic, Wells enrolled in one of Tarski's courses. An alert student, quick to spot an inconsistency on the blackboard, he often pointed out small mistakes; Wells noted that Tarski liked this because it showed that the student was paying attention to details. When Jack Silver, later to become a mathematics professor in the logic group, made an astute observation about something being wrong on the first day of a seminar and proceeded to uncover more errors the following week, Tarski grabbed Henkin and asked, "Who is that?"

Although Wells's corrections were not as dramatic as Silver's, Tarski asked him to be his teaching assistant the following year. Then, after discovering that Wells knew Russian (he had studied it as an undergraduate at MIT), Tarski asked him to write a review of a paper by Mal'cev for the *Journal of Symbolic Logic* in which Wells would point out that Tarski had found an error in one of Mal'cev's proofs as well as a way to fix it. Tarski wanted Wells to write the review so that it would not look as if he, Tarski, was criticizing Mal'cev and then trying to publish a result of his own "under the table." He said it was not really worth writing a paper about it, but he thought his correction ought to be in print.[20]

Wells worked on the review and agonized over what to say, staying up nights and accomplishing little while Tarski kept asking, "How are you doing?" Finally he wrote something and brought it to Tarski, who read it and said, "OK, this is good! I am very happy with it. Now I'm going to tell you exactly what to write," and he dictated a whole new review. "I was pleased," Wells said. "I was off the hook, and what he wrote was so much better than what I had written, but at the same time I was upset because I had spent so much time on it. It was clear to me, after the fact, that I could have come to him with almost anything because he was going to rewrite it anyhow."

During that period, Wells wrote a few more reviews at the editor's request for the *Journal*. He was now officially a Tarski advisee but, he said, "Even though I had passed the qualifying exams, I didn't know what it meant to do research. Tarski would explain mathematical ideas but I just couldn't get off the ground with them. I would think, what am I supposed to do with this? Why can't I launch forward from this stuff that is given so clearly?"

At a summer logic institute in Montreal in 1966, Wells met Andrzej Mostowski, who was now the leading logician in Poland. Gracious and inviting, Mostowski gave the impression of being less dictatorial or at least more helpful than Tarski, so Wells liked the idea of going to Warsaw for a year to study with him – as others, such as John Addison, ten years earlier, and Peter Hinman, more recently, had done. He wrote Tarski asking for advice and support, both moral and financial. The actual letter Tarski wrote in reply is missing, but Wells reconstructed the content in his own version of Tarski's style:

Warsaw is a wonderful place to visit. I love it .... The people are congenial and the women among the most beautiful. [Judith Ng's recollection is that Tarski told Wells, "Not all Polish women are nuns!"] You will certainly enjoy yourself studying there. But by this very virtue, I must caution you. You have proven to me that you are easily distracted from scientific pursuits. You claim that you will focus on your research under Mostowski's direction in the rich tradition of Polish logic, but you have not lacked for scientific opportunity, atmosphere, and guidance in Berkeley. Time and again you have shown me that you would rather do many things, perhaps *anything*, besides your work. In short, I would support your studying in Warsaw but I do not believe the argument that you will work more seriously there. I may be wrong – I hope I am – but without a broad reorientation of your goals and deep understanding of the demands of research, I am afraid this may be another method of avoiding what you must do in order to complete your degree.

Nevertheless, a year later, with deepened resolve and an intensive course in Polish under his belt, Wells went. As Tarski feared, nothing much happened at first on the level of research. Mostowski, concerned about Wells's lack of focus, kept asking him what he might tell Tarski of his progress. The answer was vague until the day Mostowski, aware that Wells knew Russian, introduced him to the owner of North-Holland Publishing Company who was looking for someone to translate a book of selected articles by Mal'cev. Wells was hired and, as he became engaged in the painstaking work of translation, he came to understand the problems involved; this in turn stimulated him to work on his own problems, whose solutions would – after many starts and stops – eventually lead to his dissertation.[21]

When Wells returned from Warsaw in 1968, the student revolution of the sixties was in full swing and "it was amazingly distracting." In Berkeley it had begun in 1964 with the free-speech movement. A crisis had arisen when the university administration banned on-campus recruitment of student activists for the civil rights movement and voter registration in

Benjamin Wells, modeling for a fashion photo, Warsaw, 1968.

the South. Students across the political spectrum protested vigorously. There were marches, speeches, confrontations, sit-ins, arrests, and investigations. On the heels of that, the war in Vietnam spawned the antiwar movement and its many forms of protest. The campus did not settle down for years. In his memoir *The Gold and the Blue*, Clark Kerr, president of UC from 1949 to 1967, wrote that "the second greatest administrative blunder in UC history was the decision to revoke permission for political advocacy tables at the south campus entrance on Telegraph Avenue."[22] (The greatest blunder, he said, was Robert Gordon Sproul's proposing the loyalty oath during the "red scare" in 1949.)

In varying degrees, most liberal professors sided with the student protestors of the sixties and the seventies, but Tarski was not among them. Wells was taken aback when he brought him a petition to sign and Tarski said, "I'm not going to sign it. I don't agree with it. I am a socialist but I am not a pacifist." As far as he was concerned, fighting communism, whether it was in Vietnam or elsewhere, trumped everything else. Furthermore, he thought students like Wells lacked understanding of world affairs; to help educate him, Tarski gave him a year's subscription to a magazine that was in line with his more conservative views. In retrospect,

Wells agreed with Tarski about the students' naiveté. On the other hand, Tarski himself was naive in his faith that the United States was always on the path of virtue.

Given Tarski's political views, Wells was amazed when, in the mid-seventies, Tarski showed him a scrapbook in which he kept extensive newspaper clippings of three women associated with revolutionary movements of the period whom he called "my heroes": Angela Davis, Bernardine Dohrn, and Patricia (Patty) Hearst. Davis, an African-American, had been an instructor of philosophy for one year (1969/70) at UCLA; Montague was among those who had opposed her reappointment, while Kalish had been for it. Davis, a former student of Herbert Marcuse, was a member of the Communist Party and supported the radical Black Panthers. In 1970, a gun registered to her was used in an attempted courtroom escape in which a judge and others were killed. Davis went into hiding but was caught after a few months and then tried on charges of conspiracy, murder, and kidnapping. She was later acquitted of these charges after being held for several months in prison; in later years she ended up on the faculty of the History of Consciousness Department of the University of California at Santa Cruz.[23]

Bernardine Dohrn was a founding member of the Weather Underground, or Weathermen, a militant revolutionary offshoot of the Students for a Democratic Society (SDS). Todd Gitlin, in *The Sixties. Years of Hope, Days of Rage*,[24] described Dohrn as "the object of many an SDS male's erotic fantasy ... she fused the two female images of the moment: sex queen and street fighter." The Weathermen advocated the overthrow of the United States government and the capitalist system, and they carried out a campaign of riots and bombings. Three Underground members died in an accidental explosion while preparing a bomb in a New York City safe house, forcing the group – Dohrn included – to indeed go "underground" for a number of years. In later life she became director of the Children and Family Justice Center and a law professor at Northwestern University.[25]

Patty Hearst, a granddaughter of William Randolph Hearst, the flamboyant founder of a publishing empire, was kidnapped in 1974 by members of the small Symbionese Liberation Army (SLA), a group of escaped convicts and student radicals. They extorted six million dollars in ransom money from the Hearsts to be used for food for the poor. In captivity, Hearst announced that she was joining the SLA and had changed her name to Tania; later she was filmed taking part in a bank robbery with the group. She was captured in 1975 and sentenced to jail for seven years for that crime, though when Tarski showed Wells his scrapbook, with clippings following each episode, she had not yet been caught. Hearst's sentence was commuted after two years by President Jimmy Carter, and she ended up marrying and raising a family.[26]

Wells was never sure how to take Tarski's statement that these three women were his heroes, nor could he account for his intense interest in them; at first he thought he was being "put on," but there was nothing in Tarski's demeanor to suggest that he was joking or being ironic. Perhaps he was fascinated by their self-assurance, their willingness to speak out and take risks to the point of putting their bodies on the line. It was the same thing he had said about Maria, years back. Moreover, all were very handsome women, and the erotic element cannot be discounted. Judith Ng, who had not seen the scrapbook, offered the opinion that Tarski "was a bit obsessed with what he considered 'strong women who had gone astray'." Even so, the reasons for his actually keeping that sort of scrapbook remain mysterious.

Wells dropped out of doing research mathematics for many years. First, seeking spiritual enlightenment, he went on a long voyage to India. After that, from 1971 to 1980 he taught in a special mathematics education program designed to increase knowledge and achievement in ghetto high schools, then later in a private high school. However, he didn't totally abandon his attachment to advanced mathematics and even while teaching he continued to translate Russian articles into English for the American Mathematical Society.

In the spirit of the zeitgeist of the seventies, while Wells was dropping out of the Ph.D. program, C. C. Chang had also gone on a spiritual quest and was soon to give up doing any serious work in mathematics. Chang, as a professor at UCLA, had done important research in a number of the fields of interest inspired by Tarski, and with Jerome Keisler he had written the first comprehensive expository text on model theory. Tarski was extremely upset at Chang's openly announced decision to forsake mathematics in favor of a deep exploration of and commitment to the Buddhist path. Although Chang was sure he was doing the right thing, Tarski thought it was a terrible mistake, "a waste of talent and purpose," and he let Wells hear exactly how he felt. "I hope you are not going to be so foolish," he said. "He was respectful of my interest in spiritual things as long as it didn't interfere with my math," Wells said.[27]

Tarski and Ng

The year Pete Wells was in Warsaw, Judith Ng visited him and other students who were there and found the idea of a year based in Poland very appealing. Unintimidated by the harsh climate, the harsh regime, and the lack of standard Western comforts and consumer goods, a number of adventurous young logicians were drawn to Poland partly by the idea that going back to "the source" would offer inspiration for their own work. What the Poles lacked in material things they made up for by being extremely hospitable; also, by the time Judith went, a regular exchange program between Warsaw University and Stanford was in effect, with students receiving a stipend that was more than enough to live on. Tarski, as he had done with Wells, tried to discourage Judith by describing all the negative aspects of life in Poland, but undeterred she went anyway and considered it a wonderful experience. "I didn't do so much in my field," she said, "but I gained a sense of independence. I also traveled elsewhere in Europe for several months." No doubt part of the attraction was simply to get out from under Tarski's thumb for a while.

When Ng came back from Warsaw, she returned to UCLA and from 1969 to 1971 worked with Alonzo Church on the Reviews section of the *Journal of Symbolic Logic*; owing to her earlier work as Tarski's research assistant, she was indeed a well-trained editor. When Church had reached mandatory retirement age at Princeton in 1968, he obtained a position at UCLA and moved there, taking the entire operation of the *JSL* Review section with him. With funding from the National Science Foundation and additional funding from UCLA, he was able to have a separate office and hire an assistant devoted to that work. This was the enterprise that Ng was involved in for two years. Church continued as editor of the Reviews until 1979, continued teaching at UCLA until 1990, and continued writing until his death in 1995 at age 92 – an altogether remarkable career. Because of Church's leadership, the Reviews section of the *Journal* was and remains a major resource for the field.[28] It is currently under the editorship of Herbert Enderton.

Exchange between UCLA and UC Berkeley was easy, via a one-day drive or a one-hour flight, and Tarski was a frequent visitor during Judith's

Tarski hiking in Joshua Tree National Park, 1967.

stay in Southern California. On one occasion after he had been the invited speaker at the logic colloquium, she prepared a large Turkish dinner for the party afterwards, a gesture that touched Alfred deeply. The next day a group went to the Huntington Library, Art Gallery, and Botanical Gardens in nearby San Marino, where – as they walked through the extensive gardens – Tarski reveled in discussing his favorite plants and then made a point of getting everyone to go inside to see the famous paintings by Thomas Gainsborough and Thomas Lawrence. In her recollection of that occasion, Ng emphasized "the delight Tarski took in introducing people to experiences that were new or foreign to them. He loved to experiment and to get others to do the same." This she thought was one of the most positive aspects of Papa Tarski's character.

Maria had accompanied Alfred to UCLA on that trip, and on their return they included Judith in their plans to take a circuitous route back home via Mexico, the Sonoran Desert, and Joshua Tree National Park. This took many days and offered Judith more new experiences including driving, because now she had her license. They hiked in the Joshua Tree area, and Tarski insisted that they stop afterwards in Indio so that Judith could have her first "date shake" – a milkshake made with the local desert

Maria with palm frond, 1967.

dates and ice cream. But the story he told over and over again whenever the opportunity arose (Ng described it as one of his embellished memories) was: "How I saved Judith from staying in a bug-infested hotel in Mexico."[29]

Montague's Death

While working on the *JSL* Reviews with Church, Judith continued to think about her thesis problem; but, as she had advised others, "finishing was the hard part." She made frequent trips to the San Francisco Bay area to see her family and to keep Tarski up to date on her progress. As it happened, she was present in early March of 1971 when Richard Montague came to discuss some details of the set-theory book that he, Tarski, and Dana Scott were writing. The plan was for her to drive Richard to the airport for his flight home, and she remembered him standing in Tarski's study, nervously shifting from one foot to another, talking on and on, evidently reluctant to leave, while she worried that he might miss his flight. Finally he turned to her and said, "Okay, let's go."

A few days later, Alfred telephoned Judith to tell her that Richard had been murdered.[30]

Richard Montague, at UCLA, c. 1967.

It is not clear how much Tarski knew about the risks Montague courted in his private life in Los Angeles. When Richard was his student in Berkeley, Tarski became well aware of his tendency to get into trouble and, along with others, had helped Montague avoid serious legal consequences when he was accused of seducing a minor. However, once Richard left Berkeley it is not likely that he confided in Tarski on private matters. On the other hand, Donald Kalish, Richard's longtime friend, said he knew as much as anyone did about the catastrophes Richard brought upon himself. According to Kalish,

On campus Richard was a model of decorum but off campus there was no risk that was too great for him. He would go to bars, "cruising", and bring people home with him. One time he picked up someone who brought along three "friends". These guys proceeded to tie Richard up and strip the house of everything of value. This was nothing "kinky" or sexual. It was just a straight old robbery and the guys knew he was not going to call the police to explain how it happened.[31]

Kalish said he and his wife – a friend of Richard's since the Berkeley days – pleaded with him to be more cautious; but he wouldn't listen. On the day of his death, he brought three or four people home with him for some sort of soirée. A friend found him in his shower, strangled by a bath towel. It was not a robbery this time – his wallet was on the soap dish – and the circumstances were ambiguous. The "visitors" escaped in his car, the beautiful Bentley, which they crashed into a telephone pole and then set on fire. The crime was never solved.

Shocked and grief-stricken, Tarski went to Los Angeles to attend Richard Montague's funeral. Besides his parents, the mourners included three distinct groups: the academic friends, the homosexual friends, and the real estate people – each apparently surprised by the presence of the others. The pastor of the church where Richard played the organ officiated at the ceremony. Herbert Enderton, a member of the UCLA faculty, had the feeling that the pastor sensed he was addressing a group of nonbelievers and thus confined himself to praising Montague's musical contributions. In addition, Montgomery Furth, then chairman of the philosophy department, spoke of Montague the scholar and his many contributions.[32]

Three weeks to the day after Montague's death, Tarski delivered a memorial lecture at UCLA in honor of Rudolf Carnap, who had died the year before at the age of seventy-nine. In his talk Tarski recalled his long association with Carnap, beginning with their first meeting in Vienna in 1930 and continuing with their interaction during the Unity of Science period of the following decade. Tarski's influence on Carnap had been decisive in his turn to semantics, while Carnap had been instrumental in bringing Tarski's theory of truth to a wider audience of philosophers. But what surprised members of the audience most was when Tarski spoke of Carnap as a generous man, without pettiness or jealousy, and then with evident emotion expressed the wish that he were possessed of those virtues.[33] Inevitably, Tarski must have thought of the contrasts between the saintly Carnap, who had lived a long, quiet, and calm academic life, and Montague: intense, obsessively careful and in perfect command of every detail in his work, yet unable to control the demons that drove him to take the wild risks that resulted in his death. Some may find it hard to imagine that those who deal in abstractions – mathematicians, physicists, philosophers, those who seem to lead a life only of the mind – lead an all too awfully "real" life as well. Might it not have been just those demands for precision and perfection that made it necessary for Montague to seek some way to release that tension? Or might it have been the other way around? In any case, Tarski probably understood as well as anyone what it was that led to Richard's tragic death.

INTERLUDE VI

Algebras of Logic

WITH THE PUBLICATION OF "On the Calculus of Relations" in the *Journal of Symbolic Logic* in 1941,[1] Tarski single-handedly revived a significant part of the algebraic approach to logic that had lain dormant since the end of the nineteenth century. From then until the end of his life, relation algebras and still richer algebras of logic were to be one of his main research projects – and one for whose realization he enlisted many students and colleagues. Throughout his career, the algebraic way of thinking in mathematics had great appeal to Tarski because of its elegance and wide applicability. He pursued it in pure (so-called) universal algebra dealing with constructions on, relations between, and decompositions of general algebraic systems. In set theory he pursued it with the development of the arithmetic of cardinal and ordinal numbers on the basis of suitable general algebraic laws. But it was first and foremost as a logician that Tarski was to exploit the algebraic way of thinking to the fullest.

Boole's Algebra of Logic and Classes

The algebraic approach to logic had its origins in the mid-nineteenth century with the work of the English mathematicians Augustus De Morgan and George Boole, who turned a subject that had been the province of philosophers since the days of Aristotle – and in which little progress had been made in the intervening 2200 years – into one that increasingly drew the attention of mathematicians and made astonishing progress in the following century. Boole, the son of a poor shoemaker in Lincolnshire with a passion for science and mathematics, was largely self-educated, but through his native brilliance he became the leading figure in that transition with his publication of two treatises, *The Mathematical Analysis of Logic* (1847) and *An Investigation of the Laws of Thought* (1854). Modelling his approach to logic on ordinary algebra, Boole used combinations of variables by means of the symbols $+$, $\cdot$, $-$, 0, and 1 to express general laws that could be interpreted in various ways – in particular, within propositional logic and the calculus of classes. For the former, the variables are interpreted to range over propositions that are either definitely true (have value 1) or definitely false (have value 0), while $+$, $\cdot$, and $-$ are interpreted (respectively) as the operations of disjunction, conjunction, and negation of propositions. Some of the laws (not all as given by Boole), such as $a + 0 = a$ and $a \cdot 1 = a$, are the same as for ordinary algebra; but others are at first sight unexpected, such as $a + (-a) = 1$, $a \cdot (-a) = 0$, $a + a = a$, and $a \cdot a = a$ – though all are perfectly justified by the given explanation of the symbols. The second interpretation takes the variables to range over arbitrary subclasses of a given non-empty class U, the "universe" of the interpretation, where 1 is interpreted to be the universal class U and 0 the empty class, while $+$, $\cdot$, and $-$ are interpreted (respectively) as the operations of union, intersection, and complement (relative to U); again, all the laws of Boole's algebra are satisfied. Years later, a general algebraic approach to Boole's concrete calculi of propositions and classes led to an abstract theory of Boolean algebras, which admits many interpretations. Much of Tarski's work in the 1930s was devoted to the study of Boolean algebras both in logical and set-theoretical terms. Incidentally, one interpretation of Boolean algebra – in terms of transmission of currents through switches in an electronic circuit – became fundamental for the construction of present-day computers; this was realized by the mathematician Claude Shannon (the father of statistical information theory) in his master's thesis at MIT in the late 1930s.

Peirce's Algebra of Relations

Boole's approach to logic had permitted him to recast the traditional theory of syllogisms in algebraic terms and then verify them by simple calculations, but the resulting logic went little beyond the traditional theory in scope. On the face of it, the statements in the language of syllogisms relate simple *unary* predicates such as "x is a man", "x is mortal", "x is blue-eyed", or the corresponding classes of all men, mortals, and blue-eyed beings. An essential step forward was made in 1870 by the remarkable American scientist, mathematician, and philosopher Charles Sanders Peirce, who surmounted the limits of what could be expressed in the syllogistic language by dealing with *binary predicates* or *relations*, such as: "x is a child of y", "x is a mother of y", "x is a father of y", "x is the same as y", and so on. Other relations may be defined in terms of these by means of the propositional operations of disjunction, conjunction, and negation together with the new operations of *converse* and *composition*. For example, the converse of the relation "x is a child of y" is defined to hold between given x and y when y is the child of x; that is, the converse of "is a child of" is the relation "is a parent of". And the composition of the two relations "x is a father of y" and "x is a parent of y" is defined to hold for given x and y when there exists a z such that x is the father of z and z is a parent of y; that is, the composition of "is a father of" and "is a parent of" is the relation "is a grandfather of".

Abstractly, in addition to the operations $+$, $\cdot$, and $-$ of Boolean algebra with its distinguished elements o and 1, the algebra of relations stemming from the work of Peirce employs two new symbols, $\smile$ for converse and ; for composition, together with two new distinguished elements: $1'$ for the identity relation and $0'$ for the diversity relation. As with the interpretation of Boolean algebra in terms of arbitrary subclasses of a non-empty class U, the algebra of relations has an interpretation in terms of arbitrary subclasses of the class $U^2 = U \times U$ consisting of all pairs of elements of U. Each subclass R of U^2 may be considered as a binary relation between elements of U; membership of a pair (x, y) in R is read as saying that x is in the relation R to y. For this interpretation, $R^\smile$ is defined to be the class of pairs (x, y) such that (y, x) belongs to R, and R ; S is defined to be the class of pairs (x, y) such that there exists a z in U with (x, z) in R and (z, y) in S. Under these definitions, if U is the class of human beings and R is the relation "is a father of" and S is the relation "is a child of," then $S^\smile$ is the relation "is a parent of" and R ; $S^\smile$ is the relation "is a grandfather of". As before, the operations $+$, $\cdot$, and $-$ are interpreted (respectively) to be the operations of union, intersection, and complement (relative to U^2) of binary relations between elements of U, with o taken to be the empty relation and 1 the universal relation (i.e., U^2 itself). Finally, $1'$ is defined to be the identity relation – that is, the class of all pairs (x, y) for which x, y are in U and $x = y$ – and $0'$ is taken to be $-(1')$. Abstractly, using variables a, b interpreted as ranging over arbitrary binary relations between elements of U, one has $a^{\smile\smile} = a$, $(a + b)^\smile = a^\smile + b^\smile$, $(a\,;b)^\smile = b^\smile\,;a^\smile$, and $a\,;1' = 1'\,;a = a$ in addition to the equations from Boolean algebra.

Relation Algebra from Schröder to Tarski

In the last decade of the nineteenth century and following Peirce's initial efforts, the German logician Ernst Schröder explored the calculus of relations in a systematic way; in particular, he observed that many elementary properties of binary relations can be re-expressed as equations in Peirce's algebra. For example, the statement that a binary relation R between elements of U is transitive – that is, that for all x, y, z in U, if (x, y) is in R and (y, z) is in R then (x, z) is in R – can be expressed by the equation $(R\,;R) + R = R$. Similarly, but with more complicated equations, we can express that R is a function – that is, a many-to-one relation – and even that it is a one-to-one function from U onto U. Schröder conjectured that every elementary property of binary relations can be expressed by an equation, but that was proved wrong in 1915 by Leopold Löwenheim using a counterexample due to A. Korselt. In modern logical terms, elementary properties are those expressible in the first-order predicate calculus. It was realized later that only properties of binary relations expressible by means of at most three distinct variables within the language of first-order logic could be re-expressed as equations of relation algebra. The work within *full* first-order logic initiated by Löwenheim overtook the calculus of relations, and it was not until Tarski's 1941 article that interest in the latter was revived. Tarski proposed an elegant and relatively simple finite axiomatization of the equations of that system, and he raised the question of whether each equation that is valid for every interpretation in the algebra of binary relations between elements of a non-empty class U is derivable from his axioms. Tarski had verified that every one of the hundreds of such equations that had been recognized by Schröder is derivable in this way.

The Tarskian School of Relation Algebras[2]

The various realizations of Tarski's axioms are called *relation algebras*, and one way in which those axioms could be shown to be complete would be by means of a *representation theorem* according to which every abstract relation algebra is representable as a concrete algebra of binary relations between elements of a universe U. This would be analogous to the important representation theorem for Boolean algebras, established in the 1930s by Marshall Stone, according to which every abstract Boolean algebra is isomorphic to a concrete algebra of subclasses of some class U with the usual set-theoretic operations of union, intersection, and complementation. In the mid-1940s, after settling in at the University of California in Berkeley, Tarski renewed his work on this and other problems that he had raised about relation algebras, and he drew the attention of students and colleagues – both at Berkeley and elsewhere – to the field. In 1950, Roger Lyndon, an algebraist at the University of Michigan, strikingly proved the existence of relation algebras that are not representable as a concrete algebra of binary relations and at the same time showed that there are equations true in all relation algebras that are not derivable from Tarski's postulates. The latter result was strengthened in 1964 by Tarski's student, Donald Monk, who showed that no finite set of postulates suffices to axiomatize the set of all true equations in this sense.

An Application of Relation Algebra: Set Theory without Variables

In contrast to the situation for Boolean algebras, many of the results concerning relation algebras in general turned out to be negative or were mixed. Nonetheless, Tarski had long recognized that a special class of such algebras had very nice properties. The point of departure in logic for those was given by formal axiomatic systems – such as axiomatic arithmetic or set theory – in which the notion of an element x representing an ordered pair (y_1, y_2) of two elements y_1 and y_2 can be defined. In these cases there are two binary *projection* relations that hold between x and y_1 and between x and y_2 just in case y_1 is the first term and y_2 the second term of the ordered pair represented by x. The significance of such projection relations is that, via them, many-placed relations can be reduced to binary relations. Tarski convinced himself that *any* formal system in which projection relations for a pairing operation can be defined using at most three distinct variables can be equivalently formalized within a version of the calculus of relations without variables or quantifiers or even sentential connectives; its only rule is the substitution of equals for equals. In 1972 he initiated a plan to prepare that work for publication, mainly with the assistance of Steven Givant. Because the detailed verification of the proof sketched by Tarski turned out to involve many technical difficulties and because of the multiple ramifications of the ideas involved, that plan turned into a major project that stretched practically to the end of Tarski's life. The manuscript was finally completed before Tarski's death in 1983, but the actual publication by Tarski and Givant, *A Formalization of Set Theory without Variables*, did not appear until four years later.[3] Because its language and single rule of inference are so simple, Tarski's system has been used by computer scientists working on programming computers to try to prove mathematical theorems automatically.[4]

First-Order Logic and Cylindric Algebras

Another approach to the algebraization of logic was also to occupy Tarski's Berkeley days, off and on, to the end of his life. This was in terms of what he called *cylindric algebras*, intended to encompass full classical first-order logic. Tarski's ideas for that were initially developed in the period 1948–1952 in collaboration with his students Louise Chin and Frederick B. Thompson.

In the first-order predicate calculus, if the variables are interpreted as ranging over some universe U and if the basic operation and relation symbols are interpreted as operations on and relations between elements of U, then we can define relations of any finite number of arguments by formulas F (or sentential functions) generated using disjunction, conjunction, negation, and existential and universal quantification. However, there are certain anomalies in this way of thinking. For example, if F defines an n-ary relation and G an m-ary relation, then $F \vee G$ is perfectly well-defined as a formula but does *not* define an n-ary or m-ary relation if n is different from m. This suggests putting the relations defined by formulas on a par by adding "don't care" clauses to the nonmatching arguments. For example, if $F(x, y, z)$ defines a ternary relation and $G(x, y)$ defines a binary relation, we replace the latter by the formula $G(x, y) \wedge z = z$, which we denote $G'(x, y, z)$. Then $F \vee G$ is replaced by $F \vee G'$ so as to define a ternary relation. If, continuing on, we want to deal solely with ternary relations, another anomaly comes up with quantification. For example, if $F(x, y, z)$ defines a ternary relation, then $(\exists z)F(x, y, z)$ defines a binary relation. In this case, to form the corresponding ternary relation $F'(x, y, z)$ we rewrite it as $(\exists u)F(x, y, u) \wedge z = z$.

Just as with binary relations and sets in the plane U^2, ternary relations R between elements of U can be treated as subsets of U^3 and viewed geometrically as volumes in 3-dimensional space that consist of all points

(a, b, c) in U^3 that are in the relation R. If R is defined by $F(x, y, z)$ then the relation R' defined by the F' in the preceding paragraph is obtained by *stretching* R along the z-axis, yielding a kind of *cylinder*, and the operation of passing from F to F' is referred to as *cylindrification* along the z-axis. More precisely, the cylindrification of R along the z-axis consists of all (a, b, c) in U^3 such that, for some u in U, (a, b, u) belongs to R. We can similarly define cylindrification of R along the x-axis or the y-axis; this corresponds to treating $(\exists x)F(x, y, z)$ and $(\exists y)F(x, y, z)$, respectively, as defining a ternary relation.

There is also a geometric interpretation of universal quantification. The largest cylinder along the z-axis contained in R corresponds to treating $(\forall z)F(x, y, z)$ as a ternary relation. Since $(\forall z)F(x, y, z)$ is logically equivalent to $\neg(\exists z)\neg F(x, y, z)$, this operation can be defined in terms of cylindrification and complement.

For another example, the binary relation of equality, $x = y$, can be treated as a ternary relation defined by $x = y \wedge z = z$; geometrically, this can be vizualized as a *diagonal plane* in the 3-dimensional space U^3. It is unchanged by cylindrification along the z-axis, whereas cylindrification along the x- or y-axis yields the set of all points in U^3.

In general, when dealing algebraically with the relations determined by formulas of first-order logic having any number of free variables, to put them all on a par we treat them (for simplicity) as defining relations in a space U^∞ of an infinite number of dimensions. By definition, U^∞ consists of all points $a = (a_1, a_2, \ldots, a_n, \ldots)$ such that each a_n is in the class U. Given a formula F, the relation R defined by F under a given interpretation of its atomic formulas in U simply consists of all a in U^∞ that satisfy F, as explained by Tarski's theory of satisfaction and truth. Thus, if the relations R and S are defined by the formulas F and G, respectively, then the relation defined by $F \vee G$ is the union of R and S, the relation defined by $F \wedge G$ is their intersection, and the relation defined by $\neg F$ is the complement of R in U^∞. Finally, the relation defined by $(\exists x_n)F$ is taken to be the cylindrification of R along the nth axis – that is, the set of all a such that there exists an a' in R that agrees with a at each position i different from n – and the relation defined by $(\forall x_n)F$ is taken to be that defined by $\neg(\exists x_n)\neg F$. Note that, for each relation R in U^∞ defined by a formula F in this way and for all but a finite number of n, the cylindrification of R along the nth axis equals R. This corresponds to the fact a sequence a's satisfying F depends only on the terms a_n for which the corresponding variable x_n is free in F.

Abstracting from these concrete interpretations, Tarski defined *cylindric algebras* to be Boolean algebras with additional operations C_n of cylindrification satisfying suitable laws that reflect properties of existential quantification in the first-order predicate calculus – for example, that $C_n(a + b) = C_n(a) + C_n(b)$, $C_n(C_n(a)) = C_n(a)$, $C_n(0) = 0$, and $C_n(1) = 1$. In addition to the new abstract operations of cylindrification, for the algebraic representation of the first-order predicate calculus with equality he also needed "diagonal" constants $d_{n,m}$, corresponding to the class of all sequences a in U^∞ such that $a_n = a_m$; motivated by the axioms that hold for the equality relation in logic, he added further axioms to the algebra such as the equations $C_m(d_{n,m}) = 1$ for any n, m. More generally, he defined cylindric algebras of any dimension k (an ordinal number), which may be finite or infinite, where the indices n and m range over the numbers less than k.

Again, the central problem for cylindric algebras of dimension k is that of *representability* as algebras of k-ary relations R between elements of some non-empty set U, that is, subsets R of U^k. In general, for both finite and infinite dimensions k there exist nonrepresentable cylindric algebras. However, for an important class of infinite-dimensional algebras, the *locally finitely dimensional* ones, there is a positive solution to the representation problem; this was established first by Tarski and later in a strengthened form by Henkin. An algebra is said to have this property if, for each element a of the algebra, $C_n(a) = a$ for all but a finite number of n; as noted before, this condition holds for the cylindric algebras of relations defined by formulas F of the first-order predicate calculus. The representation theorem for locally finite-dimensional cylindric algebras of infinite dimension k is an algebraic analogue of the celebrated completeness theorem obtained for the first-order predicate calculus by Kurt Gödel.

In addition to Louise Chin, Frederick Thompson, and Leon Henkin, many students and colleagues of Tarski were involved in the development of the theory of cylindric algebras over the years, including Donald Monk and two young Hungarian mathematicians: Hajnal Andréka and István Németi. This body of work led to the publication (by Henkin, Monk, and Tarski) of the major exposition of the subject, *Cylindric Algebras*, in two parts: the first appearing in 1971 and the second not until two years after Tarski's death, in 1985; a few years earlier, the three had collaborated with Andréka and Németi on a monograph going into other aspects of the subject. Tarski participated actively in the writing of all of Part I of *Cylindric Algebras*; his participation in the project after that was not active, though he was of course kept abreast of it all along the way.[5]

An Application of Cylindric Algebras to Computer Science

It is interesting to note that, starting in the early 1970s, cylindric algebras found an important application in the branch of computer science devoted to methods of effectively handling large data bases – such as the catalogue of a university library or the list of employees of a major corporation – in which extensive amounts of data must be stored, queried, retrieved, and updated. The initial work in this direction was by the computer scientist Edgar F. Codd, who developed a form of cylindric algebra for dealing with what are called relational data bases, apparently independently of the work of Tarski and his school. It was only later that the connections were realized, and once more the potential breadth and value of interpretations of algebraic logic was confirmed.[6]

14

A Decade of Honors

BEGINNING IN JUNE 1971 with a grand conference celebrating his seventieth birthday, Tarski enjoyed a decade of tribute that included honorary degrees from two foreign universities and ended with the award of the Berkeley Citation – UC's highest honor – and the renaming of the Logic Common Room as the Alfred Tarski Room. At the same time that all this was going on, he conducted business as usual in his own inimitable fashion. Tarski had been required to retire in 1968 but was recalled to teach for the next five years, and he continued to do research and advise students until 1982.

The 1971 eight-day symposium in Berkeley of hundreds of participants was devoted to Tarski's work and to the wealth of ideas he had inspired others to explore – particularly in the fields of model theory, set theory, algebraic logic, geometry, and the theory of truth.[1] The organizing committee's first list of topics had been even longer; looking at that list in awe, they even wondered at one point whether to expand the symposium from a week to a month.[2]

The celebration was originally to have been a surprise, but Dale Ogar, the administrative assistant for the Logic and the Methodology of Science program, thought that would be impossible. When Robert Vaught, head of the organizing committee, closed the door to her office in a very hush-hush way to tell her of the proposal, she said, "I'll try, but knowing Tarski, I don't see how we can keep it from him because the concept is huge and he's always around."

Straight away, organization for the symposium began to take up an enormous amount of time and space. Papers were everywhere, correspondence was everywhere, phone calls were constant, and in the middle of all this, Tarski would walk into Ogar's office, demand something and start riffling through the papers on her desk while she tried to hide them

Tarski disagreeing with Roland Fraïssé, at the Tarski Symposium, 1971.

with her body. "It was a nightmare," she said, "trying to contact people all over the world within the context of an eight-hour day in California. Inevitably, Tarski got wind of the plans, which was good at first because I didn't have to hide things, but then he got involved and became even more demanding."[3]

Officially Ogar was responsible to the entire logic group, but Tarski was the leader and so, de facto, she became Tarski's personal secretary. This was her first real job after leaving college. She was young and insecure and, because Tarski in his prima donna mode would often yell at her, she thought he was unhappy with her work. In fact, she was good at her job and loved the challenge of putting the meeting together, building something from scratch, and dealing with all the important people who were invited to speak, such as Quine, Popper, Erdös, and Gödel (who, in the end, did not come). The planning took a year.

Unlike Sarah Hallam, Tarski's longtime devoted secretary who had become the administrator for the entire mathematics department, Dale Ogar was never under Tarski's spell. She preferred working with Leon Henkin, Bob Vaught, and Ralph McKenzie, and she liked being involved with the students, whereas with Tarski she felt like a servant who could not teach her master to behave properly. More than twenty-five years later, she vented her frustration:

I would open his mail and try to find a way to give him the things that needed his attention without putting them on his desk, because his desk was already

Tarski listening, at the Tarski Symposium, 1971.

littered with papers and half-smoked cigars; so I started putting folders with important papers on his chair; thinking, "if I could just train him to look at the chair before he sat down"; but he wouldn't. Instead he sat on the folders and hollered, "Where is this? Where is that? I thought everything would be done by now." I'd say, "I put it on your chair." He would get up and find everything along with the piles of phone messages I had left there too. I told a friend, "One day I am going to get one of those spindles and put that on the chair with the messages and see how long it takes him to figure it out then."[4]

"I was an agitator," Ogar said. "I asked the research assistants why they felt they had to go up to Tarski's house to work at night after having worked all day." But soon she found herself caught in the same trap, succumbing to *his* needs first. "All of the other things I was supposed to do like keeping track of the students, planning the weekly colloquium, furnishing the room which eventually became the Tarski library – all of that was part of my job but as far as he was concerned that just interfered with my ability to be at his beck and call." As a result, she was forced to become super-organized, not just for the symposium but for all her work with him as boss. She couldn't "train" him, but he trained her; ultimately she considered her six years with him as great preparation for her future positions because he had pushed her to the limit with his constant demands and insistence on high standards.

Only at the closing banquet of the symposium did Tarski finally thank Dale for all she had done "far beyond the call of duty." "He was in tears" she said; "It was all so incredibly emotional and when he said 'Call me Alfred' I just lost it completely and I burst into tears myself."

Tarski was at his most expansive during the entire conference and was no doubt especially pleased with the *San Francisco Chronicle*'s article headlined "The 'Einstein of Mathematics' at UC".[5] The piece, complete with photographs of Tarski and the audience, emphasized his brilliance as a researcher and a teacher and quoted him saying, ruefully, "Yes, it is a kind of satisfaction"; and, after a meaningful pause, "but it's not the most pleasant thought to realize that you are so old." His activities and attitude in the following decade belied those words.

Almost on the heels of the Tarski birthday symposium, the fourth international Logic, Methodology and Philosophy of Science meeting was held in Bucharest, Romania, in the summer of 1971. As always, he was a central figure in the organization beforehand and at the conference itself. It was there that he was presented with the Georgian translation of *Introduction to Logic* that Tarski found extraordinarily lovely for the beauty of the script and production. Almost immediately he wrote to Wanda Szmielew, "That was the greatest pleasure for me!" Once it was in his collection, he never missed an opportunity to show it to anyone who indicated the slightest interest.

Steven Givant

As the star of his seventieth birthday symposium, Tarski was naturally expected to contribute a paper to the published proceedings. But he decided to write on a subject quite different from the one on which he had lectured concerning the foundations of geometry. Instead he chose as his topic "A Formalization of Set Theory without Variables," and, typically, he had a great deal to say. The article and its revisions became so long and took so much time to write that it held up publication of the symposium volume and finally, by mutual agreement with the editors, it was decided that Tarski would publish his material separately. After working on it for a year with still no end in sight, he concluded that he could use some assistance.

Enter Steven Givant, one of Robert Vaught's advisees, a personable and energetic graduate student who had taken many of Tarski's courses and was enthralled by his lectures. His interest and enthusiasm did not escape Tarski's attention and, in his most winning way, he asked Steven if he could work with him for "just two months" to help him finish this seemingly endless paper. Givant had already considered offering his services on that or any other paper Tarski wished to complete but had restrained the impulse because he knew his own Ph.D. work had to be done without distraction. Nevertheless, when Tarski himself asked for his help, he was so flattered to be considered capable of the job that he found the invitation irresistible.

Their working relationship began in June of 1972, but by August it was clear that they would not finish until October at the earliest. Givant was now so deeply involved in the work that he canceled a September trip to Europe (planned earlier but never mentioned to Tarski because he thought the paper would be done). In addition, with regret he turned down an attractive job offer to organize and direct a large SEED program for disadvantaged students in Oakland, part of the same program that Benjamin Wells and Judith Ng had been involved in. This was a disappointment to the leaders of the group and caused more remorse on Givant's part.

At the start of the new year, the paper was farther than ever from completion because Tarski had decided to rewrite the whole thing. Vaught, knowing Tarski's habits all too well, urged Givant to resign as research assistant and to put all his efforts into his own work, emphasizing how much more important it was for him to finish his thesis and to begin his own career than it was for Tarski to add one more paper to his already hugely impressive bibliography. There were others who could help Tarski, Vaught said. Although Vaught revered Tarski and felt grateful to him for his compassion and help during his periods of illness, he was very upset to see him using a student to further his ends – apparently without a thought of what the consequences were for the young person. And, perhaps, an unspoken rivalry for Givant's attention was part of the equation.

Givant knew Vaught was right about what he ought to do but could not bring himself to resign, he said, because of his dedication to and affection for Tarski. No doubt there were other reasons, conscious or not: stubbornness, not wanting to be called a quitter, and liking to be thought of as essential, as the best person for the job; after all, hadn't Tarski chosen *him*? So, he continued with the project without telling Tarski how this commitment was affecting his life or of the pressure he was getting from Vaught – all in the naive assumption that the paper would actually be finished soon.

Instead, the paper turned into a monograph that became longer and longer; as summer approached there were seven chapters, the last only half written. It was now clear to Givant that he could not possibly finish his dissertation by June and that Tarski's treatise would not be finished, either. Desperate for a break, he again made arrangements to go to Europe to visit friends and relatives; when Tarski asked "What are your summer plans?" Steven had to tell him. "Papa" did not like the answer at all. This time, however, Givant stuck to his plans and Tarski, first depressed and then furious, blamed *him* (most unfairly) for the uncompleted manuscript. He wrote to Wanda Szmielew, "There was a crisis in my relationship with St[even] Givant .... It's difficult to be objective but he behaved like *s*.... It cost me a lot – I would have finished my work a long time ago."[6] In anger, he hired someone else as his research assistant without informing Givant, just as he done when Dana Scott had reneged on his commitment to improve the Woodger translations in the 1950s.

When Givant returned from Europe in the fall, Tarski scarcely spoke to him. It was only after Steven wrote a long letter laying out the history of the project thus far – and explicitly stating that he thought Tarski's annoyance and hurt stemmed at least in part from the fact that "you care a great deal for me"[7] – that the air was eventually cleared and they resumed their seemingly interminable work on the monograph about the formalization of set theory without variables. Givant finished his Ph.D. with Vaught in 1975, but it was not until 1987, fifteen years after it was begun and four years after Tarski's death, that the book was finally published with Tarski and Givant as co-authors.[8]

"It was not easy to be close to Tarski," Givant wrote. "He rarely opened up or showed affection. It seemed much easier for him to criticize than to compliment."[9] That was indeed Tarski's way but, in spite of the difficulties, he and Givant had an extraordinarily tight relationship; Alfred knew he could count on him, and in the end Steven was one of Alfred and Maria's most loyal friends.

Plus ça change plus c'est la même chose

If mortality crept into Tarski's consciousness after his seventieth birthday, he kept it well hidden; to all appearances he seemed as energetic as ever. He continued the program he had already embarked upon to complete long-term projects and to ensure that the ground he had cultivated and the seeds he had sown many years earlier would bear fruit with his label on it. For this, the collaboration of others was essential. After the shock of Montague's death, the planned book on set theory with Dana Scott as third author went into limbo. Another book that had been in

Tarski lighting up in his garden, 1972.

progress since the 1960s was with Wanda Szmielew on the foundations of geometry; after the first part had been written,[10] Szmielew turned to another, quite different approach of her own. Part II of the treatise with Leon Henkin and Donald Monk on cylindric algebras was another book still in the works.[11] In addition, John Corcoran, a philosopher at the State University of New York and a great admirer of Tarski, had convinced him that there ought to be a second, revised edition of *Logic, Semantics, Metamathematics*, which had long been out of print. Thus Tarski as co-author was involved in the writing of four or five books at the same time and also (dreamer that he was) thinking of two more, one on universal algebra and the other on decision methods. These last two never got off the ground and, though none of the others appeared before his death, he had the satisfaction of knowing that an impressive body of work would be left for posterity.

As before, the flow of scholars continued and collaborators came to Berkeley to work with him and afterwards to correspond with him about their joint projects. He also kept on teaching and working with his regular students as well as visitors, presenting them with the usual difficulties. His method of attack was always the same; he didn't "mellow." As he would later tell Judith Ng, "Most people say, the older they get the more religious they are. As for me, the older, the less religious." And he might just as well have said, "the older, the tougher." Certainly in the seminar room he was as exigent as ever.

Giovanni Sambin, a twenty-three-year-old Fulbright student from the University of Padua who came to study logic at Berkeley in 1971, provides a case in point. While expressing concern that his background might be insufficient, Sambin asked Tarski if he could take part in his seminar on universal algebra. "Not a problem," said Tarski with his welcoming smile, "you have only to read my introduction to that subject in Part I of *Cylindric Algebras,*" which Sambin did. But immediately there *were* problems: first, the paper he was assigned to present a few weeks later was, according to Sambin, "the longest and hardest one assigned to any student. Maybe he forgot that I was just a first year student." Next, along with the difficulty of understanding the new material, Sambin was terribly homesick. California was a long way from mama's Italian cooking and other familiar comforts. In Berkeley he lived at the International House, which served hot dogs that he considered inedible. Although he had genuinely wanted to get away from home, the adjustment preoccupied him so much, he literally could not work. Knowing he was unprepared to do an adequate job of exposition, he made an appointment with Tarski and told him, "I have some problems."

"Which problems?" Tarski asked.

"Personal problems," Sambin answered, uncertain of what else to say.

"All right," said Tarski, "how long do you think it will take to solve these problems? Fifteen days?"

Sambin knew the depth of his anxiety hadn't been understood but felt he couldn't explain further, so he plunged ahead as best he could and, after a brief reprieve, began his lengthy report in the seminar. It went on for weeks. Many years later the ordeal was still fresh in his memory:

In my first lecture he corrected my pronunciation. If I said, "In this case you take a *finnit* number of factors," he would say "*figh night.*" So then I said, "in the *infighnite* case," and he said, "No, *infinnit.*"

During my next lecture Tarski scolded me in front of the class, "You should have taken a good course on set theory and mathematical logic," in spite of my having told him what my background had been. In any case, I understood that he meant that I had to do things his way, and write "for all algebras A" and so forth, so in the next lecture I was very formal and wrote everything in symbols and he liked that, saying, "It seems you learn fast."

Sambin soldiered on with his presentation, but as he neared the end he realized that he was unable to give applications of the main theorems to Abelian groups because, as he told Tarski, he didn't know enough about that subject – to which Tarski replied, "It is enough that you read the book by Fuchs." Sambin agreed but said he needed more time. Thus, in the middle of his ninth hour, instead of presenting the applications he said, "I will speak about them next week after reading the book by Fuchs," and the seminar ended half an hour early. The vein in Tarski's head turned purple but he said nothing. A few days later he telephoned Sambin and said, "It is not necessary that you give the report next week. I will do it. Goodbye." And although Sambin continued to attend the seminar, Tarski never spoke to him again. Describing the lasting effects of his painful experience with Tarski, Sambin wrote:

I never saw him or wrote to him after that year in Berkeley. I learned a lot about the fact that also very famous people can have their human defects. For years I worked in mathematical logic trying to show myself and him that I was able to do something valuable. In the end, I became a declared constructivist, maybe partly due to the fact that I was sure not to meet Tarski in that field.[12]

Chile and an Honorary Degree

One of the voyages that meant most to Tarski in the 1970s was his extended visit to Chile and Brazil in 1974/75, which was arranged and sponsored by Rolando Chuaqui, a UC Berkeley student of the 1960s. Chuaqui, a Chilean of Syrian ancestry, had an unusual background: he had studied medicine and had become a physician. At the same time, "just for fun," he and a cousin had studied logic on their own; they read Tarski's *Introduction to Logic* and were captured by the assertion set forth in the preface that logic might provide "a unified conceptual apparatus which would supply a common basis for the whole of human knowledge." Chuaqui, a young man of action, found this idea so exciting that he decided he would try to apply logic to medicine and would go to the source to learn how to do it. Wasting no time, he sought and received financial support from a

Latin American foundation, applied and was admitted to UC Berkeley, and was provisionally accepted (since he had no formal background in the subjects) as a graduate student in the logic and the methodology of science program. He took many courses from Tarski and others, and as his knowledge deepened he became even more interested in logic and mathematics for its own sake and gave up the idea of doing anything further in medicine. In the process, he developed an interest in probability theory and completed a Ph.D. dissertation with David Blackwell, a leading scholar in that subject.

After receiving his degree, Chuaqui returned home to begin his academic career at the Pontificia Universidad Católica de Chile in Santiago, where within a few years he created a program in logic and the first doctoral program in mathematics in all of Chile. During the next decade he returned frequently to the United States, partly because he had married Kathleen Henderson, an American who wanted to visit her family often, and also because he wanted to maintain his contacts with Tarski and Blackwell and others who had nourished his intellectual interests. According to Kathleen, "Roli admired Tarski so much and spoke of him so often that when I was first introduced to him I felt like I was meeting Schweitzer or Einstein or someone like that." Rolando's wife was enormously impressed with Tarski's wide interests and knowledge, and, happily, she considered that her husband was brilliant, too, and thought "their friendship was nice for Tarski because Roli was a person who could keep up with him in everything ... history, and, of course, medicine and science. Also, Roli was easygoing and relaxed and so they got along extremely well."[13] Others confirm Tarski's respect for Chuaqui. When he would make a comment in class or seminar, Tarski would say, "Dr. Chuaqui is completely right."[14]

In Chile, Chuaqui already had a reputation as an innovative scholar and administrator eager to raise the standards of research. To further that goal he thought nothing could be better than a long visit from Tarski. By way of encouraging his presence, Chuaqui nominated him for an honorary degree at the Universidad Católica de Chile, where the postgraduate program in the exact sciences he had begun was now underway. He also invited Tarski to conduct a seminar and made arrangements for him to talk at other universities. Knowing that he loved travel, loved mountains, and loved to enlarge the audience for logic – and that he was always extremely gratified to receive acclaim and recognition for his work – Chuaqui thought there was a good chance Tarski would accept, and he was right. Tarski enthusiastically said yes to all the proposals and, once it was known that he would be in Chile for an extended stay, he was also invited to speak at two major universities in Brazil.

Surprisingly, he had not yet been awarded an honorary degree at any university in the world – and in fact would never have that honor conferred upon him in an American or Polish university. In the latter case, serious proposals had been made that were ultimately rejected because of Tarski's vociferous opposition to the communist regime.

Alfred and Maria arrived in South America in November of 1974 and rented a small apartment in Santiago. He taught a seminar on algebraic structures and a course in universal algebra and also gave other lectures. In January of 1975, to the great satisfaction of everyone involved, he was awarded his honorary degree, *Doctor Scientiae et Honoris Causa de la Pontificia Universidad Católica de Chile*. Since Chuaqui had been in Berkeley the year before, it is not unlikely that Tarski helped him write a nominating speech or, at the very least, gave him suggestions for what to say. It was a grand ceremony with a large crowd of dignitaries from the academic community and the government in attendance. Tarski became something of a celebrity, a role quite to his liking; his presence was newsworthy and the media reported at length about his conferences and the honorary degree.

Later, when Tarski lectured in Valparaíso, Irene Mikenberg (one of Chuaqui's students) recalled that she had never seen so many people at an academic talk in Chile. Mikenberg had taken Tarski's course but, having heard that he was a "difficult person," had not dared engage in private conversation with him until almost the end of his stay, when she told him how much she had enjoyed his lectures. He asked her what she was working on and, true to form, he suggested a Ph.D. topic in algebra to her; accepting the prompt, she ended up writing a thesis on partial algebras. After he left Chile, they kept in touch and Tarski repeatedly urged her to visit Berkeley, which she and her husband did a few years later. They were wined and dined and given a tour of the house on Michigan Avenue, where Irene was particularly impressed with Tarski's study and desk because "his wife had installed plastic on the floor all around it so that he could smoke and throw [*sic*] the ashes around without having to worry about dirtying the floor."[15] The notoriously careless Tarski dropped his ashes everywhere. Once, a guest at an evening party at the Henkins' home was appalled to see him flick his lengthy cigarillo ash onto the beautiful white living room rug, obviously unaware of what he was doing.[16]

More important to Mikenberg was the "magnificent letter" Tarski wrote to the Association for Symbolic Logic in 1978, when the logicians in Chile were organizing a Simposio Latinoamericano de Lógica Matemática while a group within the ASL was urging withdrawal of sponsorship in protest of the brutally repressive government headed by Augusto Pinochet. Tarski made the point that he believed in the independence of science from politics, that it was important not to isolate countries under dictatorships, and that he would resign from the ASL if they did not sponsor the symposium. The letter had an impact; the meeting was held.

The Pinochet regime had been in power for just a year when Tarski visited in 1974/75. A military coup had ousted the duly elected president Salvador Allende, who had been attempting to implement a socialist program. Following the coup, Allende disappeared; it was officially stated that he had committed suicide but widely believed that he had been murdered. As a consequence of these events, many people fled the country in fear or dismay. Chuaqui himself had been criticized by some of his colleagues for remaining in Chile. So the very fact that Tarski had come at that time made all the Chilean academics who were still there – carrying on as best they could – grateful to him, because many foreigners had refused invitations and had tried to discourage others from accepting. As a matter of principle, Tarski rarely refused his presence or his recommendation based upon political differences. The notable exceptions were due to egregious anti-Semitic behavior.

Recreational travel, as one would expect, was an important part of Tarski's stay in Chile. Guided by the Chuaqui family, including their three young children, Alfred and Maria toured many of the high points of the country both in the north and the south – not to climb but at least to view the mountains and do some easy hikes. They drove to the coast and there, on Valentine's Day, Tarski was invited to tea at the home of Eduardo Frei Montalva, the president who had preceded Allende. Tarski felt much honored by that invitation. In his own country, former presidents had

Tarski with Kathleen and Rolando Chuaqui in Southern Chile, 1975.

not invited him to tea. (In better times to come, Frei's son Eduardo Frei Ruiz would lead the country after Pinochet was ousted.)

Love of poetry led the group to the valley of Elqui, the rural home of Gabriela Mistral, the first Latin American poet to win the Nobel Prize (1945); then they traveled far south to the Chilean lake district, to Llanquihue and Todos Los Osos, the spectacular emerald-colored lake surrounded by volcanos. Both Tarskis were unreservedly appreciative of the beauty they saw. Although Tarski's health would begin to fail after he returned to the United States, he showed no signs of fatigue or illness while in Chile except for his constant smoker's cough. The Chuaqui children's predominant memory of that trip was the billows of smoke from Tarski's cigar that engulfed them as they sat way in the back of the station wagon on their days of travel through the mountains.[17] Seven people, four adults and three children, were packed into that car. Rather, it was Maria Tarski who showed fatigue. Kathleen Chuaqui remembers her being harried with the details of the routine necessary to get Alfred started every morning. "It was a bit like getting a child ready for school. She didn't complain aloud but from her body language and facial expression I understood she had to cater to him all the time."[18]

Getting started may have been hard, but once in motion Tarski seldom stopped; his energy and enthusiasm for seeing new places and meeting new people was as indefatigable as ever. If he did feel himself running down he took doses of his Kola Astier as a pepper-upper, which was what he always did wherever he was.

The Death of Dear Friends

In August of 1975, a few months after Alfred and Maria's return from Chile, Andrzej Mostowski died of a heart attack in Vancouver, Canada, where he had gone to lecture at Simon Fraser University; he had also planned to visit his brother in Toronto. His death, at age sixty-two, was a tremendous shock to those who were close to him as well as to those who appreciated him as an academic figure. Only the week before, Mostowski and his wife had been in Berkeley, visiting with the Tarskis, going on excursions with Ina and Andrzej Ehrenfeucht and their children. There had been festive, light-hearted dinners at the Tarski home, everyone speaking Polish, everyone seemingly well and in good spirits; they were like one big happy family.

Compounding Tarski's deep personal sadness over the death of an old and treasured friend was the intellectual loss. Mostowski had become the leading logician in Poland, and he was as effective as one could be in keeping channels open between Eastern Europe and the United States while the Iron Curtain was still mostly closed. For Alfred, losing the sturdy, honorable Mostowski – who had survived the war in Poland – meant losing a major link to his past as well as a positive voice for logic in the future. "He never stopped grieving his untimely death," Jerzy Łoś said years later.[19]

Then, another blow: just one year after Mostowski's death, Wanda Szmielew died of cancer at the age of fifty-eight. Her death was not sudden; she had been diagnosed six years earlier. Tarski was aware of her illness, but that did not make her loss less painful for him. She had been a student, a lover, a colleague, and a friend for almost forty years. Wanda had returned to Berkeley many times after the first year she spent there, always supported by funds Tarski secured; he had visited her in Poland and they met often at conferences in other cities of Europe. He was in constant touch with her, writing long hand-written letters to *Kochana Wandziu* [dearest little Wanda] and signing them *Twój Fredek* [your Freddy], keeping her abreast of what he was doing, what he was thinking, where he was going, how hard he was working. "I have never worked so hard in my life and I don't see the end of it"; this he wrote to everyone, but to her he also named those he blamed for thwarting his efforts.

Tarski had an unusually deep attachment to Wanda's daughter Aleksandra (Olenka), and she reciprocated his affection. She was six when she first met him in 1960, and she remembers being excited by his visits: "He always came to see me at my house whenever he was in Poland and he always brought me little presents, American Indian jewelry made with turquoise and silver and very nice books about animals. I thought of him as a kind uncle, and I thought he was very funny because he smoked these little cigars that had little plastic tips and when he finished a cigar, he gave the tip to me."[20]

Olenka was the rare child who actively liked Tarski's smoking; she had never known anyone who smoked, and his cigarillos especially were a novelty to her. At her mother's urging she wrote him letters and he, in his letters to Wanda, always had a message for her to tell him about her studies and her other activities. She was very aware that he kept her in mind. "Wherever he went, Greece or Mexico, or Paris, or when he and my mother met somewhere, he always sent me something."

Aleksandra described her mother as "authoritarian, towards me, towards everybody, and she never worried much about what other people thought of her." Wanda was bold, determined, self-assured, and independent, as well as beautiful, and Tarski admired all these attributes. One gets the impression that once she was launched on her own, Tarski was the one trying to please her and not the other way around. He courted her favor and was most solicitous of her until her death. While she was ill he helped her buy a car and sent her money for a vacation in Italy, and after her death he worked through considerable bureaucracy to ensure that Aleksandra received money from Wanda's bank account in the United States, perhaps adding some of his own.

In the early stages of their work on geometry beginning in the late 1950s, Szmielew followed Tarski's approach and in the following decade worked vigorously on drafts of a book with him, but over the years she gradually changed her views on the foundations of geometry and began to develop her ideas along quite different lines. In 1974 Szmielew decided to write an exposition of this work, even though she was already debilitated from the effects of her cancer. According to her friend and close colleague Maria Moszyńska, "she was no longer able to carry on her university duties. During the next two years she worked intensively. The monograph soon began to live its own life, to develop beyond the initial framework. But in May 1976 it became clear that Wanda Szmielew had neither enough time nor strength to complete the work." After her death, Moszyńska finished the typescript following her detailed plan: "I tried to change only what was absolutely necessary, so as to preserve as far as possible the original text."[21] The resulting volume was published posthumously in Polish in 1981 and in English translation (by Moszyńska) as *From Affine to Euclidean Geometry* in 1983. In a review of the book a few years later, the Romanian geometer Victor Pambuccian wrote:

This book leaves the reviewer with a feeling such as one experiences watching the throwing of the javelin at the Olympic Games. Every movement seems simple, natural, the way it has to be. One almost forgets that behind this apparent ease and grace there are years of hard work, of sweat and tears.... [It] is a life's work.[22]

Tarski's own last article, "Metamathematical Discussion of Some Affine Geometries," which was written jointly with Szmielew's former student L. W. Szczerba, appeared in the *Fundamenta Mathematicae* volume for 1979; in it they presented in full the work that they had announced at the Jerusalem Logic and Methodology conference fifteen years earlier.[23] Implicitly, it was a further tribute to Szmielew.

≥ ≥ ≥

The death of his two dear friends did not cause Alfred to say to himself, "there but for the grace of God go I," even though they were much younger. He continued to drink, smoke, pop his stimulant pills, and stay up all night working on the projects he was determined to see finished. Considering these lifelong habits, Tarski remained in remarkably good health until the last few years of his life, when he was diagnosed with congestive heart failure and emphysema. This combination of ailments eventually led to serious breathing difficulties that caused him great pain and discomfort. Perversely (or perhaps the addiction was impossible to undo), he refused to stop smoking, but he did reluctantly adopt the recommended salt-free diet and took potassium pills – not without constant complaint about the tastelessness of the food and the bitterness of the pills.

Tarski's last lecture at the Berkeley logic colloquium, given in April 1978, drew an unusually large audience. He had not taught any classes since 1973. Facing a packed lecture hall, he began dramatically, outlining clearly the problems he was going to speak about. Lecturing in his customary classroom style without notes, he worked in detail through the proofs of his results. Atypically, he faltered along the way and could not carry through one of the proofs, even after many attempts. The talk dragged on for two hours, and along the way a good part of the audience lost interest and left the room. It was an embarrassing and painful experience for Tarski and also for his close colleagues. Feferman recalled asking himself, "How does one learn when to stop giving talks?"[24]

A French Doctorate

After his long stay in South America, Tarski moderated his travel somewhat; he declined many invitations and went only to places that were especially attractive to him. France was one such place. In September 1978, championed by his colleagues Roland Fraïssé and Anne Preller, he was to be the first recipient of the honorary degree of Doctor Honoris Causa at the new Université d'Aix-Marseille II.

By then Preller was teaching at the university in Marseille; she was married to Laurent Ellison, the man she had met in Berkeley. They bought a crumbling old farmhouse in Dolgue, a hamlet thirty kilometers north of Montpellier, and lovingly restored it stone by stone. The complex of buildings included an independent apartment where, at Preller's invitation, Alfred and Maria spent six weeks prior to the honorary degree ceremony. This was the Tarskis' second visit to Preller's home; in 1970, after the International Congress of Mathematicians in Nice, they had come for a shorter stay. At that time the farmhouse did not yet have electricity or running water, but Alfred and Maria had roughed it, uncomplainingly, and spent their days making excursions to nearby attractions with Preller's friends. On one trip to Mont Aigoual, the highest mountain of the area, Preller recalled driving down a narrow winding road while Tarski loudly conceded that women could be good drivers "but in mathematics they are less gifted than men." Anne and Monique Janin (her colleague in Marseille) argued vehemently that women were discouraged from even becoming interested in the field by cultural attitudes. Challenged to name a top mathematician who was a woman, Preller named Emmy Noether and was exasperated when Alfred laughed and said, "Look, you know, she was not really a woman." The argument became heated and Anne so agitated that it affected her driving and she accepted Monique's offer to replace her at the wheel. Maria Tarski changed the subject and they made it to their destination restaurant safely. A few years later, Tarski had the grace to acknowledge Preller's point when he presented her with the memoirs of Sofia Kovalevskaya (1850–1891), a Russian-born mathematician. The beautiful and gifted Kovalevskaya had overcome enormous cultural barriers in order to study – first by being allowed to have a tutor at home and then by gaining permission to travel to France, Germany, and Sweden, where ultimately she became the first woman in Europe to hold a chair in mathematics.

Much had changed between 1970 and 1978, the time of Alfred and Maria's second, longer visit with Anne Preller. In Anne's view, Tarski had lost his spark and she saw his physical energy diminished. "His illness and worry had taken away his ability to reach out to other people, which was one of the most charming things about him." Also, Maria complained that he left her alone to do the housework but that he himself was not being seriously productive, so her old rationale that she took care of everything while he exercised his genius was losing its validity. Aside from a few walks and excursions to the nearby Roman and Greek ruins, they stayed close to home. In the evenings everyone ate together, listened to music, and read poetry aloud. Anne had significantly less free time for those activities – what with her teaching, her two children, her husband, and "all that cooking without salt which excluded the ready-made dishes." Conscientious to the nth degree, Anne put so much effort into preparing healthy meals for Tarski that he returned to the United States exclaiming how much better French food was – even without salt! He seemed to be completely unaware of how much time, energy, and ingenuity it took in Preller's busy life to satisfy his craving for tasty food.

As nominator, Preller took great care preparing her laudatio for the forthcoming honorary degree ceremony; unavoidably, she did it with Tarski by her side providing biographical facts and advice about what to say. One evening Maria did not appear at dinner and Tarski later explained that they had had a fight about Wanda Szmielew (who had been dead two years). "You know," he said, "Maria does not like intellectual women." It wasn't clear to Anne how she was supposed to take that remark.

According to Preller, the honorary degree ceremony is the only one at which "ordinary French professors wear the toga and the ermine but at

Marseille only the dean had such a garment so the fanciest part of the ceremony was the champagne." Nevertheless the experience was moving. At the last sentence of her laudatio, Anne found that her voice failed her and she realized, once more, how much he meant to her. Tarski's voice, however, did not fail. Anne had been worried that he might sound feeble, but he had prepared well for his acceptance speech and his later talk by "swallowing his Kola Astier to assure that his voice and presentation would be strong."[25]

The following day the Tarskis left for America. At the airport, when they were already at the other side of the gate, Alfred turned back toward Anne to say a last goodbye. "Do you know Shakespeare's ...?" Over the noise of the airport commotion Anne could not hear which play or sonnet was being evoked, but she heard the rest: "You should read it to understand my feelings for you." A sweet parting, a touching farewell. Afterwards, they corresponded and occasionally talked on the phone, but it was the last time they saw each other.

Without being blind to his faults, Preller's positive feelings for Tarski were such that, when asked why some students regard Tarski like a god, her answer was, "because he cares so much about us." This was hardly a universal view, but it wasn't unique. Another admirer also saw him as godlike – albeit in a quite different style. The philosopher John Corcoran, who worked closely with Tarski in his last years, put it this way: "He was like a Greek God, not like a Jewish or a Christian God. He had a lot of flaws; he was exasperating and stingy and very suspicious, and he was such a glory hound it was embarassing." Nevertheless, Corcoran saw him as omnipotent and revered him for the world of thought he had created.

Corcoran encountered Tarski by way of the 1956 Oxford Press edition of a selection of his papers from the 1930s, *Logic, Semantics, Metamathematics*, which had been translated and edited by J. H. Woodger. Struck by the importance of the papers in that text, he had wanted to use it for a seminar he taught in the mid 1960s – only to discover, to his disappointment, that it was out of print. A decade or so later he would make it his mission to edit a new and improved edition of the book, but in the interim he got to know Tarski firsthand during a year's visit to Berkeley and later, when Tarski was "the star in billing and performance" at a 1974 conference that Corcoran organized at the State University of New York in Buffalo, his home base. Ever more enamored with Tarski, Corcoran

John Corcoran, 1970.

mounted a campaign to nominate him for an honorary degree. He figured it would be a "shoo-in" because the logician John Myhill – formerly of Chicago, Berkeley, and Stanford – was also a professor there who, Corcoran assumed, could be counted on as an ally. But much to his astonishment, "Myhill nixed it. He let it be known that he would fight the nomination tooth and nail."

The reason for Myhill's vehement opposition was never clear, at least not to Corcoran. He was known for being eccentric and he may have harbored a grudge against Tarski (many did) or against Corcoran himself that was registered in this way. Corcoran remarked, "Had I brought it up on a different day in a different way, Myhill might have been all for it, but once he said no, he would not change his mind and moreover he would produce reasons that followed from the laws of logic for why I was a jerk for even thinking of it." A few years later Corcoran successfully put forth Alonzo Church as a candidate, and in 1990 Church was honored by Buffalo in the way Tarski should have been.[26]

It is odd that Tarski, distinguished as he was, received no such awards in the United States. The explanation may lie partly in the system itself:

unanimity is required once a name is put forth, and unless there is a chorus of enthusiastic recommendations and an absence of dissenting voices, it is a lost cause from the outset. In addition, logic is a difficult field. Straddling mathematics and philosophy, the candidate is in competition with scholars in both fields and may not be recognized as stellar in either. Still, the fact remains that Quine, Church, and Gödel – logicians all – were recipients of honorary degrees in the United States, so the mystery remains. While there is no record of Tarski saying he felt slighted, undoubtedly he did.

The Ershov Affair

In May of 1980, a troublesome event occurred in the mathematics and logic community in the United States. Yuri Ershov, the brilliant young logician from Novosibirsk, was making a lecture tour of several universities, one of which was UC Berkeley. This visit, his first to America, was sponsored by the Fulbright Foundation. Of all the young logicians that Tarski had met during his 1966 stay in the former Soviet Union, Ershov had impressed him the most by his combination of mathematical talents (he had obtained important results in model theory, algebra, and the theory of computation) and personal liveliness. Stocky, athletic, convivial, and quick-witted, an enthusiastic vodka drinker with a good sense of humor – all these attributes appealed to Tarski. Stories were told of the boisterous drinking contests between Alfred and Yuri; in Ershov, Tarski had met his match. Since that time, in addition to his research achievements, the ambitious young Ershov had become a leading administrator in Soviet mathematical circles. After Mal'cev's death in 1967, he took charge of the seminar in algebra and logic in Novosibirsk, becoming director of the Mathematical Institute there and a corresponding member of the Soviet Academy of Sciences. Eventually, he would become rector of the University of Novosibirsk.

Like Mal'cev, Ershov was a member of the Communist Party – an essential step in the career of anyone wishing to rise in the echelons of power in the Soviet Union – and as such he followed the party line. Inevitably this led to his involvement in the widespread official and semi-official anti-Semitic practices that were well known in the academy. Evidence for this and its effects on Soviet mathematics came from a variety of sources, including reports of emigrés, visitors to the Soviet Union, and organizations of concerned scientists. Among these was a 1978 *samizdat* essay by Grigori Freiman, "It Seems I Am a Jew," and a report by emigré mathematicians entitled "The Situation in Soviet Mathematics."[27] These articles documented the denial of admission of bright young Jewish mathematics students to universities, denial of advancement to graduate study for those who had passed the first hurdle, limitation on the publication of Jewish mathematicians' research in leading journals, limitation of invitations to Soviet conferences, impossibility of travel to international conferences, rejection of Ph.D. theses, and, finally, rejection of the higher Doctor of Science thesis needed to secure a university chair.

Ershov was a member of the central committee (VAK) that made final judgments as to the acceptability of dissertations for the Doctor of Science degree. Almost without exception, theses submitted to VAK by Jewish mathematicians, even when supported by leading non-Jews, were rejected; there were also other, more specific stories that were damaging. Thus, when Ershov came to the United States he was greeted with open hostility and protests by a significant number of – though by no means all – distinguished American mathematicians and logicians.

Ershov's first talk at UCLA went off without incident but, by the time he arrived in Berkeley, a movement had been mounted to boycott his lecture of May 9. A number of the leading logicians and mathematicians in Berkeley wrote an open letter to him protesting his "extraordinary role as one who consistently interferes with the careers of Soviet mathematicians for reasons having nothing to do with the character of their work."

Terribly disturbed by the allegations and the impending boycott, Tarski invited Ershov to his home on the day before his lecture. As a friend and colleague, he wanted to hear his side of the story. Ershov firmly denied the accusations against him, but in the end Tarski, unsatisfied with his explanations, told him that he would not go to his lecture. Futhermore, despite his general principle of keeping politics and mathematics separate, he joined Leon Henkin, Julia and Raphael Robinson, John Kelley, Steven Smale, and a number of others in signing the letter of condemnation, which concluded: "Yuri Ershov, we protest that you have dishonored our profession. For shame!"[28]

Only a few people attended Ershov's lecture, and to do so they had to pass through a picket line that included a sign-carrying Julia Robinson.

In the seminar room, a banner reading "We protest the presence of the anti-Semite Yuri Ershov" was displayed for a few moments before his talk. Two of Tarski's closest colleagues, John Addison and Bob Vaught, did attend the lecture. Vaught felt a strong obligation to the guest speaker, whom he had invited on behalf of the group in logic and the methodology of science following its vote (by a narrow margin) in favor of the invitation. Terribly disturbed by the boycott, he resigned from the group in spite of Julia Robinson's plea not to do so; she said that "our differences are minor compared to thirty years of friendship."

As for Ershov, he continued to deny all complicity in anti-Semitic actions and wrote Tarski the day after his lecture, reiterating that the accusations were unproven and unjust: "As every man, I make mistakes but never in my activity was I guided by any racial or national prejudices whatever."[29] Ershov had a few defenders at the University of Chicago, the next stop on his tour, where again he spoke to a very small audience amidst vociferous public disapproval. Saunders Mac Lane was among those who said the evidence against him was insufficient, but the following week – in Cambridge, Massachussets – Willard Quine, Burton Dreben, Hilary Putnam, Aki Kanamori, Hartley Rogers, Sy Friedman, Richard Shore, and others signed a letter to Ershov that detailed reports of "a general pattern of misuse of your professional and academic position for political and anti-Semitic ends .... In the absence of convincing evidence [to the contrary] we cannot in good conscience offer you either our personal hospitality or professional collaboration." In addition, most of these distinguished scholars were signatories of a letter scolding the Fulbright Foundation for selecting Ershov in the first place. The protest made sufficient noise for the *Boston Globe* to report on the affair.

These protests were a rare event in the mathematics and logic community, where under most circumstances politics is light years away from the abstract work in which scholars are immured. The brouhaha was particularly painful to Tarski – not only because of his affection and regard for Ershov but also because of his own experience in Poland, where anti-Semitism had hampered his career. Ironically, in a replication of Tarski's experience, many of the Jewish mathematicians and logicans who were persecuted eventually made their way out of the Soviet Union to find opportunities in other countries that were far better than those in their native land.

"The Right Stuff"

John Corcoran, determined to make good on his commitment to help prepare a new edition of *Logic, Semantics, Metamathematics,* proposed a contract with the Hackett Publishing Company, which was headed by his friend William Hackett. The original publisher had been Oxford University Press, but Tarski had had disagreements with them. "In this case, Tarski worked the financial angle to his satisfaction," Corcoran said. Hackett had come to San Francisco in connection with a meeting of the American Philosophical Association and, having heard endless accounts from Corcoran about what a great man Tarski was, expressed a desire to meet him. Through Corcoran, Hackett was invited to Tarski's house in Berkeley. He was thrilled; it was an hour and a half he never forgot. "Tarski had obviously turned on the charm and was spectacularly scintillating – as he could be when he wanted to win someone over," Corcoran said. "He sweet-talked Hackett into terms much more favorable than anyone else ever got – not even Nelson Goodman. After the meeting Bill called me up and gushed about how wonderful Tarski was and how he now understood why I was so devoted to him. Bill had just finished reading Tom Wolfe's book, *The Right Stuff*, about the astronauts, and he said Tarski was a guy who had the 'right stuff.' For the next few weeks that was all he could talk about."[30]

Devotion is the apt term. For two summers in a row, 1981 and 1982, Corcoran traveled from Buffalo to Berkeley to work on the book, and for three years he spent half of his research time on it and other Tarski projects. Although Hackett offered to pay him as editor, Corcoran refused any remuneration with the rationale that this would somehow keep him "pure." It was his notion that, no matter how small an amount he took, Tarski would think it was greedy or cheap of him to accept.

While in Berkeley, Corcoran went to the Tarski home three or four nights a week to work on the revised edition and on other articles that Tarski asked him to edit or translate. He arranged his sleeping schedule to match Tarski's and even gave up his accustomed pipe in favor of the little Robert Burns cigarillos that Tarski smoked. "We would sit there with the lovely bay view in front of us all the time. Sometimes, he would be very talkative, but mostly we were working, not schmoozing."

As compensation for his selfless dedication, Tarski let Corcoran know that he was pleased with his company. This feeling of being close to Alfred was the gratification sought by many of the acolytes who went way beyond normal bounds to help him. "Maria told me that I was *good* for him; that whenever I was around him, his health would improve a lot, that he would sleep better, eat better and that he was in a better mood and that he would look forward to my coming." Since those were the years of Tarski's decline, hearing this made Corcoran very happy. And then, "When Tarski told me there was only one other philosopher he enjoyed talking to and that was Richard Montague, I really felt flattered, because of course Montague was his student."[31]

In fact, there had been many philosophers Tarski enjoyed talking to over the years: Carnap, Popper, McKinsey, Quine, and Suppes, to name only a few. Although flattery and exaggeration were not part of his regular stock in trade, Tarski knew what to say when it served his purpose; it was easy for him to make Corcoran feel appreciated – and he did. The intellectual rewards of working intimately with the man he revered were as rich as he had anticipated.

At eighty, the pattern of Tarski's evening sessions with Corcoran was only slightly different from the previous forty years of all-nighters spent with other colleagues and assistants: formulating phrases, reading and rereading them until he was satisfied. The pace, however, was significantly slower. Like others, Corcoran recalled Tarski asking him to go over a translation sentence by sentence and then paragraph by paragraph so that he could make sure the whole thing sounded right. "As I read," said Corcoran, "his eyes would droop and his head would drop. If I stopped reading, he would look directly at me and say 'please continue.' I thought everything had shut down but I was wrong: his mind was still going; he actually made some clever remarks about what I had read while he seemed to be asleep. Even if he fell asleep for a moment, he was still in his groove, looking for the optimum expression, trying this, trying that, and loving the process. His patience and enthusiasm for this type of endeavor seemed limitless." Contradicting this account, Steven Givant – who had worked with Tarski for many years and was still meeting with him during this period – remembers Tarski falling asleep for much longer than a moment. Givant found the evenings of those last years very trying because progress was slower than ever.

As with his other final projects, Tarski would not live to see the new edition of *Logic, Semantics, Metamathematics* in print (it was published in 1983). Even so, he did approve the final form of every change that was made and also wrote a postscript in which he expressed his "most genuine and cordial gratefulness to John Corcoran":

> The new edition ... is almost exclusively due to his [Corcoran's] initiative.... as editor he fulfilled his task with great fervor and enthusiasm. What is more valuable, he has provided the volume with his own introduction. While I do not always agree with Corcoran's judgments expressed in that introduction, I sincerely enjoy having it included in the volume and I feel confident that it will prove interesting and stimulating to many readers of my work.[32]

Here, amazingly, is one of the *very* rare occasions when Tarski did not insist on having absolutely everything said exactly his way.

At the end of his own introduction, Corcoran wrote that "no serious reader will fail to be profoundly impressed by Tarski's mastery of the discipline he virtually creates in these pages; but neither can the serious reader fail to be encouraged in his own research by the example of Tarski's steadfast, sober, and deeply human determination to come to grips with the same problems that perplex everyone who studies logic. *Tarski is an explorer who inspires explorers.*"[33] [emphasis added]

An Eightieth Birthday, the Berkeley Citation, and the Alfred Tarski Room

On 13 January 1981, the eve of his eightieth birthday, a group of Tarski's local colleagues, all close friends, gathered at Leon and Ginette Henkin's home in Berkeley to celebrate the occasion with a quiet dinner.[34] Alfred, impeccably groomed, wore a dark suit and a carnation in his lapel; no longer robust – he had heart trouble and emphysema – he and Maria had asked that the party be small.

Besides the cake, the centerpiece of the evening was a beautiful calligraphic document, presented with a flourish and read by Leon Henkin: "These volumes of his collected papers are presented to Alfred Tarski in celebration of his eightieth birthday with affection, admiration and gratitude by colleagues, students and friends from around the world." Attached was a list of more than a hundred people who had contributed to the gift of the four volumes of his collected papers, to be beautifully

Tarski at his eightieth birthday party, 13 January 1981.

bound in hand-decorated leather and printed on fine paper. The ornate document was, in effect, a promissory note because the books were still in production. Tarski was told that he would receive the real thing in a few months upon being awarded the Berkeley Citation, which was to be a continuation of the birthday celebration. At the Henkin party, Tarski was also given a print by the Polish artist Stefan Mrozewski, a master of the art of the woodcut.[35]

Visibly touched by both gifts, Tarski expressed his profound gratitude to all and especially to Steven Givant and Ralph McKenzie, who with enormous effort had assembled and edited the four volumes of his works. A few days later he wrote a general letter to all the signatories: "I have been overwhelmed by the outpouring of good wishes.... I have also been deeply moved by your generous gifts."

Tarski received many individual letters of congratulation for his eightieth birthday from students, friends, and colleagues from all periods of his life. One of the most interesting came from Sammy Eilenberg:

Kochany Fredziu,
... It is exactly 50 years ago that I took my first course with you.... I was 17 years old. On one occasion I was the only one in class to turn in a perfect homework. You called out my name and asked me to come forward. I pretended that I wasn't there. After class I approached you and explained that

Leon Henkin at the Berkeley Citation ceremony for Tarski, UC 1981.

the reason I didn't answer was that you might ask me to write on the board. The class was on Saturday morning and I was observing the Sabbath. That is when we actually spoke to each other for the first time.
With much admiration and love,
       Sammy

Eilenberg's amusing, affectionate greeting was double-edged, since it reiterated – between the lines – his lifelong criticism of Tarski for having abandoned his Jewish origins.

    ે& ે& ે&

The Berkeley Citation, the highest honor awarded by the University of California, is given for extraordinary achievements in a field and for outstanding service to the university. Lauding Tarski as one of the two giant figures in twentieth-century logic, Chancellor Ira Michael Heyman presented the award to him on 29 May 1981 at a ceremony in the Mathematics Common Room on the tenth floor of Evans Hall – named after Griffith Evans, the former chairman, who had hired Tarski in 1942 and was in large part responsible for creating the great mathematics department in Berkeley. Although the building itself is cold and institutional, the spacious common room is blessed with a breathtaking view of the San Francisco Bay and its surrounding mountains, bridges, and cities.

After accepting his prize, Tarski, small and now gaunt, addressed the audience briefly. Somehow through his stance and body language he gave off a tremendous intensity as he spoke of what it means to be a mathematician and of how one can never tell in advance who has what it takes and who does not. As illustration he offered his oft-told apocryphal tale of how he himself had received a "B" in logic as a student in the *gymnasium*, with the obvious implication: "and look what happened."

After Tarski's moving speech, Chancellor Heyman invited the assembled audience to follow him three floors down to 727 Evans Hall for another ceremony in the library and common room for the group in logic and the methodology of science. Acting for the regents of the university, Heyman – with Tarski by his side and a crowd squeezed around them – unveiled a bronze plaque and a framed photograph of the eponymous professor and then officially renamed the space "The Alfred Tarski Room."

The bronze plaque cited Tarski as a "great logician and inspiring teacher." Many of Tarski's students' dissertations are housed in this room along with an extensive collection of thousands of reprints of work by Tarski, his students, his colleagues, and visitors. In addition, there are course notes, project reports, and a small collection of key texts, monographs, and journals available for use by faculty and graduate students. Especially pleasing to Tarski, his "Room" also has a beautiful view of the bay.

Tarski was thrilled beyond expectation with the events of the day, which included presentation of the fine leather edition of his collected works that had been promised in January.[36] Still, there was a small thing that bothered him, and a few days later he quietly let Leon Henkin know that he would really prefer it if his favorite photograph of himself, the one taken in the Soviet Union in 1966, could replace the one that was initially hung – which was fine, but he really did like the other one better. The change was made, of course.

15

The Last Times

Now Tarski had to face the fact of his deteriorating health, but still he didn't lose his appetite for intellectual engagement and his enthusiasm for doing the things he loved when his strength permitted. No one saw this more dramatically than Verena Huber-Dyson, his companion of the 1960s, who had become a professor at the University of Calgary in Canada. She had made peace with Alfred after his reconciliation with Maria, "after he had become old and docile" (an evaluation that would have offended him). On her visits to Berkeley she always saw them both and developed an especially warm relationship with Maria, which continued after Alfred's death.

In Calgary, Verena nominated Tarski for an honorary degree – the last he would receive – and the proposal was successful. Because the academic mills grind slowly, it happened that the degree was to be granted in June 1982, while Verena was on an extended sabbatical leave in Australia. She had sponsored the award, and it was important to her and the university that she be present. "So," she said, "I found myself whisked back to Calgary and snatched off a severely delayed flight by my teaching assistant, who threw an academic gown over my head and drove me, sirens howling and lights flashing, to the convocation ceremonies which had already begun and where I saw a frail figure climbing up to the podium":

I was shaken to see him so transformed; Alfred, simply shrunk down to essentials, almost a spirit, like a wizened old goblin from an ancient German fairy tale. But there he was, clad in the academic regalia of a Polish University that someone in Calgary had taken the trouble to dig up. It was probably his last public appearance and he did not talk about specific philosophical or mathematical issues but instead discussed and defended the original *raison d'être* of academia. He stood up for academic freedom in the pursuit of knowledge and the rights of those whose foremost concern is their research, and admonished the university to treat those concerns with due respect.[1]

Maria and Ina had accompanied Alfred to Calgary to be present while he was being honored. At the president's dinner following the ceremony, Verena voiced the opinion that Mrs. Tarski also deserved an honorary degree and she knew there were others who felt the same way. Alfred, meanwhile, unfurled his social charm, making a special point of engaging all the ladies in conversation and asking everyone about their scholarly pursuits. After learning that interest in Wittgenstein ran high, the following day he gave Verena a serious moral lecture about her duty to protect the students from bad influences such as Wittgenstein's.

On the weekend, a small group escorted the Tarski family to the Canadian Rockies. Tarski, the lover of mountains, could not pass up the opportunity to see Banff, Lake Louise, and the Victoria glacier. It was like an old Tarski road trip; sick as he was, he insisted on smoking his cigarillos and keeping the car windows closed – to the major discomfort of everyone including himself. But once they were in the luxuriously warm atmosphere of the Chateau Lake Louise looking up at the glacier, Tarski came to life. "He was overwhelmed by the grandeur and the purity of that sight," Huber-Dyson said, "and all his pain seemed to fall away. That was the last time I saw Alfred and that is how I want to remember him."[2]

The Last Students: The Same Old Story

Toward the end of the 1970s, Benjamin ("Pete") Wells had come to Tarski with a very direct appeal. "What I want most in life now is a thesis with your name on it. I want to finish my Ph.D. and I want your help," he said. "I will do research on anything you want me to do. You pick it." He was still teaching high school and had done no serious mathematical research for seven years. Pleased to have him back in the fold, Tarski proposed that Wells continue his work in equational logic on pseudo-recursive varieties, a topic he had abandoned in favor of a subject of his own choosing. Wells's frank first reaction was "Not that backwater!" but soon enough he said, "Okay, you've suggested what I hoped you wouldn't, but I'll do it."[3]

Time passed; Wells brought what he thought was a solution to Tarski, who in turn described the work to Ralph McKenzie, an expert in equational logic, and asked for his opinion. McKenzie responded with words dreaded by every Ph.D. student: "I think I have already solved this problem." Hearing this, Wells quickly met with McKenzie to compare their proofs in detail. Immediately afterwards he telephoned Tarski to say, "Good news! Ralph is convinced that his proof has holes, that mine is sound and that it gives sharper results, and he is not going to think about it any more."

Agitated, Tarski said, "What do you mean he is not going to think about it? You think you can stop him from thinking? You can't turn off a mind!"

"Right," said Wells, "but he's going climbing in Yosemite for two weeks."

"Okay. So loook, Peete," said Tarski, "you've got two weeks to write it up," meaning McKenzie would indeed be thinking about it but couldn't complete it while he was in the mountains.

Two weeks turned into two years. Wells wrote up the thesis quickly enough but, in response to Tarski's criticisms, it went through several incarnations. Eventually he brought what he considered a nearly final manuscript to Tarski and waited for his approval. In the interim, however, Tarski's health had taken a serious downturn. Because of his congestive heart failure and emphysema, he had been in and out of the hospital several times; at home he tired easily and his ability for sustained concentration diminished. Yet true to form, whenever he read the thesis he inevitably asked for more. Wells recalled his utter frustration: "I got stern with Tarski. I went to him and said, 'I want this signed. I don't want to go through a long laborious refinement. You said my thesis is good so why don't you just go ahead and approve it? I have a new baby. I can't go on messing around so let's just do it and get it over with.'" No other student of Tarski's ever put the matter so bluntly. Wells's bantering insistence came naturally to him, and Tarski, sick as he was, responded positively to his nerviness because he liked him personally and did think the work was basically good. Still, this did not mean immediate approval.

Tarski's rejoinder to Wells's plea was a story about a female faculty member who expected special consideration in a tenure decision because she was pregnant. "I wouldn't support it because that's not a reason to make a scientific judgment, just because someone is pregnant." In effect he was saying to Wells: "You have a baby? So what? I wouldn't approve it even if *you* were pregnant."

Their contretemps continued into 1982, following Tarski's return from the honorary degree ceremony in Canada. After first refusing to make any further revisions, Wells took on the manuscript again and came up with new results of the sort Tarski had requested, but by then Tarski was even sicker. Time was running out for Wells. If the thesis was not signed by November 12, he would have to re-enroll for another year of graduate work, at considerable expense, and there was little chance that Tarski would be the main signatory. In a last-minute attempt, he left a final version at the door of the house on Michigan Avenue on a day when he wasn't sure whether Tarski was at home or in the hospital. A few hours later, Maria called McKenzie and told him that Tarski had signed the thesis. Wells delivered it to the appropriate administrative office at the eleventh hour, just as the doors were closing. A few days later, Leon Henkin told him that Tarski had said, innocently, "I didn't know Wells had a deadline"; Steven Givant added, "He said, 'You know if I felt better I would have kept him working longer.'" For Wells, achievement of the Ph.D. degree made a significant difference in his employment, for after a year as instructor at San Francisco State University he succeeded in obtaining a regular position in mathematics and computer science at the University of San Francisco.

Judith Ng's endgame with Tarski played out differently. Like Wells, she too had been working with him since the sixties but had not made it over the last hurdle of completing her doctoral dissertation on relation algebras with transitive closure. Needing to earn a living, she stopped being a full-time student and taught in short-term positions at various universities including California State Northridge, San Francisco State, Mount Holyoke, and UC Santa Cruz. Then, realizing "I couldn't be an academic migrant worker" forever, she accepted a permanent position at California State University, Sacramento. Although much less intensely, she worked on her thesis and returned to Berkeley during vacations and summers to consult with Tarski. Often, when the Tarskis were traveling, she would stay in their house and take care of the garden.

She was now fully into the role of surrogate daughter to Alfred and Maria and was very much a part of their family activities. It was a relationship she greatly valued, though she knew it was not to her advantage academically. The two aspects of her interaction with Tarski were in separate compartments and, close as they were, she always called him "professor" until the last six months of his life, when he begged her to call him "Alfred." Even then, she said, she never felt quite right about it because she took seriously his general rule about first-name usage being accorded only *after* the Ph.D. had been officially granted (or at parties when he was tipsy).

Rebellion was necessary for Ng when finally, after the long years of effort, she had what she and other colleagues thought were sufficiently important results for a thesis and Tarski, in his maddening fashion, asked for more. "At that point," she said, "I balked." It wasn't in her nature to make a scene and make demands the way Wells had done. Instead, she told Tarski she wasn't going to do any more; and she didn't. Taken aback by Ng's refusal, Tarski relented and said, "Okay, you have a thesis," explaining that he hadn't approved it earlier because he was worried people would think he was being lenient toward her since she was his friend, so he had to lean the other way.[4]

Perversely undermining her case, Ng then insisted she wanted to do more; she wanted a thesis that would satisfy him completely. However, by the time the "more" was done, Tarski was unable to read it with his usual attentiveness. He approved the work verbally, including the last additions, but asked Ralph McKenzie to take over. In the end, she completed her dissertation after Tarski's death under the supervision of McKenzie. The work is dedicated "To the memory of my teacher, Alfred Tarski" and in her acknowledgments she writes: "Words cannot express my gratitude and indebtedness to my teacher, Alfred Tarski, originator of the theory of relation algebras."

Rhododendrons and the Rangpur Lime

In his last year, as long as he felt up to it, Alfred continued to make the almost ritual Sunday excursions that had always been part of the family program. One day in spring of 1983, Judith Ng and Peter Hoffman drove Tarski and Maria to the Kruse Rhododendron Reserve, a beloved spot in Sonoma County, twenty miles north of the point where the Russian River meets the Pacific Ocean. The outing was to include lunch at Fiori's, a favorite Italian restaurant in the small town of Occidental, and Alfred insisted they go by way of the beautiful but tortuous coast highway with its spectacular ocean views. En route, though, he soon fell asleep and Hoffman, who was driving, decided to take the easier inland road. He was anxious about Alfred's breathing – he was already using an oxygen tank

Steven Givant with Alfred and Maria in their garden, 1981.

when necessary. When Alfred awoke and realized they were not driving along the Pacific, he was vexed and, as Judith said, "had a grumpy pout about it because he insisted that he wanted *us* to have the pleasure of the sea route." Another disappointment was that only a few rhododendrons were in bloom at the garden. Then, during lunch, he had a horrendous coughing fit, after which it was discovered that the valve on the oxygen tank had been left open and the tank was empty. Anxiety caused them to cut the trip short and hurry home.[5]

Another place Alfred wished to go was to the Four Winds nursery in Fremont, some forty miles from Berkeley, to buy a Rangpur lime tree. There were good horticultural resources nearer to home, but Four Winds specialized in dwarf citrus. He had bought plants from them since the late 1950s, and he *had* to have the best specimen to replace one that had recently died. It was Judith's impression that he wanted the tree not for himself – that he knew he would not be harvesting its fruit – but for the integrity of the garden. He never ceased loving plants of all varieties, in his own garden and elsewhere; it was an essential ingredient of his enthusiastic appreciation of the diversity of the natural world. Given any opportunity, he rarely failed to mention that his first intellectual love had been biology and botany; Steven Givant recalled that several times in his later years Tarski expressed regret "at having abandoned his first love, and jokingly vowed to return, in his next incarnation, as a biologist."[6] Besides rhododendrons, fuchsias, and roses, he had a special love for exotic fruits and citrus such as the Rangpur lime.

Judith arrived at the house early on the day they had planned to go to Four Winds. Only then did she really realize how very sick Alfred was. He was in the small room next to the kitchen with the view of the bay, lying in a hospital bed with his head elevated to make breathing easier. Although he was awake and alert, his face was extremely puffy; there was no question of his going along on this trip, but he was still fully committed to the purchase. Planting a tree that would bear fruit for others to harvest cannot be compared to the ramifications of his intellectual efforts; still something in the impulse is the same.

ɞ   ɞ   ɞ

Tarski died on October 27, 1983, in a Berkeley hospice, feverishly mumbling lines of poetry from Heine. Maria was by his side through all his trials, supported by his "family" of loyal students, friends, and colleagues. A few days before his death, Steve Givant went with Maria to see him.

"He took my hand and said, 'Ach, Steve. Good boy,' and I thought, here I am, forty years old, and Alfred still thinks of me as a boy – but he said it with such tenderness. It was very touching."[7]

Two memorials were held to honor Tarski. At Stanford, Patrick Suppes, Solomon Feferman, and Jon Barwise (Feferman's former student, thus Tarski's "grandstudent") emphasized Tarski's scientific achievements and his personal importance to them.[8] In Berkeley, on December 16, John Addison introduced seven speakers – Benson Mates, John Kelley, Robert Vaught, Dana Scott, Czesław Miłosz, Steven Givant, and Leon Henkin – who reminisced about Tarski's life from his earliest days there to the end. Miłosz, who had known Tarski only by reputation in Warsaw, told of their friendship in Berkeley, their shared love of poetry, and "a literary debt" – Tarski had alerted him to a nineteenth-century translation of the Bible in Hebrew and Polish. It seemed very appropriate, then, that he ended by reading the twenty-ninth Psalm in Polish.

Leon Henkin, speaking last, emphasized "that supreme absence of self-doubt" which set Tarski apart from most people and described him with an alliterative series of adjectives: "proud, penetrating, persistent, powerful, passionate." He also described Alfred's sociability and conviviality, the characteristics that joined him to the rest of mankind. Leaving the podium, Henkin closed the ceremony with a tribute to the valiant Maria, walking over to her and kissing her hand. The gesture also reminded the audience of Alfred at his most romantic and charming, playing (and knowing he was playing) the *galant*, slightly awkward, smiling, blushing – and wanting to please both the lady whose hand was being kissed and himself.

And it was that Alfred, the romantic, who brought forth the lines from Heine as he lay dying:

> *Es ist eine alte Geschichte,*
> *Doch bleibt sie immer neu;*
> *Und wem sie just passieret,*
> *Dem bricht das Herz entzwei.*

ଈ ଈ ଈ

Maria Tarski continued to live in the house on Michigan Avenue, visited by her many friends and her children and grandchildren. Jan returned to the United States permanently and lived with his mother until her death on January 8, 1990.

Coda

At the end of May 2001, one hundred years after Tarski's birth, a conference was held in Warsaw to celebrate his life and work.[9] A glittering reception in the Kazimierzowski Palace of the university was part of the program. It was the first celebration in his honor in Poland, where – for the first thirty-eight years of his life – the cultural and intellectual milieu had shaped his spirit and his way of understanding the world.

He spent the rest of his life in the United States. In Berkeley, with the combination of human talent, financial resources, and the freedom available to him, he built an empire; but his core Polishness never left him, and Poland never stopped counting him as theirs. Hence the centenary conference and, in conjunction with it, an exhibition in the philosophy department library of his papers – especially his early works. Included among these was his old friend Witkacy's personal copy of "Pojęcie Prawdy" [The Concept of Truth], complete with his notorious off-color annotations.

The most dramatic and visible recognition of Tarski is in the handsome new university library built along the Vistula River. At the main entrance,

Statues of Jan Łukasiewicz and Alfred Tarski at the University of Warsaw Library.

on the broad staircase, his statue stands atop a high column, engraved with his name and a fragment from the manuscript of his masterwork on truth. As students enter the library they see him, life-sized, flanked by statues of Poland's other famous philosopher-logicians: Kazimierz Twardowski, Jan Łukasiewicz, Stanisław Leśniewski, each on his own column. The architect's intended effect is a reference to classic Greek architecture and the origins of logic and philosophy in ancient Greece. Although at first there was skepticism about the concept, the statues have become a symbol of the library.[10] The sculptor, Adam Myjak, rector of the Academy of Fine Arts, has juxtaposed the figures in an arresting fashion, none gazing directly at any of the others. Tarski – erect, legs apart, arms crossed – has finally been accorded his rightful position in Poland.

Authors' Note and Acknowledgments

Although we didn't think so then, we were young and innocent when we became acquainted with Alfred Tarski at the beginning of the 1950s. Solomon was his Ph.D. student and his teaching and research assistant; Anita was preparing to teach in elementary school.

Because Tarski was a gregarious man, we saw him frequently. Students were an important part of his social as well as his intellectual life; he needed a coterie as an extension of himself, for academic reasons and for his personal well-being. Solomon spent long evenings working with Tarski at his house, and both of us were invited to lively parties and weekend excursions with Alfred and Maria. At those times, Tarski liked to let his hair down, if one can say that about a man without much hair.

We moved to Stanford in 1956 and continued to see Tarski often at meetings – and both Alfred and Maria at the parties that followed the regular Stanford–Berkeley logic colloquia. We were no longer close, but our association was long. We exchanged holiday greetings and kept them abreast of our children's activities. On Alfred's eightieth birthday we joined a group of friends to celebrate the occasion at the Henkins' house in Berkeley.

In 1993, ten years after Tarski's death, Anita had just finished a biography of Jean van Heijenoort and Solomon was approaching the end of his work as editor-in-chief of the collected works of Kurt Gödel. It occurred to us that Tarski would be an interesting subject for a biography – *he* obviously thought so because he once said to a friend, "you make me so angry I will make sure you will not be in my biography." If we were to write one, we had to start while people who knew him in his early days were still alive. Many things besides our personal acquaintance were in our favor. Tarski's papers were housed in the Bancroft Library at UC Berkeley, only an hour's drive from Stanford, and a core group of his colleagues and friends were still in the San Francisco Bay area. Most importantly, his son and daughter – whom we had known when they were young children – lived in Berkeley, and Jan was still living in the Tarski house on Michigan Avenue.

So we began and soon became aware of how much we did not know about the man we thought we knew. To expand our knowledge we made several trips to Warsaw to acquaint ourselves with Tarski's home ground, and we traveled to Zakopane to see the mountains he loved. Our research drew on archival material concerning Tarski as well as on books and articles by or about him and about his cultural, intellectual, and historical milieu. Of equal if not greater importance were the long interviews and casual conversations we had with those who knew him in varied circumstances. The story of Tarski's life could not have been told the way we wanted to tell it without the unstinting help of the many, many good people who generously offered us their firsthand recollections, observations, and insights. We feel privileged, grateful, and indebted to all of them for enriching our understanding and thus his biography.

Our first thanks must go to Ina (Tarski) Ehrenfeucht and Jan Tarski, who provided us with family history and photographs that only they could have given; they answered our endless questions with honesty and unfailing good humor, no matter what the topic. It would be impossible to overemphasize the value of their contribution. Leon Henkin and John Addison, who worked with Tarski as partners from the 1950s to the end of his life, shed light on his academic achievements and their multiple projects and also gave us leads to other sources. Steven Givant, Judith Ng, and Benjamin (Pete) Wells were particularly helpful about the later period of Tarski's life; Givant and Ng were like members of the Tarski family.

From Henry Hiż and Peter Hoffman we learned details of Tarski's earlier years and his milieu in Warsaw. Michał Krynicki in Warsaw amazed us by finding all of Tarski's student records and other documents assumed to have been destroyed in the war. Also, in Warsaw he was our guide to Tarski's neighborhood and some of his favorite haunts. Jan Woleński answered every question we asked about the history of Polish logic and philosophy as well as many we didn't ask. Jan Berg unearthed documents – that we did not know existed – from the Swedish State Archives revealing how Maria, Ina, and Jan Tarski managed to leave Poland in 1945. Eckart Förster found and provided us with a rare copy of Hugo Gerlach's collection *Die vom Hinterhaus* containing the short story *In Letzter Stunde* described in Chapter 1. Herbert Enderton supplied us with an audiotape of his March 1978 conversation with Tarski.

Through interviews and correspondence with the following people, many of whom were close to Tarski at one time or another, we enlarged the picture of his life and work. In most cases the specifics of their contributions are evident in the text or the notes. In others, it may be less visible but valuable nonetheless. Our deepest thanks to: Ernest Adams, Mary Ann Addison, Irene Applebaum, Kenneth Arrow, Stefan Bauer-Mengelberg, Arianna Betti, David Blackwell, Zofia Borsuk, Jean Butler, Chen-Chung Chang, Oswaldo Chateaubriand, Charles Chihara, Tomás Chuaqui, Nino Cocchiarella, George Collins, John Corcoran, William Craig, Donald Davidson, John W. Dawson, Jr., John Doner, Andrzej Ehrenfeucht, Samuel Eilenberg, Herbert Enderton, Paul Erdös, Leo Esakia, John Etchemendy, Jens Erik Fenstad, Paulette Février, Dennis Flanagan, Dagfinn Follesdal, Thomas Frayne, Haim Gaifman, Leonard Gillman, Andrzej Grzegorczyk, Haragauri Gupta, Sarah Hallam, Lee Hause, Olaf Helmer, Henry Helson, Diane Hempel, Kathleen Henderson (Kathleen Chuaqui), Joseph Hersch, Peter Hinman, Wilfrid Hodges, Verena Huber-Dyson, Hidé Ishiguro, Aleksander Jędrzejczak, Richard Jeffrey, Bjarni Jónsson, Donald Kalish, Hans Kamp, David Kaplan, H. Jerome Keisler, John Kelley, Donald Knuth, Antoni Kosiński, Manfred Krifka, Boris Kushner, Erich Lehmann, Leonid Levin, Azriel Levy, June Lewin, Roger Maddux, Menachem Magidor, Ruth Barcan Marcus, Karol Martel, Benson Mates, Yuri Matiyasevich, Ralph McKenzie, Trevor McMinn, Irene Mikenberg, David Miller, Czesław Miłosz, John Mitchell, Donald Monk, Madeline S. Moore, Delos Morel, Maria Mostowska, Marcin Mostowski, Adam Myjak, Jan Mycielski, Lensey Chao Namioka, Dale Ogar, Hiroakira Ono, Leszek Pacholski, Charles Parsons, Volker Peckhaus, John Perry, Donald Pigozzi, Vaughan Pratt, Anne Preller, Hilary Putnam, Willard Van Orman Quine, Michael Rabin, Helena Rasiowa, Constance Reid, George Reisch, Gonzalo Reyes, Jean Rubin, Gabriel Sabbagh, Giovanni Sambin, Boris Schein, Dana Scott, Irene Schreier Scott, Andrzej Skowron, Hans Sluga, Marianne Freundlich Smith, Frits Staal, Patrick Suppes, Lesław Szczerba, Aleksandra Szmielew, William Tait, Michael Taitslin, Frederick B. Thompson, Boris Trakhtenbrot, Anna-Teresa Tymieniecka, Adam Ulam, Françoise Ulam, Johan van Benthem, Paul van Ulsen, Marilyn Vaught, Robert Vaught, Anita Wasilewska, Celina Whitfield, Richard Wiebe, Joseph Wisnovsky,

Wiktoria Witkowska, Elizabeth Wolgast, Richard Wollheim, Urszula Wybraniec-Skardowska, Janina Zawadowska, and Wacław Zawadowski. Our apologies to anyone we inadvertently omitted.

Lanier Anderson and Phokion Kolaitis read portions of the manuscript in its early stages, as did the members of the Biographers' Seminar at Stanford. Herbert Enderton, Branden Fitelson, Michael Friedman, and Paolo Mancuso read the entire manuscript. All provided useful suggestions and comments. Emilie Mycielski, Jan Mycielski, Grigori Mints, and Tadeusz Slawek helped with translation of a variety of documents. We thank them all.

We used the following archival sources and wish to thank the institutions where they are housed and most especially the people who helped us access the material there, offering us time and effort far beyond our expectation: Susan Snyder, David Farrell, Jessica Lemieux at the Bancroft Library, UC Berkeley; Michał Krynicki from the Warsaw University Archives and the Central Polish Archives, Ministry of Religious Matters and Public Education, Warsaw; Marcia Tucker at the Institute for Advanced Study, Princeton, New Jersey; Steven W. Siegel at the 92nd Street Y (Young Men's and Young Women's Hebrew Association) in New York City; Jan Berg from the Swedish State Archives, Stockholm; Wojciech Zalewski, Curator for Slavic and East European Collections at the Stanford University Library; Paul van Ulsen from the Evert W. Beth Archives, Amsterdam; Kai Wehmeier from the Heinrich Scholz *Nachlass*, Münster; and G. Thomas Tanselle at the John Simon Guggenheim Foundation.

For encouragement at the beginning of our project and for help in seeking a publisher we thank Donald Albers and Peter Renz.

Finally our warmest appreciation to Lauren Cowles, our editor at Cambridge University Press, who has been patient, kind, and unflappable; she also read the entire manuscript and offered many valuable suggestions. Our thanks as well to Katherine Hew for her dedicated attention to final details and to Matt Darnell for his invaluable work as copy editor.

The last words of an acknowledgment are traditionally reserved for "the person without whose patience, kindness, forebearance and understanding this book could not have been written." So permit us to say: it wasn't easy but we couldn't have done it without each other.

Tarski's Ph.D. Students[1]

Andrzej Mostowski, *On the Independence of Finiteness Definitions in a System of Logic*, 1938
Bjarni Jónsson, *Direct Decompositions of Finite Algebraic Systems*, 1946
Louise Hoy Chin Lim, *Distributive and Modular Laws in Relation Algebras*, 1948
Julia Bowman Robinson, *Definability and Decision Problems in Arithmetic*, 1948
Wanda Szmielew, *Arithmetical Properties of Abelian Groups*, 1950
Frederick Burtis Thompson, *Some Contributions to Abstract Algebra and Metamathematics*, 1952
Anne C. Davis Morel, *A Study in the Arithmetic of Order Types*, 1953
Robert Lawson Vaught, *Topics in the Theory of Arithmetical Classes and Boolean Algebras*, 1954
Chen-Chung Chang, *Cardinal and Ordinal Factorization of Relation Types*, 1955
Solomon Feferman, *Formal Consistency Proofs and Interpretability of Theories*, 1957
Richard Merritt Montague, *Contributions to the Axiomatic Foundations of Set Theory*, 1957
H. Jerome Keisler, *Ultraproducts and Elementary Classes*, 1961
James Donald Monk, *Studies in Cylindric Algebra*, 1961
Haim Gaifman, *Two Contributions to the Theory of Boolean Algebras*, 1962
William Porter Hanf, *Some Fundamental Problems Concerning Languages with Infinitely Long Expressions*, 1963
Robert Earl Bradford, *The Axiom of Choice in the Arithmetic of Cardinals*, 1965
Haragauri N. Gupta, *Contributions to the Axiomatic Foundations of Geometry*, 1965

John Doner, *An Extended Arithmetic of Ordinal Numbers and its Metamathematics*, 1968
Donald Pigozzi, *Amalgamation and Interpolation Properties of Cylindric Algebras*, 1970
George McNulty, *The Decision Problem for Equational Bases of Algebras*, 1972
Charles Martin, *Decision Problems in Equational Logic*, 1973
Roger Maddux, *Topics in Relation Algebras*, 1978
Benjamin Franklin Wells III, *Pseudorecursive Varieties and Their Implications for Word Problems*, 1982
Kan Ching (Judith) Ng, *Relation Algebras with Transitive Closure*, 1984 (completed under the direction of Ralph McKenzie)

Tarski also had a significant influence on the Ph.D. dissertations of the following students: Adolf Lindenbaum, Leonard Gillman, Dana Scott, and Steven Givant.

Credits

Front cover: pastel portrait of Alfred Tarski by Witkacy, 1934, courtesy of Jan Tarski
Back cover: pastel portrait of Maria Tarski by Witkacy, 1938, courtesy of Ina (Tarski) Ehrenfeucht
Frontispiece: photo courtesy of Ina (Tarski) Ehrenfeucht

PAGE
6 photo by E. Stumman studio; courtesy of Ina (Tarski) Ehrenfeucht
7 photo courtesy of Ina (Tarski) Ehrenfeucht
11 photo by J. Tyras Polski studio; courtesy of Ina (Tarski) Ehrenfeucht
12 photo courtesy of Ina (Tarski) Ehrenfeucht
14 reproduction of manuscript; used by permission of the Bancroft Library, University of California at Berkeley
15 reproduction of manuscript; used by permission of the Bancroft Library, University of California at Berkeley
19 photo courtesy of Ina (Tarski) Ehrenfeucht
36 photo courtesy of Ina (Tarski) Ehrenfeucht
40 reproduction of document; used by permission of the archives of the University of Warsaw
58 photo by Solomon Feferman; courtesy of Solomon Feferman
65 photo courtesy of Ina (Tarski) Ehrenfeucht
80 photo courtesy of Douglas Boynton Quine
82 photo by Maria Kokoszyńska-Lutmanowa; courtesy of Ina (Tarski) Ehrenfeucht
86 photo courtesy of Douglas Boynton Quine
89 photo courtesy of Ina (Tarski) Ehrenfeucht
90 photo by Maria Kokoszyńska-Lutmanowa; courtesy of Ina (Tarski) Ehrenfeucht
91 photo courtesy of Ina (Tarski) Ehrenfeucht
105 photo courtesy of Ina (Tarski) Ehrenfeucht
107 photo courtesy of Ina (Tarski) Ehrenfeucht

PAGE

125	photo courtesy of Françoise Ulam
129	photo courtesy of Douglas Boynton Quine
145	photo by Steven Givant; courtesy of Steven Givant
146	photo by Steven Givant; courtesy of Steven Givant
155	photo courtesy of Bjarni Jónsson
169	photo courtesy of Ina (Tarski) Ehrenfeucht
174	photo courtesy of Constance Reid
178	photo courtesy of Ina (Tarski) Ehrenfeucht
187	photo by Steven Givant; courtesy of Steven Givant
197	photo by Wanda Szmielew; courtesy of Delos Morel
202	photo by Steven Givant; courtesy of Steven Givant
207	photo courtesy of Willard Van Orman Quine
212	photo by Anita B. Feferman; courtesy of Solomon Feferman
216	photo by Steven Givant; courtesy of Steven Givant
226	photo by Verena Huber-Dyson; courtesy of Verena Huber-Dyson
227	photo by Verena Huber-Dyson; courtesy of Verena Huber-Dyson
233	photo courtesy of Richard Wiebe
234	photo courtesy of Paulette Février
236	photo by Jean-Louis Destouches; courtesy of Paulette Février
237	photo by Jean-Louis Destouches; courtesy of Paulette Février
240	photo courtesy of Ina (Tarski) Ehrenfeucht
241	photo by Steven Givant; courtesy of Steven Givant
253	photo by Steven Givant; courtesy of Steven Givant
258	photo by Maria Kokoszyńska-Lutmanowa; used by permission of the Institute for Advanced Study, Princeton, New Jersey
259	photo by Steven Givant; courtesy of Steven Givant
261	photo by Steven Givant; courtesy of Steven Givant
263	photo by Steven Givant; courtesy of Steven Givant
267	photo by Steven Givant; courtesy of Steven Givant
289	photo courtesy of Ina (Tarski) Ehrenfeucht
289	photo courtesy of Jan Tarski
307	photo used by permission of the Bancroft Library, University of California at Berkeley
311	photo courtesy of Judith Ng
319	photo by Steven Givant; courtesy of Steven Givant
321	photo courtesy of Donald Pigozzi
326	photo courtesy of Benjamin Franklin Wells III
330	photo by Judith Ng; courtesy of Judith Ng
331	photo by Judith Ng; courtesy of Judith Ng
332	photo courtesy of Richmond Thomason

PAGE

344 photo by Steven Givant; courtesy of Steven Givant
345 photo by Steven Givant; courtesy of Steven Givant
349 photo by Steven Givant; courtesy of Ina (Tarski) Ehrenfeucht
355 photo courtesy of Kathleen Henderson (Chuaqui)
362 photo by Lynn Corcoran; courtesy of John Corcoran
369 photo by Christine Suppes; courtesy of Christine Suppes
370 photo by Judith Ng; courtesy of Judith Ng
377 photo by Steven Givant; courtesy of Steven Givant
380 photo used by permission of the University of Warsaw Library

Polish Pronunciation Guide[1]

The Polish Alphabet

a, ą, b, c, ć, d, e, ę, f, g, h, i, j, k, l, ł, m, n, ń, o, ó, p, r, s, ś, t, u, w, y, z, ź, ż (no q, v, or x)

Pronunciation

This is shown mainly for letters and letter combinations whose pronunciation in Polish differs from that in English.

a	as in f**a**ther
ą	as in c**on**, n**on** (French, nasal)
b	as in **b**ig
bi	as in **b**eautiful
c	as in ca**ts**
ć	as in **ch**eek
ci	as in **ch**eek (used before vowels)
ch	as in lo**ch** (Scottish)
cz	as in **ch**alk
d	as in **d**o
dz	as in be**ds**
dż	as in **j**aw
dź	as in **j**eans
dzi	as in **j**eans (used before vowels)
e	as in **e**ver
ę	as in s**en**se, **un** (French, nasal)
g	as in **g**irl
gi	as in bu**g y**ou
h	as in **h**all

391

i	as in feet
j	as in yes, boy
ł	as in wag, bow
mi	as in harm you (used before vowels)
ń	as in canyon
ni	as in canyon (used before vowels)
o	as in poke
ó	as in toot
r	as in arriba (Spanish, trilled)
rz	as in pleasure
ś	as in sheep
si	as in sheep (used before vowels)
sz	as in shut
u	as in boot
w	as in vat
y	as in ill
ż	as in pleasure
ź	as in azure
zi	as in azure (used before vowels)

The *stress* in Polish falls on the next-to-last syllable.

Spelling

Polish spelling is followed for names of individuals with minor exceptions – for example, when they have taken up residence in the West or as shown in certain documents. Female names in Poland end in 'a' when the corresponding male name ends in 'i'; for example, 'Maria Tarska' in Poland is changed to 'Maria Tarski' in the United States. Similarly, 'Krystyna Tarska' is the Polish usage versus 'Kristina Tarski' or 'Ina Tarski' in the United States.

Polish place names have been Anglicized throughout: 'Warsaw', 'Cracow', 'Lvov', and so forth. Some Polish words are Anglicized – for example 'slivovitz' instead of 'śliwowica'.

Notes

Chapter 1: The Two Tarskis

1. Elizabeth Wolgast interview with ABF, November 1997.
2. Trevor McMinn letter to SF, May 1996.
3. Dale Ogar interview with ABF, September 1995.
4. Bjarni Jónsson letter to ABF, May 1995.
5. Anna-Teresa Tymieniecka interview with ABF, November 1997.
6. John Corcoran interview with ABF and SF, May 1997.
7. In some biographical articles, Tarski's birthdate has been given incorrectly as 14 January 1902.
8. Filip Friedman (1930), pp. 36–7. Reymont (in his 1904 novel *The Promised Land*) also cites the Prussak owners as the first to use steam in their textile factories. Some information about the Prussak family is available on the website "Jews of the Old Lodz," ⟨http://www.lodzjews.com/indexus.htm⟩.
9. Peter Hoffman interview with ABF, September 1994.
10. Genealogy dated 1995 provided by Marc Hersch (son of Joseph Hersch, Tarski's Swiss cousin).
11. Nowicki (1992), pp. 15–16.
12. Peter Hoffman interview with ABF, September 1994.
13. Ibid.
14. Ibid.
15. A story in Gerlach (1894).
16. In his translation, Alfred substituted Warsaw for Berlin.
17. Davies (1982), pp. 378–92.
18. Ibid., p. 389.

Chapter 2: Independence and University

1. Andrzej Grzegorczyk interview with ABF and SF, September 1993.
2. Peter Hoffman interview with ABF, September 1994.
3. Davies (1982); Zamoyski (1988), Ch. 14ff.
4. Zamoyski (1988), p. 261.
5. Ibid., p. 321.

6. Roos (1966), p. 14.
7. Zamoyski (1988), pp. 332–3.
8. Nowicki (1992), p. 7.
9. Davies (1982), pp. 393–4.
10. Abramsky et al. (1986), p. 7.
11. Ibid., p. 3.
12. Davies (1982), p. 240.
13. Ibid., pp. 241–2.
14. Ibid., p. 25: "Assimilated Jews ... were noted for their tendency to become more Polish than the Poles."
15. Garlicki (1982), p. 134.
16. Tarski's course register for the University of Warsaw, 1918–1924, from the University of Warsaw archives.
17. Tarski (1986a), Vol. 1, pp. 1–12 (in Polish); it is listed as a report to Leśniewski's seminar.
18. Garlicki (1982), p. 55. Total enrollment in the School of Philosophy in 1918/19 was 1096 students and for 1919/20 was 1361 students. The School of Philosophy included among its divisions Philosophy, History, Philology (Languages and Literature), and the Exact Sciences, of which the last in turn had further subdivisions into Mathematics, Physics, Chemistry, Botany, Zoology, and Geology. Figures for the enrollment in these separate divisions and subdivisions are not available.
19. Kuratowski (1980), pp. 163ff.
20. Ibid., p. 165.
21. Henry Hiż letter to ABF, April 1995.
22. Cf. Smith (1994).
23. Woleński (1989), p. 3, quoting Władysław Witwicki.
24. Ibid., p. 2.
25. Polish Philosophy Page, ⟨http://www.fmag.unict.it/~polphil/PolPhil/LvovWarsaw/LvovWarsaw.html#anchor36290⟩.
26. Ibid.
27. Later, Łukasiewicz was to be elected rector of the university twice, in 1922/23 and in 1931/32.
28. Translation in Tarski (1983a) as "Logical Operations and Projective Sets," pp. 142–51.
29. Bocheński (1994), pp. 1–2.
30. Developed independently by the American logician Emil Post; cf. ⟨http://plato.stanford.edu/entries/logic-manyvalued/#His⟩. For Łukasiewicz's papers on the subject, see Łukasiewicz (1970).
31. Bell (1934).
32. Bocheński (1994), p. 4.
33. From Łukasiewicz's memoir for 9 May 1949, translation by Owen LeBlanc and Arianna Betti in ⟨http://www.fmag.unict.it/~polphil/PolPhil/Lesnie/LesnieDoc.html#lukdiary⟩.

34. Letter of 9 December 1913 from Kotarbiński to Twardowski, translated in ⟨http://www.fmag.unict.it/~polphil/PolPhil/Lesnie/LesnieDoc.html#TK1913⟩.
35. Cf. "Tadeusz Kotarbiński and the Lvov–Warsaw School," in Woleński (1999), pp. 36–44.
36. Kazimierz Pasenkiewicz, "Recollections about Alfred Tarski from the Years 1921–1925" for the Historical-Philosophical Department of the Jagiellonian University in Cracow (translation by Jan Tarski). Tarski Archives, Bancroft Library, University of California, Berkeley.
37. Ibid.
38. Letter of 24 November 1922 from Marian Borowski to Twardowski, ⟨http://www.fmag.unict.it/~polphil/PolPhil/Lesnie/LesnieDoc.html#Borowski⟩.
39. Dana Scott interview with ABF, May 1994.
40. John Corcoran interview with ABF and SF, May 1997.
41. Aleksander Jędrzejczak interview on behalf of ABF by Michał Krynicki, July 1997.
42. From Łukasiewicz's memoir for 17 May 1949, communicated by Arianna Betti.
43. University of Warsaw Archives.
44. Henry Hiż, "Remarks on Leśniewski," notes of a lecture delivered to the Polish Semiotic Society in Warsaw, 22 May 1992.
45. The dedication for the second edition (1983a) of *Logic, Semantics, Metamathematics,* following Kotarbiński's death (and before Tarski's own death), reads: "To the memory of his teacher TADEUSZ KOTARBINSKI. The author."
46. Sundholm (2003); Woleński (1999), p. 67.
47. Jan Mycielski interview with ABF, December 1997.
48. Śrzednicki and Rickey (1984) and Sundholm (2003).

Interlude I: The Banach–Tarski Paradox, Set Theory, and the Axiom of Choice

1. Wagon (1985) provides an exposition of the paradox, its proof, and related developments.
2. Moore (1982) gives a history of this development and the ensuing controversies.
3. Cf. Gödel (1990), pp. 1–101.
4. According to Gödel's second incompleteness theorem, no proof of the consistency of the Zermelo–Fraenkel axioms can be given within that system if it is consistent; this puts strong limitations on any possible absolute proof of consistency. Cf. Gödel (1986), pp. 126ff.
5. Cf. Cohen (1966).
6. Solovay (1970).
7. Levy (1988), p. 2.
8. Charles Chihara letter to ABF, June 1994.
9. Unpublished extemporaneous remarks (preserved on tape) during the discussion period for a symposium chaired by Tarski on the philosophical implications of Gödel's incompleteness theorems, held in Chicago at a joint meeting of the

Association for Symbolic Logic and the American Philosophical Association, 29–30 April 1965.
10. Elsewhere Tarski has said more specifically that he subscribed to the philosophical views of his teacher and hero, Tadeusz Kotarbiński, called *reism* or *concretism* – a kind of physicalistic nominalism.
11. Oswaldo Chateaubriand interview with SF, August 1998.

Chapter 3: *Polot!* The Polish Attribute

1. Samuel Eilenberg interviews with ABF, April–May 1995.
2. Nowicki (1992), pp. 83–4, and Zamoyski (1988), p. 326.
3. Karol Martel letter to ABF, 15 January 1996.
4. Ibid.
5. Miłosz (1994), p. 72.
6. Gerould (1992), p. 272.
7. Ibid., pp. 243ff.
8. Ibid., pp. 239–42.
9. Ibid., p. 253.
10. Anne Preller letter to ABF, May 1996.
11. Tarski letter to Verena Huber-Dyson, 18 March 1963.
12. Henry Hiż, "Remarks on Leśniewski."
13. Henry Hiż letter to ABF, November 27, 1993.
14. Irena Grosz conversation with Jan Tarski in Warsaw, mid-1960s.
15. Celina Whitfield interview with ABF, April 1995.
16. Woleński (1999), pp. 61–2.
17. Polish Philosophy Page, ⟨http://www.fmag.unict.it/~polphil/PolPhil/Chwi/Chwistek.html⟩.
18. Betrand Russell letter dated 29 December 1929 to the Dean of the Faculty of Mathematics and Natural Sciences of the University of Lvov, quoted in Estreicher (1971), p. 202, and in Woleński (1999), p. 61.
19. Paolo Mancosu letter to SF, 6 October 2003; according to Christopher Pincock, there is circumstantial evidence that Frank Ramsey and others in Russell's circle supplied the additional references in the 1925 edition of *Principia Mathematica*.
20. Kac (1985), p. 50 footnote.

Interlude II: The Completeness and Decidability of Algebra and Geometry

1. Zygmunt (1991).
2. Tarski and Givant (1999), p. 214.

Chapter 4: A Wider Sphere of Influence

1. Menger (1994), p. 146.
2. Ibid., pp. 146–7.

3. Stadler (1997), Feigl (1969), and M. Friedman (1999).
4. Feferman (1986), pp. 28ff.
5. Menger (1994) p. 147.
6. Gödel (1986), pp. 44–125.
7. Richard Jeffrey letter to ABF, 24 January 2000.
8. Carnap (1963), p. 30.
9. Menger (1994), p. 147.
10. Gödel (1986), pp. 126–45.
11. Givant (1991), p. 25.
12. Tarski (1983a), pp. 93–4.
13. Tarski (1983a), p. 277.
14. Feferman (1999), pp. 55–6, 60.
15. Givant (1991), p. 25.
16. Aleksander Jędrzejczak interview on behalf of ABF by Michał Krynicki, July 1997.
17. Quine (1985), p. 97.
18. David Kaplan interview with ABF, February 1995.
19. Quine (1985), p. 98.
20. W. V. O. Quine letter to Tarski, 10 January 1981.
21. Quine (1985), p. 104.
22. Creath (1990), pp. 28ff. Quine's three lectures at Harvard in November 1934 on Carnap's work and ideas are published in that volume for the first time; cf. also Quine (1985), pp. 116–17.
23. Woleński refers to her as M. Kokoszyńska-Lutman, or simply M. Kokoszyńska.
24. Tarski Archives.
25. Feigl (1969), p. 637.
26. Menger (1994), pp. 60–1.
27. Cartwright et al. (1996), p. 78.
28. Dawson (1997), pp. 68–9.
29. Stadler (1997), pp. 395–402.
30. Popper (1974), pp. 397–8.
31. Ibid., p. 399.
32. Menger (1994), p. 194.
33. Cartwright et al. (1996), p. 83.
34. Menger (1994), pp. 194ff.
35. Carnap (1963), p. 60.
36. Stadler (1997), p. 404.
37. Ibid., p. 405.
38. Ibid., pp. 402–12.
39. Carnap (1963), p. 61.
40. Roos (1966), pp. 141ff, and Nowicki (1992), pp. 176–7.
41. Rudnicki (1987), p. 254.
42. Ibid., pp. 257–9.
43. English translation in Tarski (1983a), pp. 152–278.
44. Woodger (1974), p. 482.

45. Polish Philosophy Page, ⟨http://www.fmag.unict.it/~polphil/PolPhil/Lesnie/LesnieDoc.html#Leśniewski⟩, translation by Owen LeBlanc. There is also a partial translation in Woleński (1999), pp. 68–9.
46. Ibid.
47. Woleński (1999), p. 112.
48. Henry Hiż, "Remarks on Leśniewski."
49. Woleński (1999), p. 70.
50. Tarski interview with Herbert Enderton, March 1978.
51. Crossley (1975), p. 45.
52. Moore (1982), p. 276.
53. Crossley (1975), p. 36.
54. Ibid., pp. 32–3.
55. Nowicki (1992), p. 231.
56. Henry Hiż, "Remarks on Leśniewski."

Interlude III: Truth and Definability

1. Cf. Kirkham (1992) and Lynch (2001).
2. Tarski (1983b) is an English translation of the *Wahrheitsbegriff*.
3. Cf. Hodges (1985/1986) and Feferman (2004).
4. Cf. pp. 54–5 of Kripke's "Outline of a Theory of Truth," reprinted in Martin (1984).
5. Tarski (1983b), p. 153.
6. Here, in contrast to Interlude II, lowercase variables are used to range over classes of individuals.
7. The symbolism of sentential functions chosen here follows current practice; it also requires the use of parentheses to avoid ambiguity. In the *Wahrheitsbegriff*, Tarski used what is called "Polish notation," which does not require parentheses; he wrote Ixy for xIy, NF for $\neg F$, AFG for $F \vee G$, and ΠxF for $(\forall x)F$.
8. Cf. Gödel (1986), pp. 102ff.
9. Tarski and Vaught (1957).
10. Cf. Tarski (1983a), Article VI, pp. 110–42.
11. Unbeknownst to Tarski, a similar explanation of the notion of definability in a structure had been proposed by Hermann Weyl in an article in 1910. See Feferman (1998), pp. 258–9.
12. Tarski (1983a), p. 110.
13. Ibid., p. 111.
14. Tarski (1983a), Article XV, pp. 401–8.
15. Tarski (1944).
16. Cf. Martin (1984).
17. Cf. Montague (1974).
18. Cf. Portner and Partee (2002).
19. Cf. Mosses (1990).

Chapter 5: How the "Unity of Science" Saved Tarski's Life

1. Ulam (1976), p. 114. The shipboard photo showing "Adam, Tarski, and me" did not appear in Ulam's memoir. The authors were able to obtain a copy from Françoise Ulam, his wife. On close inspection, there is evidence of a fourth party at the table, a woman whose face has been scratched out.
2. Ibid., p. 56.
3. Adam Ulam interview with ABF, December 1998.
4. Ulam (1976), p. 115
5. Ibid., p. 119
6. *New York Times*, 3 September 1939.
7. Carnap (1963), p. 35.
8. In the later publication of the proceedings, Tarski's lecture is listed as "New Investigations on the Completeness of Deductive Theories"; see Stadler (1997), p. 432.
9. Quine (1985), p. 140.
10. Ibid., pp.141-2.
11. W. V. O. Quine letter to U.S. Consul in Havana, November 1939; Tarski Archives, Bancroft Library, UC Berkeley.
12. Marshall Stone, quoted in Quine (ibid.).
13. Quine (1985), p. 149.
14. Leon Henkin interview with ABF, December 1993.
15. Stefan Bauer-Mengelberg interview with ABF, July 1996.
16. Kenneth Arrow interview with ABF, November 1994.
17. Ibid.
18. Dewey and Kallen (1941), p. 18.
19. Ibid., p. 20.
20. Russell (1951), p. 343.
21. Issued 22 May 1940.
22. Tarski letter to Józef Bocheński, 20 April 1940.
23. Tarski letter to Heinrich Scholz, 7 January 1941; Heinrich Scholz Archives of the University of Münster.
24. Swedish State Archives, Stockholm.
25. John Simon Guggenheim Foundation Archives.
26. Carnap (1963), p. 36; cf. also Quine (1985), p. 150.
27. Ernest Nagel and Albert Hofstadter promoted the engagement at the 92nd Street YMHA.
28. 92nd Street YMHA archives.
29. Olaf Helmer interview with ABF and SF, March 1999; Diane Hempel interview with ABF, May 2000.
30. Mary Ann Addison story told at Tarski's eightieth birthday party, 13 January 1981.
31. Ruth Barcan Marcus letter to ABF, April 1997.
32. J. C. C. McKinsey letter to B. A. Bernstein, 25 October 1941; B. A. Bernstein Archives, Bancroft Library, UC Berkeley.

33. Givant (1991), p. 30.
34. Dawson (1997), pp. 149–51.
35. Ibid., p. 34.
36. Ibid., p. 153.
37. Paul Erdös letter to "Feferman family," September 1997.
38. Ibid.
39. Rider (1989), p. 294.
40. Tarski file, Institute for Advanced Study, Princeton.
41. Ibid.

Chapter 6: Berkeley Is So Far from Princeton

1. Reid (1982), p. 181.
2. Institute for Advanced Study Archives.
3. Tarski (1999), letter to Kurt Gödel, 12 December 1942. There is no record of Tarski's ever having taught a course in cryptography in the list of courses taught by him in the UC Berkeley Mathematics Department for 1942–1946.
4. Tarski (1999), letter to Adele Gödel, 4 January 1943.
5. Ibid., letter to Adele and Kurt Gödel, 9 December 1943.
6. Stadtman (1970), pp. 312–15.
7. Course listings from the catalogues of the University of California at Berkeley for 1942–1946.
8. Bjarni Jónsson letters to ABF, May 1994.
9. Ibid.
10. Guggenheim Foundation Archives.
11. There is a 24-page report by Bernstein on this matter in the archives of B. A. Bernstein in Bancroft Library, UC Berkeley.
12. Sarah Hallam interviews with ABF, January 1994.
13. Ibid.
14. Ibid.
15. Tarski (1999), letter to Adele and Kurt Gödel, 9 December 1943.
16. Ibid., letter to Kurt Gödel, 12 February 1944.
17. Ibid., letter to Gödel, 9 March 1944.
18. Ibid., postcard to Gödel, 4 April 1944.
19. Later, Montana State University.
20. Davies (1982), p. 480.
21. Jan Tarski interview with ABF, January 1994.
22. Ina (Tarski) Ehrenfeucht interview with ABF, October 1993.
23. Quine (1985), p. 190.
24. W. V. O. Quine interview with ABF and SF, March 1997.
25. Swedish State Archives, Stockholm.
26. Jan Tarski interview with ABF, January 1994.
27. Benson Mates interview with ABF, May 1994.
28. *Oakland Tribune*, 6 January 1946.

29. Tarski letter to Heinrich Scholz, 21 October 1946; Heinrich Scholz Archives, University of Münster.
30. Ibid.

Chapter 7: Building a School

1. SF personal recollection, March 1995.
2. Benson Mates interview with ABF, May 1994.
3. Tarski letter to Heinrich Scholz, 26 June 1948; Heinrich Scholz Archives.
4. Jean Butler interview with ABF, February 1996.
5. Reid (1996), p. 47.
6. Ibid., p. 53.
7. Ibid., p. 51.
8. Jan Tarski interview with ABF, January 1994.
9. Ina (Tarski) Ehrenfeucht interview with ABF, October 1993.
10. Ina (Tarski) Ehrenfeucht and Jan Tarski interviews with ABF, March 1994.
11. Frederick B. Thompson interview with ABF, November 1996.
12. (Eleanor) Lee Hause interview with ABF and SF, February 1997.
13. Frederick B. Thompson interview with ABF, April 1996.
14. Ibid.
15. Kelley (1989), pp. 486–7.
16. Stadtman (1970), pp. 319–39, provides a history of the UC loyalty oath controversy.
17. John L. Kelley interview with ABF and SF, August 1995.
18. Robert Vaught interview with ABF and SF, March 1994.
19. Vaught (1986), p. 879.

Interlude IV: The Publication Campaigns

1. Tarski (1941b).
2. Cf. Givant (1986), [43] on p. 921 and [61b] on p. 925.
3. Cf. McKinsey and Tarski (1944) as well as Givant (1986), [46] and [48] on p. 922.
4. Tarski (1944); cf. also Givant (1986), [44a] on p. 921.
5. Cf. Givant (1986), [39] on p. 921 and [45] on p. 922.
6. Tarski (1994); Givant (1986), [41^m](1) on p. 932.
7. Cf. Givant (1986), [41^m](6) on p. 933.
8. Tarski (1948).
9. Tarski (1951).
10. Cf. Givant (1986), [67a^m] on p. 934.
11. Tarski's decision procedure was never implemented for a computer, except for special cases; cf. Caviness and Johnson (1998). Fischer and Rabin (1974) proved that, in general, no procedure for elementary algebra is computationally feasible.
12. Tarski (1949).
13. Jónsson and Tarski (1947).

14. Henkin, Monk, and Tarski (1971).
15. Henkin, Monk, and Tarski (1985).
16. Tarski, Mostowski, and Robinson (1953).
17. Tarski (1956a).
18. Tarski (1986b).
19. Tarski (1983a).
20. Tarski (1956b).

Chapter 8: "Papa Tarski" and His Students

1. Elizabeth Wolgast interview with ABF, November 1997.
2. Gonzalo Reyes letter to ABF, November 1995; the student was Ann Yasuhara.
3. C. C. Chang interview with ABF and SF, February 1995.
4. Sarah Hallam interview with ABF, January 1994.
5. Anne Davis letters to Tarski, 1949–1950; Tarski Archives.
6. Delos Morel interview with ABF, February 1996.
7. Anne Davis letter to Tarski, 22 February 1960; Tarski Archives.
8. Delos Morel interview with ABF, January 2004.
9. Jean Butler interview with ABF, February 1996.
10. C. C. Chang interview with ABF and SF, February 1995.
11. Ibid.
12. Ibid.
13. Benson Mates interview with ABF, July 2000.
14. Pieter A. M. Seuren interview with SF, October 2001; cf. also Staal (1965) and van Ulsen (2000).
15. Jan Mycielski letter to SF, December 2003.
16. Paulette Février letter to ABF, November 1997.
17. Reid (1982), p. 239.
18. Logic and the Methodology of Science Program records, UC Berkeley.
19. Ibid.
20. Hans Kamp letter to ABF, November 1996.
21. Nino Cocchiarella interview with ABF and SF, April 2000.
22. Leonard Gillman letter to ABF and SF, November 1995.
23. Benson Mates interview with ABF, May 1994.
24. Ibid.
25. Translator's Preface to Tarski (1956a) and (1983a).
26. Dana Scott interview with ABF and SF, May 1994.
27. Dana Scott letter to Tarski, 18 August 1955; Tarski Archives.

Chapter 9: Three Meetings and Two Departures

1. Nerode (1996), p. 83.
2. The lists of participants, speakers, and summaries of talks presented at the Cornell meeting were typescripted, dittoed, and then organized into a volume entitled: *Summer Institute for Symbolic Logic, Cornell University 1957. Summaries of Talks.*

No editor is listed. A bound second facsimile edition was put out on 25 July 1960 by the Communications Research Division of the Institute for Defense Analyses and distributed to select libraries. Cf. also Feferman (2003).
3. Moschovakis (1989), p. 346.
4. Halmos (1985), p. 215.
5. See note 11 to Interlude IV.
6. Nerode (1996), p. 83.
7. Dauben (1995).
8. Cf. Delzell (1996).
9. Hilary Putnam interview with ABF and SF, April 1995.
10. William Tait letter to SF, October 2000.
11. Monk (1990), p. 498.
12. Ibid., p. 499.
13. Ibid.
14. Odifreddi (1996), p. 75.
15. Huber-Dyson (1996), p. 52.
16. Verena Huber-Dyson interview with ABF, January 1995.
17. See note 2 in this chapter.
18. George Collins letter to SF, 29 September 2000.
19. Michael Rabin interview with ABF, April 1994.
20. Cf. Henkin, Suppes, and Tarski (1959).
21. Tarski and Givant (1999).
22. Leon Henkin interview with ABF, December 1993.
23. Cf. Henkin et al. (1959) for the full program.
24. Paulette Février letter to ABF, November 1997.
25. Ibid.
26. Irene Schreier interview with ABF, September 1995.
27. Ina (Tarski) Ehrenfeucht interview with ABF, October 1993.
28. Jan Mycielski interview with ABF, December 1997.
29. Givant (1991), p. 22.
30. Paulette Février letter to ABF, July 1998.
31. Zofia Borsuk interview with ABF and SF, September 1996.
32. Ibid.
33. Wiktor Marek interview with ABF, June 2001.
34. Ralph McKenzie interview with ABF and SF, December 1995.

Chapter 10: Logic and Methodology, Center Stage

1. Cf. Feigl (1969) and Dahms (1993).
2. Reisch (n.d.).
3. Letter of Hook to Carnap, 29 March 1949, reproduced in part by Reisch (ibid.).
4. Evert W. Beth Archives, Amsterdam; cf. also van Ulsen (2000).
5. Pilet (1977), and H. Guggenheimer letter to SF, 9 September 2001.
6. Tarski (1986a), Vol. 4, pp. 715–16.
7. Ibid.

8. Friedman (1999), p. 12.
9. Henkin, Suppes, and Tarski (1959).
10. See the usage note for 'methodology' in the *American Heritage Dictionary*, 4th ed., Houghton-Mifflin, Boston, 2000.
11. Tarski letter to Patrick Suppes, October 1959; Tarski Archives.
12. The full program is to be found in Nagel, Suppes, and Tarski (1962).
13. The programs of the Unity of Science meetings 1935–1941 are given in full in Stadler (1997), pp. 406–36.
14. Nino Cocchiarella interview with ABF and SF, April 2000.

Chapter 11: Heydays

1. John Addison interview with ABF and SF, April 1994.
2. Ibid.
3. Gaifman interview with ABF and SF, September 1995.
4. Ibid.
5. Michael Rabin interview with ABF and SF, April 1994.
6. Ibid.
7. Donald Monk letter to ABF, 26 September 1995.
8. Peter Hinman interview with ABF, July 1994.
9. Thomas Frayne letters to SF, May 1999.
10. H. Jerome Keisler letter to ABF, 4 March 1996.
11. *Encyclopedia Judaica*, Vol. 14, col. 908.
12. Michael Rabin interview with ABF, April 1994.
13. Menachem Magidor interview with ABF and SF, August 1996.
14. Czesław Miłosz interview with ABF, October 1994.
15. Witkacy's annotation is: "Czy przy pomocy metajęzyka, można zrobić metaminetę?" (Rough translation, provided by Benjamin Wells: "Is it possible with the aid of the metatongue [metalanguage] to perform metacunnilingus?")
16. Czesław Miłosz interview with ABF, October 1994.
17. Cf. Huber-Dyson (1996).
18. Verena Huber-Dyson interview with ABF, January 1995.
19. Verena Huber-Dyson letter to ABF, January 1994.
20. Tarski letter to Verena Huber-Dyson, 28 October 1961.
21. Verena Huber-Dyson letter to Tarski, 29 October 1961.
22. Verena Huber-Dyson letter to ABF, January 1994.
23. Tarski letter to Verena Huber-Dyson, 18 April 1963.
24. Verena Huber-Dyson letter to ABF, January 1997.

Interlude V: Model Theory and the 1963 Symposium

1. Vaught (1974), p. 159.
2. Cf. Gödel (1986), pp. 44ff.
3. Chang (1974), p. 173.
4. Cf. Vaught (1974), p. 164.

5. Cf. Givant (1986), [54a], [54b], [55] on p. 923.
6. Tarski and Vaught (1957).
7. Cf. Dauben (1995).
8. Cf. Givant (1986), [58] on p. 924.
9. Ibid., [58ᵃc] on p. 930.
10. Hanf (1964).
11. Keisler and Tarski (1964).
12. Ibid., p. 226.
13. Cf. Robinson (1966).
14. Cf. Addison et al. (1965).
15. Cohen (1966).
16. Solovay (1970).

Chapter 12: Around the World

1. Jan Tarski interview with ABF, February 1999.
2. Ina (Tarski) Ehrenfeucht interview with ABF, April 1997.
3. Ibid.
4. ⟨http://www.piasa.org/history.html⟩.
5. Richard Wollheim letter to SF, 13 September 2002, and Michael Martin letter to SF, 17 September 2002.
6. Schweik, the picaresque eponymous antihero in the antiwar novel by the Czech writer Jaroslav Hasek.
7. Hide Ishiguro letter to ABF and SF, April 1997.
8. Hans Sluga interview with ABF and SF, April 1995.
9. David Miller letter to ABF, January 1999.
10. Ibid.
11. Joseph Wisnovsky interview with ABF, July 1999.
12. Tarski (1969).
13. "Tarski behaved like a gangster": Charles Parsons attributed this quote to Stanley Cavell, who denied having said it; but if the words were not accurate, the sentiment was.
14. Hans Sluga interview with ABF and SF, April 1995.
15. Wilfrid Hodges letter to ABF, August 1994.
16. Albers et al. (1987), pp. 26–7.
17. Ibid., pp. 34–5.
18. Also transliterated as 'Maltsev'.
19. Cf. Givant (1986), [41ᵐ](2), p. 932.
20. In a copy of the volume held by the library of the University of California at Berkeley, there is a handwritten note, in Russian: "The introduction to the appendix is not written by me. S. Yanovskaya."
21. Essays in memory of S. A. Yanovskaya, on the centenary of her birth, by B. Kushner in *Modern Logic* 6 (1996) and B. Trahktenbrot in *Modern Logic* 7 (1997).
22. Albers et al. (1987), p. 35.
23. Yuri Matiyasevich letter to SF, 10 July 2002.

24. Leo Esakia letter to ABF and SF, August 1996.
25. Ibid.
26. Mal'cev (1971).
27. Michael Taitslin interview with SF, July 1997, and Boris Trakhtenbrot letter to SF, 14 February 2002.
28. Haraguari Gupta letter to SF, 10 June 1997.
29. Hiroakira Ono letter to SF, 9 September 2002.

Chapter 13: Los Angeles and Berkeley

1. Judith Ng interview with ABF, August 2002.
2. Ibid.
3. David Kaplan interview with ABF, February 1995.
4. Donald Kalish interview with ABF, February 1995.
5. Hans Kamp letter to ABF, November 1996.
6. Judith Ng letter to ABF, September 2002.
7. Frits Staal interview with ABF and SF, November 1993.
8. David Kaplan letter to ABF, August 2002.
9. Partee (2001). See Montague (1974) for the relevant articles on formal semantics.
10. Cf. Portner and Partee (2002).
11. Anne Preller letter to ABF, April 1996.
12. Ibid.
13. Anne Preller letter to ABF, May 1996.
14. Frits Staal interview with ABF and SF, November 1993.
15. Ralph McKenzie interview with ABF and SF, December 1995.
16. Ibid.
17. Henkin, Monk, and Tarski (1971).
18. Donald Pigozzi letter to ABF, January 1996.
19. Ibid.
20. Benjamin Wells interview with ABF and SF, March 1995.
21. The result of Wells's translation work is Mal'cev (1971).
22. Kerr (1971).
23. Cf. ⟨http://humwww.ucsc.edu:16080/HistCon/faculty_davis.htm⟩.
24. Gitlin (1987).
25. Cf. ⟨http://www.law.northwestern.edu/faculty/clinic/dohrn/dohrn.html⟩.
26. Cf. ⟨http://www.mistersf.com/notorious/notpattyindex.htm⟩.
27. Benjamin Wells interview with ABF and SF, March 1995.
28. Enderton (1998).
29. Judith Ng interview with ABF, August 2002.
30. Richard Montague died on March 7, 1971.
31. Donald Kalish interview with ABF, February 1995.
32. Herbert Enderton letter to ABF, January 2003; Montague's funeral was at the Praisewater Funeral Home, Los Angeles, on March 12, 1971.
33. John Perry letter to ABF, April 1995.

Interlude VI: Algebras of Logic

1. Tarski (1941b).
2. Tarski and Givant (1987), Preface.
3. Tarski and Givant (1987); Steven R. Givant interview with SF, February 2003.
4. See, among others: Kawahara (1988), Maddux (1996), and Formisano et al. (2000); cf. also Formisano et al. (2001) and the website ⟨http://www.tarski.org⟩.
5. Steven R. Givant interview with SF, February 2003.
6. Cf. Halpern et al. (2001), Sec. 3.

Chapter 14: A Decade of Honors

1. The Tarski Symposium, UC Berkeley, 23–30 June 1971; the proceedings were published as Henkin et al. (1974).
2. Ibid., p. xii.
3. Dale Ogar interview with ABF, September 1995.
4. Ibid.
5. *San Francisco Chronicle*, 27 June 1971.
6. Tarski letter to Wanda Szmielew, 15 December 1973, Tarski Archives; the Polish word indicated by 'g....' in the original is probably 'gówno' ('shit' in English).
7. Steven Givant letter to Tarski, 17 September 1973, Tarski Archives.
8. Tarski and Givant (1987).
9. Givant (1991), p. 30.
10. The part of that work that had already been written was eventually incorporated into the joint monograph by Schwabhäuser, Szmielew, and Tarski (1983).
11. Part II of the treatise on cylindric algebras eventually appeared as Henkin, Monk, and Tarski (1985). An offshoot of work on it – a monograph authored by Henkin, Monk, and Tarski together with Hajnal Andréka and István Németi – appeared in 1981. Cf. Givant (1986), [81^m] on p. 935.
12. Giovanni Sambin letter to ABF, July 1996.
13. Kathleen (Henderson) Chuaqui interview with ABF, January 1996.
14. Gonzalo Reyes letter to ABF and SF, November 1995.
15. Irene Mikenberg letter to ABF, October 2002.
16. William Craig interview with ABF, April 1999.
17. Tomás Chuaqui letter to ABF, October 2002.
18. Kathleen (Henderson) Chuaqui interview with ABF, January 1996.
19. Jerzy Łoś lecture to the Polish Mathematical Society, 1984.
20. Aleksandra Szmielew interview with ABF and SF, June 1997.
21. Maria Moszyńska, Preface to Szmielew (1983).
22. Victor Pambuccian, *American Mathematical Monthly* (January 1986), pp. 74–5.
23. Cf. Givant (1986), [79] on p. 926.
24. Solomon Feferman diary, 28 April 1978.
25. Anne Preller letter to ABF, May 1996.
26. John Corcoran interview with ABF and SF, January 1997.

27. Freiman (1980); the report "The Situation in Soviet Mathematics" is included as an appendix.
28. Yuri Ershov visit report, Tarski Archives.
29. Yuri Ershov letter to Tarski, 10 May 1980, Tarski Archives.
30. John Corcoran interview with ABF and SF, January 1997.
31. Ibid.
32. Tarski (1983a), p. xiv.
33. Ibid., Editor's Introduction, p. xxvii.
34. Seated around the dinner table with Alfred and Maria were John and Mary Ann Addison, William Craig, Solomon and Anita Feferman, Steven Givant, Leon and Ginette Henkin, Ralph and Katherine McKenzie, Judith Ng, Julia and Raphael Robinson, Patrick and Christine Suppes, and Robert and Marilyn Vaught.
35. Stefan Mrozewski (1894–1975), a contemporary of Tarski who was also born in Poland, lived in Walnut Creek – a few miles from Berkeley.
36. The fine leather edition was printed in limited numbers and offered to select university libraries.

Chapter 15: The Last Times

1. Verena Huber-Dyson letter to ABF, January 1997.
2. Ibid.
3. Benjamin Wells interviews with ABF, March 1995.
4. Judith Ng interview with ABF and SF, April 1997, and letters to ABF, August 2002.
5. Ibid.
6. Givant (1991), p. 28.
7. Steven R. Givant interview with SF, February 2003.
8. Suppes et al. (1989).
9. The Tarski Centenary Conference in Warsaw (May 28 – June 1, 2001) was organized by the Stefan Banach Mathematical Center of the Institute of Mathematics of the Polish Academy of Sciences.
10. Adam Myjak letter to ABF and SF, 16 October 2002.

Tarski's Ph.D. Students

1. Hodges (1986).

Polish Pronunciation Guide

1. This guide is drawn mainly from the website ⟨http://polish.slavic.pitt.edu/firstyear⟩ created by the University of Pittsburgh for the teaching of Polish over the World Wide Web.

Bibliography

Selected Works by Alfred Tarski

Tarski, Alfred (1935), "Der Wahrheitsbegriff in den formalisierten Sprachen," *Studia Philosophica* 1, pp. 261–405. [Reprinted in Tarski (1986a), Vol. 2.]

Tarski, Alfred (1941a), *Introduction to Logic and to the Methodology of Deductive Sciences*, Oxford University Press, New York.

Tarski, Alfred (1941b), "On the calculus of relations," *Journal of Symbolic Logic* 6, pp. 73–89. [Reprinted in Tarski (1986a), Vol. 2.]

Tarski, Alfred (1944), "The semantic conception of truth and the foundations of semantics," *Philosophy and Phenomenological Research* 4, pp. 341–75. [Reprinted in Tarski (1986a), Vol. 2, and in Lynch (2001).]

Tarski, Alfred (1948), *A Decision Method for Elementary Algebra and Geometry* (prepared for publication by J. C. C. McKinsey), RAND report R-109, RAND Corp., Santa Monica, CA.

Tarski, Alfred (1949), *Cardinal Algebras* (with an appendix by B. Jónsson and A. Tarski), Oxford University Press, Oxford.

Tarski, Alfred (1951), *A Decision Method for Elementary Algebra and Geometry*, 2nd rev. ed. of Tarski (1948), University of California Press, Berkeley. [Reprinted in Tarski (1986a), Vol. 3.]

Tarski, Alfred (1956a), *Logic, Semantics, Metamathematics. Papers from 1923 to 1938* (J. H. Woodger, transl.), Oxford University Press, New York.

Tarski, Alfred (1956b), *Ordinal Algebras* (with appendices by C. C. Chang and B. Jónsson), North-Holland, Amsterdam.

Tarski, Alfred (1969), "Truth and Proof," *Scientific American* 220(6), pp. 63–77. [Reprinted in Tarski (1986a), Vol. 4.]

Tarski, Alfred (1983a), *Logic, Semantics, Metamathematics. Papers from 1923 to 1938* (J. Corcoran, ed.), 2nd rev. ed. of Tarski (1956a), Hackett, Indianapolis.

Tarski, Alfred (1983b), "The concept of truth in formalized languages," translation of Tarski (1935), in Tarski (1983a), pp. 182–278.

Tarski, Alfred (1986a), *Collected Papers*, Vols. 1–4 (S. R. Givant and R. N. McKenzie, eds.), Birkhäuser, Basel.

Tarski, Alfred (1986b), "What are logical notions?" (J. Corcoran, ed.), *History and Philosophy of Logic* 7, pp. 143–54.

Tarski, Alfred (1994), *Introduction to Logic and to the Methodology of Deductive Sciences* (Jan Tarski, ed.), 4th rev. ed. of Tarski (1941a), Oxford University Press, New York.

Tarski, Alfred (1999), "Letters to Kurt Gödel, 1942–1947" (J. Tarski, ed. and transl.), in Woleński and Köhler (1999), pp. 261–73.

Selected Works by Tarski with Others

Addison, John W., Leon Henkin and Alfred Tarski, eds. (1965), *The Theory of Models. Proceedings of the 1963 international symposium at Berkeley*, North-Holland, Amsterdam.

Chin, Louise H. and Alfred Tarski (1951), "Distributive and Modular Laws in the Arithmetic of Relation Algebras," *University of California Publications in Mathematics* (N.S.) 1(9), pp. 341–84. [Reprinted in Tarski (1986a), Vol. 3.]

Henkin, Leon, Donald Monk and Alfred Tarski (1971), *Cylindric Algebras, Part I. With an introductory chapter: General theory of algebras*, North-Holland, Amsterdam.

Henkin, Leon, Donald Monk and Alfred Tarski (1985), *Cylindric Algebras, Part II*, North-Holland, Amsterdam.

Henkin, Leon, Patrick Suppes and Alfred Tarski, eds. (1959), *The Axiomatic Method with Special Reference to Geometry and Physics*, North-Holland, Amsterdam.

Jónsson, Bjarni and Alfred Tarski (1947), *Direct Decompositions of Finite Algebraic Systems* (Notre Dame Mathematical Lectures, vol. 5), Notre Dame Press, Notre Dame, IN. [Reprinted in Tarski (1986a), Vol. 3.]

Keisler, H. Jerome and Alfred Tarski (1964), "From accessible to inaccessible cardinals," *Fundamenta Mathematicae* 53, pp. 225–308. [Reprinted in Tarski (1986a), Vol. 4.]

McKinsey, J. C. C. and Alfred Tarski (1944), "The algebra of topology," *Annals of Mathematics* (2) 45, pp. 141–91. [Reprinted in Tarski (1986a), Vol. 2.]

Nagel, Ernest, Patrick Suppes and Alfred Tarski (1962), *Logic, Methodology and Philosophy of Science. Proceedings of the 1960 international congress*, Stanford University Press, Stanford.

Schwabhäuser, Wolfram, Wanda Szmielew and Alfred Tarski (1983), *Metamathematische Methoden in der Geometrie*, Springer-Verlag, Berlin.

Tarski, Alfred and Steven Givant (1987), *A Formalization of Set Theory without Variables* (Colloquium Publications, vol. 41), American Mathematical Society, Providence.

Tarski, Alfred and Steven Givant (1999), "Tarski's system of geometry," *Bulletin of Symbolic Logic* 5, pp. 175–214.

Tarski, Alfred, Andrzej Mostowski and Raphael M. Robinson (1953), *Undecidable Theories*, North-Holland, Amsterdam.

Tarski, Alfred and Robert L. Vaught (1957), "Arithmetical extensions of relational systems," *Compositio Mathematica* 13, pp. 81–102. [Reprinted in Tarski (1986a), Vol. 3.]

Works about Tarski and His Achievements

Addison, John (1984), "Eloge: Alfred Tarski, 1901–1983," *Annals of the History of Computing* 6(4), pp. 335–6.

Blok, W. J. and Don Pigozzi (1988), "Alfred Tarski's work on general metamathematics," *Journal of Symbolic Logic* 53, pp. 36–50.

Doner, John and Wilfrid Hodges (1988), "Alfred Tarski and decidable theories," *Journal of Symbolic Logic* 53, pp. 20–35.

van den Dries, Lou (1988), "Alfred Tarski's elimination theory for real closed fields," *Journal of Symbolic Logic* 53, pp. 7–19.

Etchemendy, John (1988), "Tarski on truth and logical consequence," *Journal of Symbolic Logic* 53, pp. 51–79.

Feferman, Anita Burdman (1999), "How the Unity of Science saved Alfred Tarski," in Woleński and Köhler (1999), pp. 43–52.

Feferman, Solomon (1999), "Tarski and Gödel between the lines," in Woleński and Köhler (1999), pp. 53–63.

Feferman, Solomon (2003), "Alfred Tarski and a watershed meeting in logic: Cornell, 1957," in J. Hintikka et al., eds., *In Search of the Polish Tradition. Essays in honour of Jan Woleński on the occasion of his 60th birthday*, Kluwer, Dordrecht, pp. 151–62.

Feferman, Solomon (2004), "Tarski's conceptual analysis of semantical notions," in A. Benmakhlouf, ed., *Sémantique et épistémologie*, Editions Le Fennec, Casablanca, pp. 79–108.

Formisano, A., E. G. Omodeo and M. Temperini (2000), "Goals and benchmarks for automated map reasoning," *Journal of Symbolic Computation* 29(2), pp. 259–97.

Formisano, A., E. G. Omodeo and M. Temperini (2001), "Instructing equational set-reasoning with Otter," in R. Goré et al., eds., *Automated Reasoning* (Lecture Notes in Computer Science, vol. 2083), Springer-Verlag, New York, pp. 152–67.

Givant, Steven (1986), "Bibliography of Alfred Tarski," *Journal of Symbolic Logic* 51, pp. 913–41.

Givant, Steven (1991), "A portrait of Alfred Tarski," *Mathematics Intelligencer* 13(3), pp. 16–32.

Givant, Steven (1999), "Unifying threads in Alfred Tarski's work," *Mathematics Intelligencer* 21(9), pp. 47–58.

Henkin, Leon et al., eds. (1974), *Proceedings of the Tarski Symposium*, American Mathematical Society, Providence.

Hodges, Wilfrid (1985/1986), "Truth in a structure," *Proceedings of the Aristotelian Society* (N.S.) 86, pp. 131–51.

Hodges, Wilfrid (1986), "Alfred Tarski," *Journal of Symbolic Logic* 51, pp. 866–8.

Kawahara, Yasuo (1988), "Applications of relational calculus to computer mathematics," *Bulletin of Informatics and Cybernetics* 23, pp. 67–78.

Levy, Azriel (1988), "Tarski's work in set theory," *Journal of Symbolic Logic* 53, pp. 2–6.

Maddux, Roger (1996), "Relation algebraic semantics," *Theoretical Computer Science* 160, pp. 1–85.

McNulty, George F. (1986), "Alfred Tarski and undecidable theories," *Journal of Symbolic Logic* 51, pp. 890–8.
Monk, J. Donald (1986), "The contributions of Alfred Tarski to algebraic logic," *Journal of Symbolic Logic* 51, pp. 899–906.
Moore, Gregory H. (1990), "Alfred Tarski," *Dictionary of Scientific Biography*, Suppl. II, Vol. 18, Scribner, New York, pp. 893–6.
Popper, Karl (1974), "Some philosophical comments on Tarski's theory of truth," in Henkin et al. (1974), pp. 397–409.
Sinaceur, Hourya (2001), "Alfred Tarski: Semantic shift, heuristic shift in metamathematics," *Synthese* 126, pp. 49–65.
Sundholm, Göran (2003), "Tarski and Leśniewski on languages with meaning versus languages without use," in J. Hintikka et al., eds., *In Search of the Polish Tradition. Essays in honour of Jan Woleński on the occasion of his 60th birthday*, Kluwer, Dordrecht, pp. 109–28.
Suppes, Patrick (1988), "Philosophical implications of Tarski's work," *Journal of Symbolic Logic* 53, pp. 80–91.
Suppes, Patrick, Jon Barwise and Solomon Feferman (1989), "Commemorative meeting for Alfred Tarski. Stanford University – November 7, 1983," in Duren (1989), Part III, pp. 393–403.
Szczerba, L.W. (1986), "Tarski and geometry," *Journal of Symbolic Logic* 51, pp. 907–12.
Vaught, Robert L. (1986), "Tarski's work in model theory," *Journal of Symbolic Logic* 51, pp. 869–82; "Errata," *Journal of Symbolic Logic* 52 (1987), p. vii.
Wagon, Stanley (1985), *The Banach–Tarski Paradox*, Cambridge University Press.
White, Morton (1987), "A philosophical letter of Alfred Tarski," *Journal of Philosophy* 84, pp. 28–32.
Woleński, Jan and Eckehart Köhler, eds. (1999), *Alfred Tarski and the Vienna Circle*, Kluwer, Dordrecht.
Woodger, Joseph Henry (1974), "Thank you, Alfred," in Henkin et al. (1974), pp. 481–2.

Other References

Abramsky, Chimen, Maciej Jachimczyk and Antony Polonsky, eds. (1986), *The Jews in Poland*, Basil Blackwell, Oxford.
Albers, Donald J., Gerald L. Alexanderson, and Constance Reid (1987), *International Mathematical Congresses: An illustrated history, 1893–1986*, Springer-Verlag, New York.
Bell, Eric Temple (1934), *The Search for Truth*, Reynal & Hitchcock, New York.
Bocheński, Józef M. (1994), "Morals of thought and speech," in J. Woleński, ed., *Philosophical Logic in Poland*, Kluwer, Dordrecht, pp. 1–8.
Carnap, Rudolf (1928), *Der Logische Aufbau der Welt*, Weltkreis, Berlin. [Translated by R. George as *The Logical Structure of the World*, University of California Press, Berkeley, 1967.]
Carnap, Rudolf (1934), *Logische Syntax der Sprache*, Springer, Wien. [Translated by A. Smeaton as *The Logical Syntax of Language*, Kegan Paul, London, 1937.]

Carnap, Rudolf (1963), "Intellectual autobiography," in P. A. Schilpp, ed., *The Philosophy of Rudolf Carnap* (Library of Living Philosophers, vol. 11), Open Court, La Salle, pp. 3–84.
Cartwright, Nancy, Jordi Cat, Lola Fleck and Thomas E. Uebel (1996), *Otto Neurath. Philosophy between science and politics*, Cambridge University Press.
Caviness, Bob F. and Jeremy R. Johnson, eds. (1998), *Quantifier Elimination and Cylindric Algebraic Decomposition*, Springer-Verlag, Wien.
Chang, Chen-Chung (1974), "Model theory 1945–1971," in Henkin et al. (1974), pp. 173–86.
Cohen, Paul (1966), *Set Theory and the Continuum Hypothesis*, Benjamin, New York.
Creath, Richard, ed. (1990), *Dear Carnap, Dear Van: The Carnap–Quine correspondence and related work*, University of California Press, Berkeley.
Crossley, John N., ed. (1975), "Reminiscences of logicians," in *Algebra and Logic* (Lecture Notes in Mathematics, vol. 450), Springer-Verlag, Berlin, pp. 1–62.
Dahms, Hans Joachim (1993), "The emigration of the Vienna Circle," in P. Weibel and F. Stadler, eds., *Vertreibung der Vernunft. The cultural exodus from Austria*, Löcker, Vienna, pp. 65–97.
Dauben, Joseph W. (1995), *Abraham Robinson: The creation of non-standard analysis: A personal and mathematical odyssey*, Princeton University Press, Princeton.
Davies, Norman (1982), *God's Playground. A history of Poland in two volumes. Volume II, 1795 to the present*, Columbia University Press, New York.
Dawson, John W., Jr. (1997), *Logical Dilemmas. The life and work of Kurt Gödel*, A. K. Peters, Wellesley.
Delzell, Charles N. (1996), "Kreisel's unwinding of Artin's proof," in Odifreddi (1996), pp. 113–246.
Dewey, John and Horace M. Kallen, eds. (1941), *The Bertrand Russell Case*, Viking Press, New York.
Duren, Peter, ed. (1989), *A Century of Mathematics in America*, Parts I–III, American Mathematical Society, Providence.
Enderton, Herbert (1998), "Alonzo Church and the Reviews," *Bulletin of Symbolic Logic* 2, pp. 171–80.
Estreicher, Karol (1971), *Leon Chwistek. Biografia artysty* [*Leon Chwistek. Biography of an artist*], Państwowe Wydawnictwo Naukowe, Krakow.
Feferman, Solomon (1986), "Gödel's life and work," in Gödel (1986), pp. 1–36.
Feferman, Solomon (1998), *In the Light of Logic*, Oxford University Press, New York.
Feigl, Herbert (1969), "The Wiener Kreis in America," in D. Fleming and B. Bailyn, eds., *The Intellectual Migration. Europe and America, 1930–1960*, Belknap, Cambridge, MA, pp. 630–73.
Fischer, Michael J. and Michael O. Rabin (1974), "Super exponential complexity of Presburger's arithmetic," *SIAM-AMS Proceedings* 6, pp. 27–41.
Freiman, Grigori (1980), *It Seems I Am a Jew. A Samizdat essay* (transl. and ed. with an introduction by M. B. Nathanson), Southern Illinois University Press, Carbondale.

Friedman, Filip (1930), *Żydzi w Łódzkim przemyśle włókienniczym: w pierwszych stadjach jego rozwoju* [*Jews in the Lodz Textile Industry in the First Stages of Development*]. Łódz.
Friedman, Michael (1999), *Reconsidering Logical Positivism*, Cambridge University Press.
Garlicki, Andrzej, ed. (1982), *Dzieje Uniwersytetu Warszawskiego 1915–1939*, Państwowe Wydawnictwo Naukowe, Warsaw. [Summary in English, pp. 340–8.]
Gerlach, Hugo (1894), *Die vom Hinterhaus: Novellen*, Berlin.
Gerould, Daniel, ed. and transl. (1992), *The Witkiewicz Reader*, Northwestern University Press, Evanston.
Gieysztor, Aleksander, ed. (1967), *Universitas Varsoviensis: The University of Warsaw*, Warsaw University Press, Warsaw.
Gitlin, Todd (1987), *The Sixties. Years of hope, days of rage*, Bantam, Toronto.
Gödel, Kurt (1986), *Collected Works, Vol. I. Publications 1929–1936* (S. Feferman et al., eds.), Oxford University Press, New York.
Gödel, Kurt (1990), *Collected Works, Vol. II. Publications, 1938–1974* (S. Feferman et al., eds.), Oxford University Press, New York.
Halmos, Paul R. (1985), *I Want to Be a Mathematician. An automathography*, Springer-Verlag, New York.
Halpern, Joseph Y. et al. (2001), "On the unusual effectiveness of logic in computer science," *Bulletin of Symbolic Logic* 7, pp. 213–36.
Hanf, William (1964), "Incompleteness in languages with infinitely long expressions," *Fundamenta Mathematicae* 53, pp. 309–24.
Huber-Dyson, Verena (1996), "Thoughts on the occasion of Georg Kreisel's 70th birthday," in Odifreddi (1996), pp. 51–73.
Kac, Mark (1985), *Enigmas of Chance*, Harper & Row, New York.
Kelley, John L. (1989), "Once over lightly," in Duren (1989), Part III, pp. 471–93.
Kerr, Clark (1971), *The Gold and the Blue. A personal memoir of the University of California, 1949–1967*, Vol. 2, University of California Press, Berkeley.
Kirkham, Richard L. (1992), *Theories of Truth*, MIT Press, Cambridge, MA.
Kuratowski, Kazimierz (1980), *A Half Century of Polish Mathematics. Remembrances and reflections*, Pergamon, Oxford.
Łukasiewicz, Jan (1970), *Selected Works* (L. Borkowski, ed.), North-Holland, Amsterdam.
Lynch, Michael P., ed. (2001), *The Nature of Truth. Classic and contemporary perspectives*, MIT Press, Cambridge, MA.
Mal'cev, Anatolii Ivanovic (1971), *The Metamathematics of Algebraic Systems. Collected papers: 1936–1967* (B. F. Wells III, ed. and transl.), North-Holland, Amsterdam.
Martin, Robert L., ed. (1984), *Recent Essays on Truth and the Liar Paradox*, Oxford University Press, Oxford.
Menger, Karl (1994), *Reminiscences of the Vienna Circle and the Mathematical Colloquium*, Kluwer, Dordrecht.
Miłosz, Czesław (1994), *A Year of the Hunter*, Farrar, Strauss, Giroux, New York.
Monk, Ray (1990), *Ludwig Wittgenstein. The duty of genius*, Free Press, New York.

Montague, Richard (1974), *Formal Philosophy. Selected papers of Richard Montague* (R. H. Thomason, ed.), Yale University Press, New Haven.

Moore, Gregory H. (1982), *Zermelo's Axiom of Choice. Its origins, development and influence*, Springer-Verlag, New York.

Moschovakis, Yiannis N. (1989), "Commentary on mathematical logic," in Duren (1989), Part II, pp. 343–6.

Mosses, Peter D. (1990), "Denotational semantics," in J. van Leeuwen, ed., *Handbook of Theoretical Computer Science, Volume B. Formal models and semantics*, Elsevier, Amsterdam, pp. 575–631.

Nerode, Anil (1996), "An appreciation of Kreisel," in Odifreddi (1996), pp. 81–8.

Nowicki, Ron (1992), *Warsaw: The cabaret years*, Mercury House, San Francisco.

Odifreddi, Piergiorgio, ed. (1996), *Kreiseliana. About and around Georg Kreisel*, A. K. Peters, Wellesley.

Partee, Barbara H. (2001), "Montague grammar," in *International Encyclopedia of the Social and Behavioral Sciences* (N. J. Smelser and P. B. Baltes, eds.), Elsevier, Amsterdam.

Pilet, Paul-Emil (1977), "Ferdinand Gonseth – sa vie, son oeuvre," *Dialectica* 31, pp. 23–33.

Portner, Paul and Barbara H. Partee, eds. (2002), *Formal Semantics. The essential readings*, Blackwell, Oxford.

Quine, Willard Van Orman (1985), *The Time of My Life*, MIT Press, Cambridge, MA.

Reid, Constance (1982), *Neyman*, Springer-Verlag, New York.

Reid, Constance (1996), *Julia. A life in mathematics*, Mathematical Association of America, Washington, DC.

Reisch, George (n.d.), "From the life of the present to the icy slopes of logic: Logical empiricism, the Unity of Science movement and the Cold War," in A. Richardson and T. E. Uebel, eds., *The Cambridge Companion to Logical Empiricism*, Cambridge University Press (forthcoming).

Reymont, Władysław S. (1927), *The Promised Land*, Knopf, New York.

Rider, Robin E. (1989), "An opportune time: Griffith C. Evans and mathematics at Berkeley," in Duren (1989), Part II, 283–302.

Robinson, Abraham (1966), *Non-standard Analysis*, North-Holland, Amsterdam.

Roos, Hans (1966), *A History of Modern Poland*, Knopf, New York.

Rudnicki, Szymon (1987), "From 'numerus clausus' to 'numerus nullus'," *Polin* 2, pp. 246–68.

Russell, Bertrand (1951), *The Autobiography of Bertrand Russell, 1914–1944*, Little, Brown, Boston.

Smith, Barry (1994), *Austrian Philosophy: The legacy of Franz Brentano*, Open Court, Chicago.

Solovay, Robert (1970), "A model of set theory in which every set of reals is Lebesgue measurable," *Annals of Mathematics* (2) 92, pp. 1–56.

Śrzednicki, Jan T. J. and V. Frederick Rickey, eds. (1984), *Leśniewski's Systems. Ontology and mereology*, Martinus Nijhoff, The Hague.

Staal, J. Frits (1965), "E. W. Beth, 1908–1964," *Dialectica* 19(1/2), pp. 158–79.
Stadler, Friedrich (1997), *Studien zum Wiener Kreis. Ursprung, Entwicklung und Wirkung des Logischen Empirismus im Kontext*, Suhrkamp Verlag, Frankfurt am Main.
Stadtman, Verne A. (1970), *The University of California, 1868–1968*, McGraw-Hill, New York.
Szmielew, Wanda (1983), *From Affine to Euclidean Geometry. An axiomatic approach* (ed., prep. for publication, and transl. from the Polish by M. Moszyńska), D. Reidel, Dordrecht.
Ulam, Stanisław M. (1976), *Adventures of a Mathematician*, Scribner, New York.
van Ulsen, Paul (2000), *E. W. Beth als Logicus*, ILLC Dissertation Series 2000–04, Institute for Logic, Language and Computation, University of Amsterdam.
Vaught, Robert L. (1974), "Model theory before 1945," in Henkin et al. (1974), pp. 153–72.
Whitehead, Alfred North and Bertrand Russell (1925), *Principia Mathematica*, Vols. I–III, 2nd ed., Cambridge University Press.
Wittgenstein, Ludwig (1922), *Tractatus Logico-Philosophicus*, Routledge & Kegan Paul, London.
Woleński, Jan (1989), *Logic and Philosophy in the Lvov–Warsaw School*, Kluwer, Dordrecht.
Woleński, Jan (1999), *Essays in the History of Logic and Logical Philosophy*, Jagiellonian University Press, Cracow.
Woodger, Joseph Henry (1937), *The Axiomatic Method in Biology* (with appendices by A. Tarski and W. F. Floyd), Cambridge University Press.
Zamoyski, Adam (1988), *The Polish Way*, Franklin Watts, New York.
Zygmunt, Jan (1991), "Mojżesz Presburger: Life and work," *History and Philosophy of Logic* 12, pp. 211–23.

Index of Names

Abramsky, Chimen, 394
Achilles, 44
Adams, Ernest, 210, 232, 234
Addison, John, 141, 222, 229, 259–61, 264, 285, 325, 365, 378, 404, 405, 408
Addison, Mary Ann (née Church), 141, 399, 408
Ajdukiewicz, Kazimierz, 67, 89, 93, 97, 169, 252, 253
Albers, Donald J., 405
Alexander III, Czar, 22
Allende, Salvador, 354
Andréka, Hajnal, 341, 407
Aristotle, 32, 33, 35, 44, 110, 131, 188, 334
Arrow, Kenneth J., 133, 134, 399
Artin, Emil, 224
Austen, Jane, 199
Aydelotte, Frank, 148, 149

Baldwin, James, 13
Banach, Stefan, 28, 29, 32, 41, 43, 44, 48–52, 67, 287, 395, 408
Bauer-Mengelberg, Stefan, 399
Bell, Eric Temple, 131, 394
Beltrami, Eugenio, 278–80
Bergman, Stefan, 127
Berkeley, Edmund C., 129
Bernays, Paul, 141, 207, 233, 234, 255
Bernoulli, Daniel, 300
Bernstein, Benjamin A., 142, 156, 157, 399, 400
Beth, Corry, 181, 206–8

Beth, Evert W., 181, 206, 207, 222, 249, 250, 285, 318, 404
Betti, Arriana, 394, 395
Bezhanishvili, Misha, 305
Blackwell, David, 352
Bocheński, Józef, 136–8, 207, 250, 255, 394, 399
Boltzmann, Ludwig, 231
Bolyai, János, 69, 278, 280
Boole, George, 334–8, 341
Borel, Emile, 47
Borowski, Marian, 395
Borsuk, Karol, 209, 232–6, 243–5
Borsuk, Zofia (Zosia), 243–5, 403
Bradford, Robert E., 385
Brentano, Franz, 30, 31
Brezhnev, Leonid, 300
Bridgman, Percy, 233
Broadwyn, L., 127
Broadwyn, S., 127
Brouwer, Luitzen Egbertus Jan, 77, 161, 294
Butler, Jean, 173, 200, 201, 222, 401, 402

Cantor, Georg, 29, 43, 45–6, 47, 48, 198
Carnap, Ina, 80, 85
Carnap, Rudolf, 80–3, 85–7, 92–5, 97, 98, 101, 102, 106, 122, 127, 128, 131, 139, 159, 247, 248, 251, 252, 255, 261, 262, 301, 312, 314, 333, 367, 397, 399, 403
Carter, Jimmy, 328

Cartwright, Nancy, 397
Cavell, Stanley, 405
Caviness, Bob F., 401
Chambers, Whittaker, 181
Chang, Chen-Chung, 172, 195, 198, 201–5, 211, 218, 222, 234, 266, 281, 282, 303, 312, 328, 385, 402, 404
Chao, Lensey, 175
Chao, Yuen Ren, 210
Chihara, Charles, 395
Chin, Louise H., 171, 173, 183, 191, 192, 198, 339, 341; see also Lim, Louise Hoy Chin
Chomsky, Noam, 255, 294, 314
Chopin, Frédéric, 9
Chuaqui, Kathleen Henderson, 352, 355, 407
Chuaqui, Rolando, 351–5
Chuaqui, Tomás, 407
Church, Alonzo, 70, 129, 130, 132, 141, 145, 146, 160, 193, 218, 219, 221, 222, 228, 229, 255, 259, 294, 329, 331, 362, 363
Church, Mary Ann: see Addison, Mary Ann
Chwiałkowski, Zbigniew, 57
Chwistek, Leon, 57, 61, 66–8, 97, 103, 128, 396
Cocchiarella, Nino, 213, 402, 404
Codd, Edgar, 342
Cohen, Paul, 51, 260, 286, 302, 315, 395, 405
Collins, George, 222, 229, 403
Conant, James B., 128
Coniglione, Francesco, 31
Corcoran, John, 5, 195, 349, 361, 362, 366–8, 393, 395, 407, 408
Courant, Richard, 132, 231
Couturat, Louis, 37
Craig, William, 222, 229, 259, 299, 407, 408
Creath, Richard, 397
Crick, Francis, 277
Crossley, John N., 398

Curry, Haskell B., 129–31, 141, 159, 207, 303

Dahms, Hans Joachim, 403
Dauben, Joseph W., 403, 405
Davidson, Donald, 216, 260
Davies, Norman, 18, 24, 393, 394, 400
Davis, Alan, 198, 199
Davis, Angela, 327, 406
Davis, Anne C.: see Morel, Anne C. Davis
Davis, Martin, 145, 221, 229
Dawson, John W., Jr., 400
de Broglie, Louis, 208
Delzell, Charles N., 403
De Morgan, Augustus, 334
Dennes, William, 172, 214
Destouches, Jean-Louis, 97, 208, 233–5, 242, 244, 250, 252
Dewey, John, 247, 248, 301, 399
Diener, Karl-Heinz, 299
Dohrn, Bernardine, 327, 406
Dollfuss, Engelbert, 94, 95
Doner, John, 315, 386
Dreben, Burton, 222, 365
Dyson, Freeman, 226–8

Ehrenfeucht, Andrzej, 209, 239, 240–2, 291–3, 356
Ehrenfeucht, Anya (Tarski's grandchild), 293
Ehrenfeucht, Kristina (Ina, née Krystyna Tarska), 292, 293, 356, 373, 400, 401, 403, 405; see also Tarski, Ina
Ehrenfeucht, Michael (Tarski's grandchild), 293
Ehrenfeucht, Renya (Tarski's grandchild), 293
Eilenberg, Samuel, 54, 55, 148, 159, 215, 369, 370, 396
Einstein, Albert, 77, 139, 144, 232, 346, 352
Ellison, Lawrence (Laurent), 317, 359

Index of Names

Enderton, Herbert, 103, 174, 329, 333, 398, 406
Erdös, Paul, 146, 148, 159, 189, 215, 284, 344, 400
Ershov, Yuri L., 302, 303, 307, 308, 363–5, 408
Esakia, Leo, 304, 305, 406
Euclid, 69, 231, 278, 279, 280
Euler, Leonhard, 300
Evans, Griffith C., 147, 148, 151, 158, 167, 172, 210, 234, 370

Feferman, Anita B., 303, 408
Feferman, Solomon, 171, 201, 211–14, 222, 223, 260, 272, 282, 286, 303, 359, 378, 385, 398, 403, 407, 408
Feigl, Herbert, 397, 403
Février, Paulette, 208, 209, 233–6, 242, 244, 288, 402, 403
Feydeau, Georges, 199
Feys, Robert, 207, 254
Fischer, Michael J., 401
Flanagan, Dennis, 298
Foster, Alfred, 157
Fraenkel, Abraham, 47, 51, 395
Frank, Phillip, 78, 79, 93, 94, 97, 248
Frayne, Thomas, 234, 264–6, 320, 404
Fraïssé, Roland, 315, 344, 359
Frege, Gottlob, 79, 80, 131, 188, 225, 299
Freiman, Grigori, 364, 408
Frei Montalva, Eduardo, 354
Frei Ruiz, Eduardo, 355
Freud, Sigmund, 30
Freudenthal, Hans, 253
Friedman, Filip, 393
Friedman, Michael, 251, 397, 404
Friedman, Sy, 365
Froda, Alexandre, 232
Furth, Montgomery, 333

Gagarin, Yuri, 257
Gaifman, Haim, 261, 262, 266, 292, 385, 404

Gainsborough, Thomas, 330
Galbraith, John Kenneth, 140
Galilei, Galileo, 44, 45
Garlicki, Andrzej, 394
Gates, Bill, 140
Gerlach, Hugo, 13, 17, 393
Gerould, Daniel, 396
Gillman, Leonard, 214, 215, 385, 402
Gitlin, Todd, 327, 406
Givant, Steven, 143, 262, 320, 338, 339, 346–8, 367, 369, 375, 377, 378, 382, 386, 396, 397, 400, 401, 403, 405, 407, 408
Gödel, Kurt, 1, 5, 30, 44, 51, 70–2, 80–5, 89, 90, 93, 94, 104, 119, 121, 132, 143–5, 151, 152, 157, 159, 160, 161, 174, 188, 212, 221–3, 225–8, 231, 258, 281, 283, 286, 341, 344, 363, 395, 397, 398, 400, 404
Goethe, Johann Wolfgang von, 92, 228
Gonseth, Ferdinand, 249–53
Goodman, Nelson, 141, 366
Grelling, Kurt, 103, 128
Grosz, Irena, 63, 64, 166, 209, 396
Grosz, Wiktor, 166
Guggenheimer, H., 404
Gupta, Haragauri, 264, 309, 385, 406
Gurion, David Ben, 26

Hackett, William, 366
Hahn, Hans, 78–81, 92, 94
Hallam, Sarah, 2, 157, 158, 178, 180, 198, 344, 400, 402
Halmos, Paul, 220–2, 225, 403
Halpern, Joseph Y., 407
Hanf, William P., 189, 265, 282–4, 385, 405
Harsanyi, John, 255
Hausdorff, Felix, 47, 48, 50, 155, 198
Hause, Eleanor Lee, 401
Hearst, Patricia (Patty, "Tania"), 327, 328, 406
Heidegger, Martin, 299
Heine, Heinrich, 379

Helmer, Olaf, 97, 100, 127, 134, 141, 399
Hempel, Carl Peter, 97, 100, 126, 127, 144
Hempel, Diane, 399
Henderson, Kathleen: *see* Chuaqui, Kathleen Henderson
Henkin, Ginette, 368, 369, 408
Henkin, Leon, 132, 133, 145, 172, 186, 188, 192, 206, 207, 210, 212, 222–4, 232–4, 251, 257, 259, 282, 285, 306, 310, 320, 321, 324, 341, 344, 349, 364, 368–71, 375, 378, 379, 399, 402–4, 406–8
Hersch, Jeanne (Tarski's cousin), 6, 270
Hersch, Joseph (Tarski's cousin), 6, 393
Hersch, Marc (Tarski's cousin), 393
Herzl, Theodor, 25
Heyman, Ira Michael, 370, 371
Heyting, Arend, 233, 250, 255
Hilgard, Ernest, 255
Hinman, Peter, 264, 325, 404
Hiss, Alger, 181
Hitler, Adolf, 88, 99, 103, 125, 128, 162, 271
Hiż, Henry, 63, 106, 222, 225, 394, 395, 396, 398
Hodges, Wilfrid, 398, 405, 408
Hoffman, Dustin, 140
Hoffman, Peter, 11, 12, 376, 393
Holding, Eileen (Olaf Helmer's first wife), 127, 141
Hook, Sidney, 139, 248
Hosiasson-Lindenbaum, Janina, 93, 97, 108, 129, 169
Huber-Dyson, Verena, 226–8, 272, 273–6, 293, 317, 372, 373, 396, 403, 404, 408
Hurwicz, Leonid, 255
Husserl, Edmund, 30

Ishiguro, Hidé, 295, 296, 405

James, William, 247
Janin, Monique, 359

Janiszewski, Zygmunt, 28
Jaśkowski, Stanisław, 97
Jędrzejczak, Aleksander, 395, 397
Jeffrey, Richard, 207, 397
Johnson, Jeremy R., 401
Jónsson, Bjarni, 3, 154, 155, 156, 158, 171, 191, 192, 195, 234, 385, 393, 400, 401
Jordan, Pascual, 233
Jurzykowski, Alfred, 294

Kac, Mark, 68, 396
Kafka, Franz, 296
Kalicki, Jan, 172, 204, 205, 206, 215, 235
Kallen, Horace M., 399
Kamp, Hans, 313, 402, 406
Kanamori, Aki, 365
Kaplan, David, 216, 312, 314, 397, 406
Karp, Carol, 282
Kay, Jean, 135
Keisler, H. Jerome, 189, 214, 265–8, 282–4, 297, 328, 385, 404, 405
Kelley, John L., 185, 186, 188, 218, 364, 378, 401
Kemeny, John, 145
Kennedy, John F., 300
Kerr, Clark, 326, 406
Khrushchev, Nikita S., 300
Kirkham, Richard L., 398
Kissinger, Henry, 140
Klee, Victor, 201
Kleene, Stephen C., 129, 130, 141, 145, 219, 221, 222, 228, 253, 255, 259, 260, 303
Klein, Felix, 279, 280
Knaster, Bronisław, 38, 55, 76
Kochen, Simon, 145, 221, 307
Kokoszyńska-Lutmanowa, Maria, 88–91, 94, 97, 98, 105, 243, 397
Kolakowski, Leszek, 270
Kolmogorov, Andrei N., 301, 306
Korselt, A., 337
Koschembahr-Łyskowski, Ignacy, 39

Index of Names

Kotarbiński, Tadeusz, 30, 31, 35, 36, 41, 42, 62, 67, 83, 97, 106, 169, 209, 243, 395, 396
Kovalevskaya, Sofia, 360
Kreisel, Georg, 207, 219, 222–8, 255, 260, 272, 285, 286
Kripke, Saul, 112, 286, 398
Krynicki, Michał, 397
Kuratowski, Kazimierz, 29, 32, 76, 104, 125, 194, 243, 394
Kushner, Boris, 405

Lakatos, Imre, 297
Lampe, Wiktor, 39
Langford, Cooper H., 72, 73, 121
Lawrence, Thomas, 330
Lazarsfeld, Paul, 255
Lebesgue, Henri, 47
LeBlanc, Owen, 394
Leibniz, Gottfried Wilhelm von, 96, 285
Lenzen, Victor F., 210
Leśniewski, Stanisław, 26, 27, 29–32, 34–42, 48, 63, 67, 77, 83, 87, 88, 100–2, 106, 380, 394–6, 398
Levy, Azriel, 51, 52, 268, 395
Lim, Louise Hoy Chin, 385; see also Chin, Louise H.
Lindenbaum, Adolf, 37, 93, 97, 101, 108, 155, 169, 191, 194, 195, 198, 386
Lobachevsky, Nikolai, 69, 278, 280
Łoś, Jerzy, 239, 282, 283, 290, 356, 407
Löwenheim, Leopold, 72, 103, 121, 281, 337
Łukasiewicz, Jan, 30–6, 38, 41, 67, 73, 81, 87, 88, 99, 101, 102, 194, 380, 394, 395
Lutman, R., 89
Lutman-Kokoszyńska, Maria: see Kokoszyńska-Lutmanowa, Maria
Luxemburg, Rosa, 25, 35
Lynch, Michael P., 398
Lyndon, Roger, 260, 272, 338

Mach, Ernst, 78–80, 92, 94, 95, 231
Mac Lane, Saunders, 365
Maddux, Roger, 386
Magidor, Menachem, 268, 404
Mal'cev, Anatolii I., 282, 301–3, 306–8, 324, 325, 363, 405, 406
Mancosu, Paolo, 396
Marcus, Ruth Barcan, 141, 399
Marcuse, Herbert, 327
Marek, Wiktor, 403
Markov, Andrei A., 301, 303
Martel, Karol, 56, 396
Martin, Charles, 386
Martin, Michael, 405
Mates, Benson, 167, 172, 205, 210, 214, 299, 378, 400, 401, 402
Matiyasevich, Yuri, 405
Mazurkiewicz, Stefan, 28, 29, 32
McCarthy, Joseph, 186
McKenzie, Ralph, 318–20, 344, 369, 373–6, 386, 403, 406, 408
McKinsey, John Charles Chenowith ("Chen"), 141–3, 156, 159–61, 185, 189–91, 194, 213, 216, 232, 251, 304, 367, 399, 401
McNulty, George, 386
Meinong, Alexius, 30
Menger, Karl, 76–8, 80, 81, 83, 94, 106, 235, 396, 397
Michalski, Stanisław, 39
Mickiewicz, Adam, 23
Mikenberg, Irene, 353, 354, 407
Miller, David, 297, 405
Miłosz, Czesław, 58, 269–71, 294, 378, 396, 404
Mistral, Gabriela, 355
Monk, J. Donald, 192, 263, 266, 318–21, 338, 341, 349, 385, 402–4, 406, 407
Monk, Ray, 403
Montague, Richard M., 123, 181, 201, 211, 213, 214, 216, 217, 222, 232, 282, 312–14, 327, 331–3, 348, 367, 385, 398, 406
Moore, Gregory H., 395

Morel, Anne C. Davis, 197–200, 265, 276, 283, 385, 402
Morel, Delos, 198–201, 402
Morgenstern, Oskar, 144, 145
Morrey, Charles B., Jr., 210
Morris, Charles, 93, 97
Morse, Anthony P. ("Tony"), 157, 205
Morse, Marston, 148, 149
Morse, Mary, 157
Moschovakis, Yiannis, 403
Mosses, Peter D., 398
Mostowska, Maria, 243
Mostowski, Andrzej, 104, 105, 170, 173, 174, 193, 209, 212, 239, 240, 243, 264, 324, 325, 356
Moszyńska, Maria, 357, 358, 407
Mrozewski, Stefan, 369, 408
Mycielski, Jan, 238, 239, 395, 402, 403
Myjak, Adam, 380, 408

Naess, Arne, 98
Nagel, Ernest, 93, 131, 132, 247, 254, 399, 404
Napoleon, 21
Németi, István, 341, 342, 407
Nerode, Anil, 220, 222, 402, 403
Neurath, Otto, 78, 79, 92–5, 97, 98, 122, 128, 129, 247
Newton, Isaac, 231, 277
Neyman, Jerzy, 150, 151, 210
Ng, Judith (Kan Ching), 310–13, 320, 322, 323, 325, 328–31, 347, 350, 375, 376, 386, 406, 408
Nicholas II, Czar, 22
Nietzsche, Friedrich, 299
Noether, Emmy, 359
Novikov, Petr S., 301
Nowicki, Ron, 393, 394, 396, 398

Odifreddi, Piergiorgio, 403
Ogar, Dale, 2, 343–5, 393, 407
Ono, Hiroakira, 309, 406
Ono, Katuzi, 309
Oppenheim, Gabrielle, 149
Oppenheim, Paul, 127, 149

Pacholski, Leszek, 239
Paderewski, Ignacy Jan, 9, 23, 32
Pambuccian, Victor, 358, 407
Parsons, Charles, 222, 405
Partee, Barbara, 314, 398, 406
Pasenkiewicz, Kazimierz, 37, 395
Peck, Gregory, 181
Peirce, Charles Sanders, 247, 335, 336, 337
Pepis, Józef, 169
Perry, John, 406
Peter the Great, 300
Picasso, Pablo, 1
Pigozzi, Donald, 320–3, 386, 406
Pilet, Paul-Emil, 404
Piłsudski, Józef, 22–5, 28, 64, 65, 98, 99, 102
Pincock, Christopher, 396
Pinochet, Augusto, 354, 355
Piveteau, Jean, 252
Planck, Max, 79
Plato, 139
Poniatowski, Stanisław-August, 21
Popper, Karl, 93, 94, 122, 255, 294, 297, 344, 367, 397
Portner, Paul, 398, 406
Post, Emil, 228, 394
Preller, Anne, 315–17, 359–61, 396, 406, 407
Presburger, Mojżesz, 73, 74, 169, 229
Prussak, Abraham Mojżesz, 5
Prussak, Rosa (Rachel): *see* Teitelbaum, Rosa
Prussak family, 5, 393
Putnam, Hilary, 224, 365, 403

Quine, Marjorie (Willard Quine's second wife), 207
Quine, Naomi (Willard Quine's first wife), 86
Quine, Willard Van Orman ("Van"), 63, 85–8, 106–8, 125, 129–31, 139, 141, 165, 201, 202, 207, 221, 222, 246, 247, 259, 344

Index of Names

Rabin, Michael O., 145, 219, 221, 229, 257, 262, 263, 268, 269, 401, 403, 404
Ramsey, Frank, 396
Rasiowa, Helena, 239
Reichenbach, Hans, 92, 93, 97, 103
Reid, Constance, 174, 400, 401
Reisch, George, 247, 403
Reyes, Gonzalo, 402, 407
Reymont, Władysław S., 393
Rickey, V. Frederick, 395
Rider, Robin E., 400
Robinson, Abraham, 207, 222–4, 229, 255, 261, 282, 285, 294, 306, 307, 315, 405
Robinson, Julia, 172–5, 177, 193, 222, 234, 364, 365, 385, 408
Robinson, Raphael M., 157, 173–5, 193, 210, 222, 231, 234, 364, 402, 408
Rogers, Hartley, 221, 365
Roos, Hans, 394, 397
Rosenberg, Ethel and Julius, 181
Rosser, J. Barkley, 70, 71, 73, 129, 130, 145, 193, 221, 222, 229
Rougier, Louis, 93
Royden, Halsey, 231, 260
Rubin, Herman, 216, 232, 234
Rubin, Jean, 216
Rudnicki, Szymon, 98, 397
Russell, Bertrand, 37, 43, 67–9, 71, 77, 79, 96, 97, 131, 134, 135, 136, 139, 160, 225, 294, 299, 301, 396, 399
Ryll-Nardzewski, Czesław, 239

Sambin, Giovanni, 350, 351, 407
Schayer, Władysław, 57
Schlick, Moritz, 78, 80, 85, 93, 95, 247
Schmidt, Arnold, 207, 254
Scholz, Heinrich, 97, 137, 138, 167, 170, 172, 399, 401
Schröder, Ernst, 337
Schrödinger, Erwin, 294
Schütte, Kurt, 302
Schwabhäuser, Wolfram, 410

Schweitzer, Albert, 352
Scott, Dana S., 145, 172, 214–19, 221–3, 229, 231–4, 237, 243, 257, 260, 265, 272, 282, 283, 285, 287, 315, 331, 348, 378, 386, 395, 402
Scott, Irene Schreier, 218, 237, 403
Seuren, Pieter A. M., 402
Shanin, Nikolai, 302, 303
Shannon, Claude, 335
Shearman, A. T., 294, 295, 298, 305
Shore, Richard, 365
Sierpiński, Wacław, 29, 30–2, 41, 48, 198, 214, 243, 268
Silver, Jack, 324
Singer, Isaac Bashevis, 5
Singer, Israel Joshua, 5, 7
Skolem, Thoralf, 72, 121, 281, 285
Sluga, Hans, 296–9, 405
Słupecki, Jerzy, 243
Smale, Stephen, 302, 364
Smith, Barry, 394
Sobieski, King Jan III, 98
Solovay, Robert, 51, 286, 287, 395, 405
Spector, Clifford, 222, 260
Sproul, Robert Gordon, 210, 326
Staal, Frits, 207, 313, 318, 402, 406
Stadler, Friedrich, 397, 399, 404
Stadtman, Verne A., 400, 401
Stalin, Joseph, 125, 128, 146, 209, 300, 302, 303
Steinhaus, Hugo, 28, 29, 32, 67, 68
Stone, Harlan F., 166
Stone, Marshall H., 131, 166, 234, 338, 399
Sturm, Charles-François, 75
Sundholm, Göran, 395
Suppes, Christine, 408
Suppes, Patrick, 216, 232–4, 251–4, 260, 272, 367, 378, 403, 404, 408
Szczerba, Lesław, 239, 358
Szmielew, Aleksandra (Olenka), 209, 357, 407
Szmielew, Wanda, 104, 105, 172, 177–80, 193, 194, 198, 209, 232–6,

Szmielew, Wanda *(cont.)*
243–5, 271, 290, 346, 348, 349, 356–8, 361, 385, 407
Śrzednicki, Jan T. J., 395

Tait, William, 225, 260, 272, 403
Taitslin, Mikhail, 307, 308, 406
Tajtelbaum, see Teitelbaum
Tarska, Anna (Tarski's niece), 38, 164, 209
Tarska, Krystyna: *see* Tarski, Ina
Tarska, Maria: *see* Tarski, Maria
Tarski, Alfred: *see* Teitelbaum, Alfred, for entries prior to name change
Tarski, Ina (Tarski's daughter), 91, 107, 162–4, 166, 167, 169, 176, 179, 182, 206, 238–42, 245, 289, 291, 292; *see also* Ehrenfeucht, Kristina
Tarski, Jan (Janusz) (Tarski's son), 91, 107, 162–7, 169, 176, 177, 179, 206, 238, 289–91, 379, 395, 396, 400, 401, 405
Tarski, Maria (née Witkowska), 64–6, 88, 89, 91, 92, 137, 138, 158, 162–7, 175, 176, 178–82, 187, 200, 203, 204, 206, 237–42, 244, 270, 271, 276, 292, 293, 299, 303, 304, 311, 312, 316, 317, 322, 323, 328, 330, 331, 348, 353–6, 359, 360, 367, 368, 372, 373, 375–9
Tarski, Tamara (Tarski's sister-in-law), 164
Teitelbaum, Alfred, 20, 26, 36–8, 40, 53–5, 147
Teitelbaum, Berek (Tarski's paternal grandfather), 6
Teitelbaum, Ignacy (Isaak) (Tarski's father), 6–9, 53
Teitelbaum, Rosa (née Prussak) (Tarski's mother), 5, 6, 7, 8, 10, 107
Teitelbaum, Stanisław (Tarski's uncle), 6
Teitelbaum, Wacław (Tarski's brother, prior to name change), 8, 10, 38, 164
Teitelbaum dynasty, 268

Teitelbaum family, 5, 9, 11, 17, 18, 54, 270
Thompson, Frederick B., 171, 172, 183–5, 192, 197, 201, 212, 339, 341, 385, 401
Trakhtenbrot, Boris, 307, 308, 406
Turing, Alan, 70, 145, 222, 228, 229
Tuwim, Julian, 270
Twardowski, Kazimierz, 30, 31, 34, 35, 37, 38, 67, 89, 100, 380, 395
Tymieniecka, Anna-Teresa, 393

Ulam, Adam, 124–6, 399
Ulam, Stanisław, 124–7, 399

van Ulsen, Paul, 402, 404
Vaught, Marilyn, 239, 408
Vaught, Robert L., 181, 185–8, 197, 222, 239, 257, 260, 272, 282, 302, 303, 310, 343, 344, 347, 348, 365, 378, 385, 398, 401, 404, 405, 408
Vavilov, S., 300
Veblen, Oswald, 151, 172
Vidal, Gore, 196
von Neumann, John, 93, 124, 126, 144, 228, 231

Wagon, Stanley, 395
Waismann, Friedrich, 95
Wajda, Andrzej, 245
Wajsberg, Mordechai, 37
Wang, Hao, 207
Watson, James D., 277
Wedberg, Anders, 138, 165, 166
Wedberg, Birger, 138, 165
Weinstock, Niute (Tarski's paternal grandmother), 6
Wells, Benjamin ("Pete"), 306, 323–8, 347, 373–6, 386, 404, 406, 408
Westheimer, Ruth, 140
Weyl, Hermann, 231, 398
Wheeler, John, 255
Whitehead, Alfred North, 37, 69, 71, 77, 301

Whitfield, Celina, 66, 187, 270, 271, 396
Whitfield, Frank, 270, 271
Williams, Bernard, 297
Wilson, Woodrow, 23
Wisnovsky, Joseph, 298, 405
Witkacy (Witkiewicz, Stanisław Ignacy), 58–62, 66, 67, 271, 379, 404
Witkiewicz, Stanisław Ignacy ("Witkacy"), 57–9, 66, 271
Witkowska, Józefa: *see* Zahorska, Józefa
Witkowska, Maria: *see* Tarski, Maria
Witkowski, Antoni (Tarski's brother-in-law), 182
Wittgenstein, Ludwig, 77, 80, 85, 225, 299, 373
Witwicki, Władysław, 394
Wojeński, Teofil, 56
Woleński, Jan, 31, 394–8
Wolfe, Dorothy, 233, 234, 237, 239, 241, 244, 245
Wolfe, Tom, 366
Wollheim, Richard, 295, 405
Woodger, Joseph Henry, 97, 100, 129, 194, 195, 217, 234, 235, 251, 348, 361, 397
Woolf, Virginia, 58
Wundheiler, Alexander, 127

Yanovskaya, Sofia A., 302, 405
Yasuhara, Ann, 402

Zahorska, Józefa (née Witkowska) (Tarski's sister-in-law), 64, 88, 162, 164, 293
Zamoyski, Adam, 22, 393, 394, 396
Zawirski, Zygmunt, 97, 102
Zeno, 44
Zermelo, Ernst, 43, 46, 47, 51, 395
Żeromski, Stefan, 56, 62, 63, 65, 66, 83
Zilsel, Edgar, 139
Zygmund, Antoni, 151, 167